Engineering Separations Unit Operations for Nuclear Processing

Edited by

Reid A. Peterson

CRC Press
Taylor & Francis Group
Boca Raton London New York

CRC Press is an imprint of the
Taylor & Francis Group, an **informa** business

CRC Press
Taylor & Francis Group
6000 Broken Sound Parkway NW, Suite 300
Boca Raton, FL 33487-2742

ISBN 13: 978-1-138-60582-4 (hbk)

Visit the Taylor & Francis Web site at
http://www.taylorandfrancis.com

and the CRC Press Web site at
http://www.crcpress.com

Contents

Preface

Separation processes for nuclear materials, an essential part of the nuclear fuel cycle, lie at the intersection of chemical and nuclear engineering disciplines. They have been the source of some of the greatest challenges and triumphs in either field. The first full-scale nuclear reactor produced plutonium that was recovered through separation processes; plants that separate radioisotopes from reactor products or waste streams rank among the most complex and expensive processing facilities ever built. Unfortunately, neither the chemical engineering nor nuclear engineering discipline truly embraces the processes involved. Most nuclear engineering curricula focus on the nuclear reactor itself, paying attention to post-reactor materials only from an economic perspective. Chemical engineering curricula come closer to addressing the processing of post-reactor materials but do not attend to the complex issues that arise when dealing with highly radioactive materials. Nuclear materials processing generally involves parts of the periodic table that are unfamiliar to most chemical engineers. Undergraduate course work pays little to no attention to the chemistry of actinides nor to the myriad fission products present in most post-reactor materials. In short, there is a gap between chemical and nuclear engineering. This book aims to help fill that gap.

Much of this book focuses on the development and performance of historical processes. Some of these processes like PUREX form the backbone of the current nuclear materials processing. Others like the Bismuth Phosphate Process were highly inefficient and have long been surpassed. However, it is useful to understand how they were developed and what challenges their engineers had to overcome. Hopefully, these lessons will help to ensure that some of the mistakes from the past are not repeated. The future of nuclear processing relies heavily on the ability to develop new processes to minimize waste while limiting the potential for proliferation. Addressing the engineering issues for these processes will be of paramount importance to their success.

Preface

Acknowledgments

First and foremost, this book would not have been possible without the assistance of Barb Beller. She helped coordinate the efforts of all the authors, tracked down references, and made sure we all stayed on task. Rose Perry worked diligently to make sure that all the figures in this book were consistent and met the authors intent and the publisher's guidelines. Andrew Pitman and his team of technical editors did a fantastic job formatting the manuscript and catching the grammatical errors that the engineers and chemists that authored the different chapters were apt to make. The chapter authors acknowledge the valuable critiques of the technical reviewers: Sandra Fiskum for Chapters 1 and 6, Sergey Sinkov and John Swanson for Chapters 2 and 3, Cal Delegard and Gregg Lumetta for Chapter 4, and Richard Daniel for Chapter 5.

This book was supported by the Nuclear Process Science Initiative (NPSI) at Pacific Northwest National Laboratory (PNNL). PNNL is a multi-program national laboratory operated for the U.S. Department of Energy (DOE) by Battelle Memorial Institute under Contract No. DE-AC05-76RL01830.

Acknowledgments

Editor

Reid A. Peterson currently leads a team of 25 chemists and engineers at PNNL, working on issues associated with separations and monitoring for nuclear materials processing. His work is primarily in the field of waste processing for treatment of high-level waste. He has spent much of his career taking projects from inception to pilot-scale proof of concept. Dr. Peterson holds a BS in chemical engineering from Iowa State University and a PhD in chemical engineering from the University of Wisconsin.

Contributors

Susan E. Asmussen is a chemist at PNNL in Richland, Washington. She received her PhD in chemistry at the University of Western Ontario, in London, Ontario, Canada, where her work focused on the effects of radiation on ionic liquids in contact with water and different cover gasses. Prior to joining the PNNL team, she worked at the Canadian Nuclear Laboratories as a research chemist (formerly Atomic Energy of Canada) in the Analytical Chemistry Branch. Her research interests at PNNL involve studying aspects of solvent extraction in macro- and micro-scale systems, understanding fundamental rates of transfer and complex formation using *in situ* measurement techniques, investigating and understanding solvent extraction and biphasic systems in extreme environments, studying fuel pellet dissolution using a single-pass flow-through set up, and performing research supporting long term disposal of nuclear waste.

Garrett Brown received his BS in chemistry from Western Washington University and his PhD in analytical chemistry from University of Colorado, Boulder. Dr. Brown has more than two decades of technical management focused on research, development, and technology commercialization in the areas of the production and purification of short-lived radionuclides, radiopharmaceuticals, medical devices, and nuclear waste remediation.

Amanda J. Casella earned her BS in chemical engineering in 2002 and her PhD in nuclear engineering in 2008 from the University of Missouri-Columbia. She is currently a research engineer in the Nuclear Chemistry and Engineering Group, part of the Energy and Environment Directorate at PNNL. Dr. Casella's research has focused on numerous processes relating to the back end of the nuclear fuel cycle. Her specific work has involved dissolution of used nuclear fuel for post-irradiation analysis and long-term fuel storage applications and studying fundamental properties of the coordination chemistry and interfacial transport behavior of lanthanides and actinides in various liquid:liquid extraction reprocessing schemes. This extraction work includes the advancement of real-time, online spectroscopic techniques for both enhanced control and safeguards accountability applications.

Andrew M. Casella is currently the materials development and computation team lead within the Reactor Materials and Mechanical Design Group, part of the Energy and Environment Directorate at PNNL. In his 12 years at PNNL, his research has focused on mass and thermal transport in nuclear systems, post-irradiation examination of reactor fuels and accelerator targets, and nuclear reactor analysis. He holds a BS in chemical engineering and a PhD in nuclear engineering, both from the University of Missouri-Columbia.

Calvin H. Delegard graduated from Washington State University (1970) with a BS in chemistry. He has 16 years of experience in radiochemical processing at the Hanford Site and 23 additional years associated with PNNL in Richland, Washington, specializing in Hanford tank waste chemistry and plutonium separations and processing. He also was a nuclear materials safeguards inspector for eight years at the International Atomic Energy Agency in Vienna, Austria. Delegard continues his career in process chemistry on a part-time basis.

Gabriel B. Hall received a BS in chemistry from Truman State University and a PhD in organometallic chemistry from the University of Arizona. He joined the staff of PNNL in 2015 where he began working on solvent extraction in the nuclear fuel cycle. His research interests include electrocatalysis, understanding metal-ligand covalency, and optimizing solvent extraction processes.

Amy M. Rovira is an engineer at PNNL. She received a BS in chemical engineering from Montana State University in Bozeman, MT. Rovira joined the staff of PNNL in 2017 when she began working on the Radioactive Waste Test Platform at PNNL, which is designed to support start-up and operation of direct-feed low-activity waste to the Waste Treatment Plant at Hanford. Prior to joining the PNNL team, she worked at the Waste Treatment Plant in Plant Engineering.

Philip P. Schonewill earned his BS in chemical engineering from Calvin College in Michigan and PhD in chemical engineering from the University of Notre Dame. He is a senior chemical engineer and team lead of the Experimental Fluids team, a part of the Experimental & Computational Engineering Group in the Energy and Environment Directorate at PNNL. Over a 10-year career at PNNL, Dr. Schonewill has investigated several nuclear waste treatment technologies via experimentation—ranging from bench-scale to full-scale—and modeling. Currently, he manages PNNL technology development support to the Low-Activity Pretreatment System (LAWPS) and Tank Side Cesium Removal (TSCR) systems, which includes simulant development, filtration testing, ion exchange media testing, flammable gas generation measurements, and testing to develop a flammable gas safety basis for the facility.

Michael F. Simpson is currently a professor and department chair for Materials Science and Engineering at the University of Utah. After earning a BS in chemical engineering from the California Institute of Technology in 1991 and a PhD in chemical engineering from Princeton University in 1996, he joined Argonne National Laboratory-West as a research engineer. His career at ANL-West (later Idaho National Laboratory) notably included contributing to the successful demonstration of the electrometallurgical treatment process for irradiated fuel from the Experimental Breeder Reactor and then further developing pyroprocessing technology via collaborations with researchers from the Republic of Korea and U.S. universities. He founded the International Pyroprocessing Research Conference in 2006, which has been held every two years since. He has discovered, invented, and developed a number of methods to improve pyroprocessing of nuclear materials pertaining to actinide monitoring and separation, dehalogenation, waste volume minimization, and safeguards. He joined the faculty of the University of Utah in 2013 as an associate professor in the Department of Metallurgical Engineering and immediately started building a world class academic research program in pyrochemical processing of nuclear materials and molten salt electrochemistry.

1 Overview of Process Flowsheets

Reid A. Peterson

CONTENTS

1.1 PROCESS FLOWSHEETS

Unit operations have formed the basis of chemical engineering since they were introduced by Arthur Little in 1915. This book will focus on separation unit operations and, in particular, those unit operations that are typically used on an industrial scale to post-process materials from nuclear reactors. The classical separations unit operations include humidification, gas absorption, distillation, liquid-liquid extraction, adsorption and ion exchange, crystallization, precipitation, and leaching. With the exception of humidification and distillation, all of these unit operations are used in nuclear materials processing. This book will discuss those unit operations in the order in which they were historically used. From that perspective, it is important to note that these unit operations were not deployed independently, but rather in process flowsheets linking several unit operations together. Therefore, this chapter will introduce those overall flowsheets and discuss, in general, how the unit operations interacted; relevant chapters discuss each unit operation in detail.

1.1.1 Bismuth Phosphate Process

The first nuclear process to be deployed on an industrial scale was the Bismuth Phosphate Process. While generally known for its core process of separating plutonium from uranium (the Bismuth Phosphate Precipitation step), the overall Bismuth Phosphate Process is actually a series of unit operations linked together to achieve the ultimate goal of separating, purifying, and concentrating plutonium. Figure 1.1 provides a summary flowsheet showing how these unit operations were linked

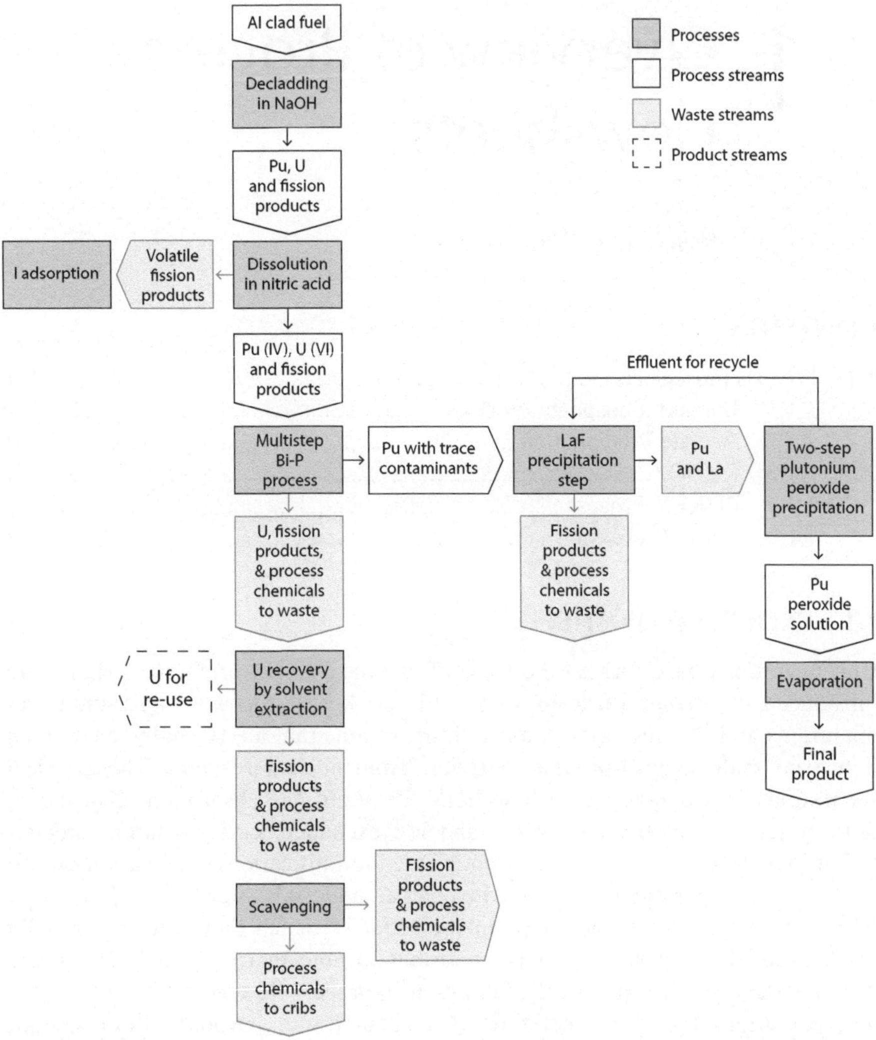

FIGURE 1.1 Bismuth Phosphate Precipitation flowsheet.

together. This figure represents the fully matured Bismuth Phosphate Process. The uranium recovery step and the scavenging step were added after the initiation of the process to address some of the overall process limitations. The process scheme starts with spent fuel from a reactor in which plutonium has been produced from uranium. The first two steps in this process separate the cladding from the fuel charge, then dissolve the fuel charge for further processing (Chapter 2). The off-gas from these steps was generally treated to recover selected volatile off-gas radionuclides to avoid their release into the environment. The dissolved fuel charge then went through two separate precipitation processes, the Bismuth Phosphate Process and the lanthanum-fluoride process (Chapter 3), to first separate the plutonium from the uranium and fission products and then to further purify the resultant plutonium product. The solution was finally evaporated to produce a purified, concentrated product. These process steps represent the initial process flowsheet for the Bismuth Phosphate Process. This process had two primary limitations: first, it did not recover the uranium from the waste stream; second, it resulted in a significant volume of waste. Subsequently, additional process steps were added to the flowsheet to recover the uranium by solvent extraction (Chapter 4) and precipitate the radionuclides (Chapter 3) from the aqueous phase to allow disposal of bulk of the aqueous phase in cribs.

1.1.2 Acetate Precipitation

The second major flowsheet deployed on an industrial scale was the acetate precipitation process used at Mayak (see Figure 1.2). Similarly to the Bismuth Phosphate Process, this process separated the cladding from the fuel by dissolution with caustic, then dissolved the fuel charge in nitric acid. The advantage of the acetate precipitation process (Chapter 3) over the Bismuth Phosphate Process was that it directly recovered uranium for reuse. The separated plutonium stream was then treated with a lanthanum fluoride process to purify and concentrate the final product.

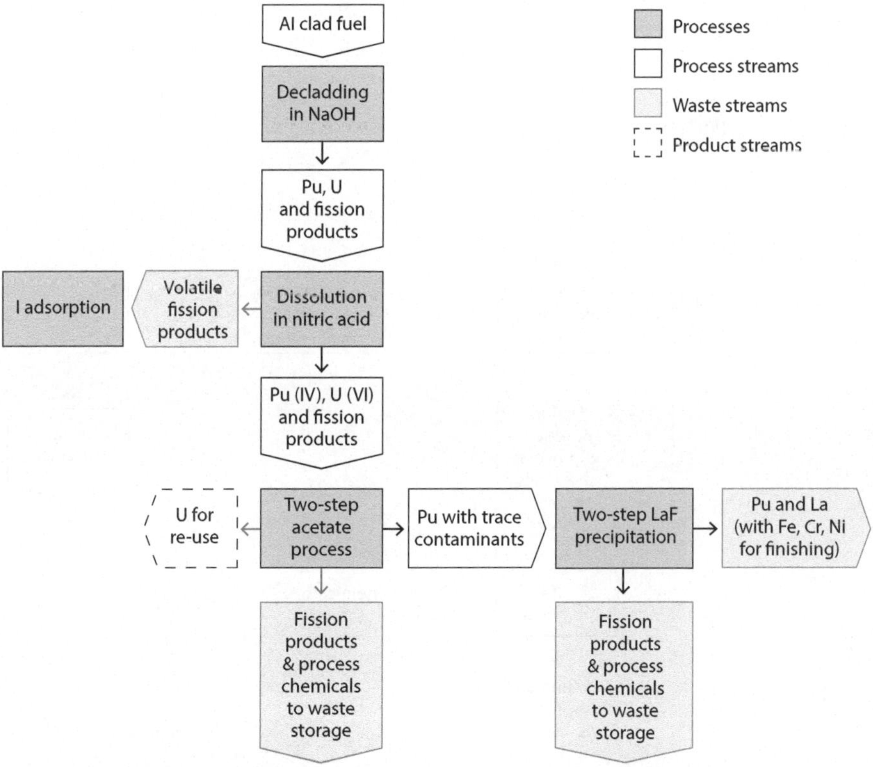

FIGURE 1.2 Mayak precipitation flowsheet.

1.1.3 REDOX

The reduction-oxidation (REDOX) process followed on the heels of the Bismuth Phosphate Process at the Hanford Site (see Figure 1.3). The front end of the REDOX process was the same as that deployed in prior processes, but the uranium/plutonium separation was carried out by a solvent extraction process (Chapter 4). Additional purification and concentration were achieved through anion exchange (Chapter 6) or evaporation. Final product concentration was achieved through a final precipitation step. The REDOX process was more efficient than earlier processes; however, it resulted in the production of waste streams with a high thermal load. This led to the self-boiling of some of the tanks. To address this issue, fission products were

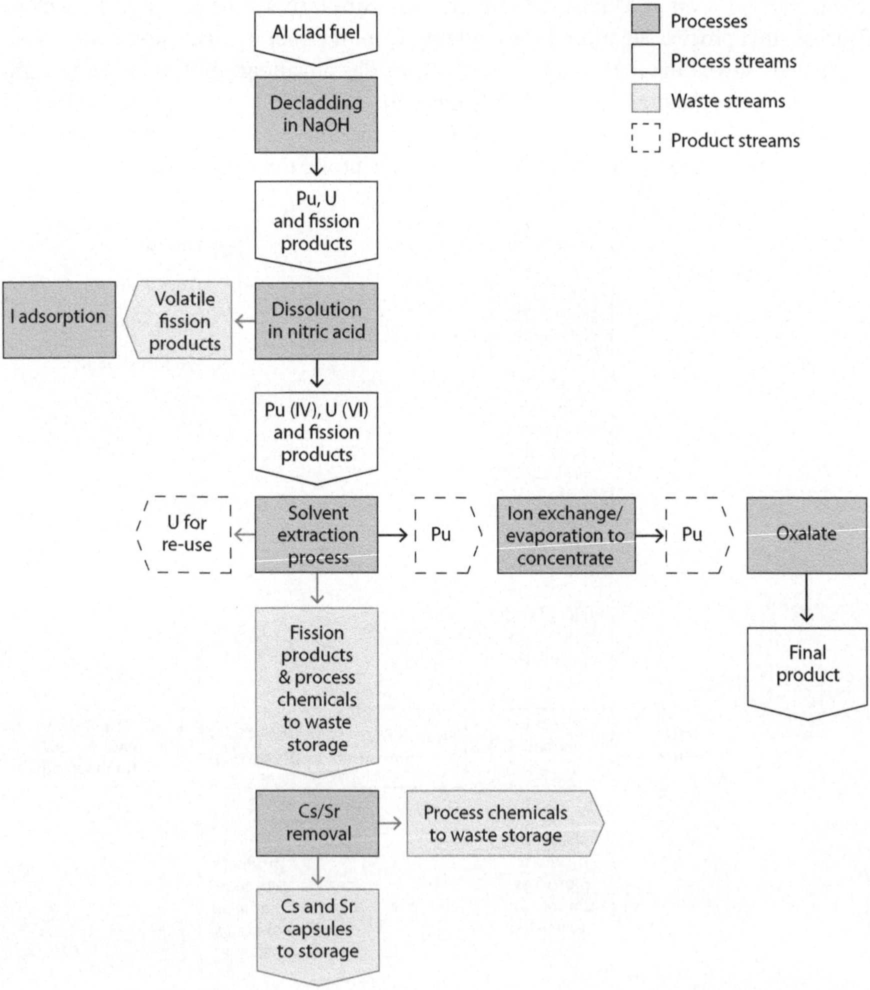

FIGURE 1.3 REDOX solvent extraction flowsheet.

removed from the waste stream by subsequent processing through solvent extraction for removal of strontium and ion exchange for removal of cesium (Chapter 6).

1.1.4 PUREX Process

Figure 1.4 shows the Plutonium Uranium Extraction (PUREX) process flowsheet. Again, the front end involves segregation of the cladding and dissolution of the fuel charge. The basic process then involves segregation of uranium and plutonium from the fission products by solvent extraction. The PUREX process forms the basis of most modern reprocessing plants and, as such, has as many different flowsheet variations as there are deployments of the PUREX process. It is worth discussing some of the earliest PUREX flowsheets, as these formed the basis for most subsequent process flowsheets. The first deployment of the PUREX flowsheet was at the Hanford site.

Figure 1.5 provides a flow diagram for this flowsheet. The primary differences here are the deployment of anion exchange for concentration of the plutonium product and removal of the fission products from the waste stream by solvent extraction

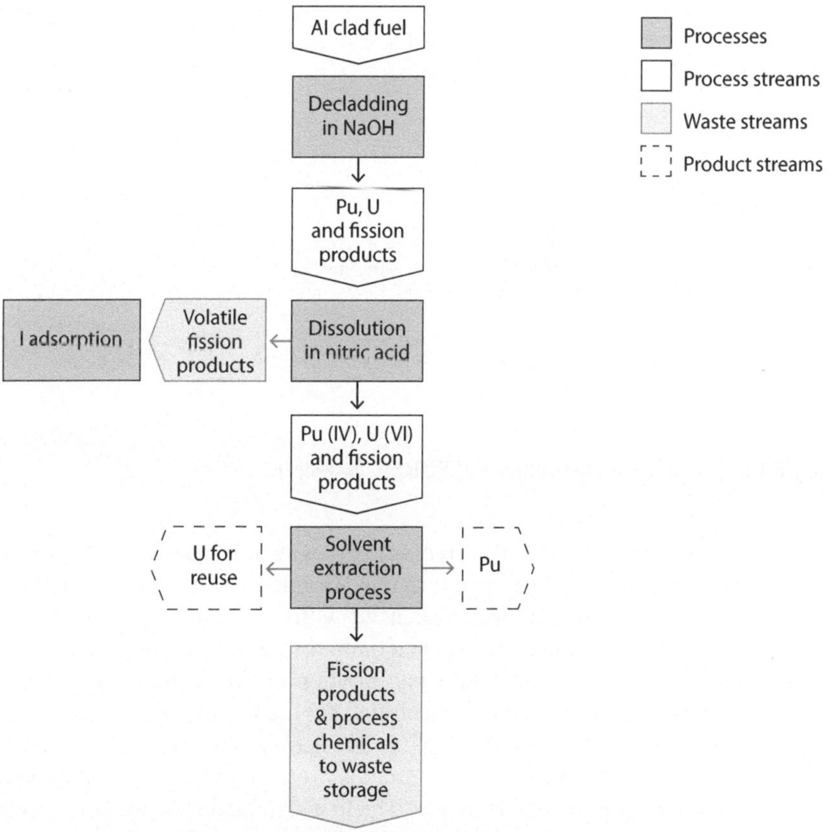

FIGURE 1.4 Generic PUREX solvent extraction flowsheet.

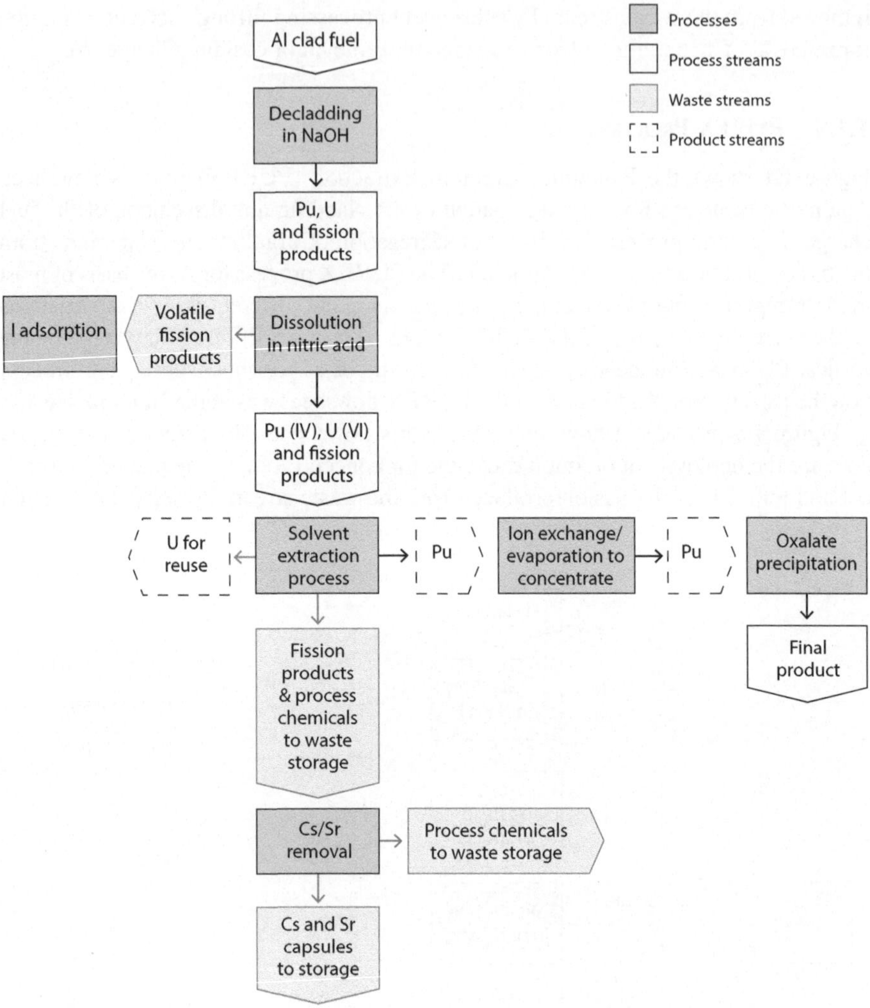

FIGURE 1.5 Hanford implementation of PUREX flowsheet.

and ion exchange. Shortly after the Hanford deployment of the PUREX process, the Savannah River Site deployed two different variations on the process. The first of these processes (Figure 1.6) was associated with recovery of plutonium from depleted uranium targets. Since the depleted uranium was of marginal value, it was not recovered as part of this process. Concentration of the product was achieved through cation ion exchange, which concentrated but did not significantly purify the product. At the same time, the Savannah River Site deployed the H-Canyon Modified (HM) PUREX process (Figure 1.7). This process involved the recovery of uranium and neptunium from enriched fuel charges. The first significant change here was the simultaneous dissolution of the cladding and the fuel charge. The other significant change was associated with the recovery of neptunium in place of the recovery of

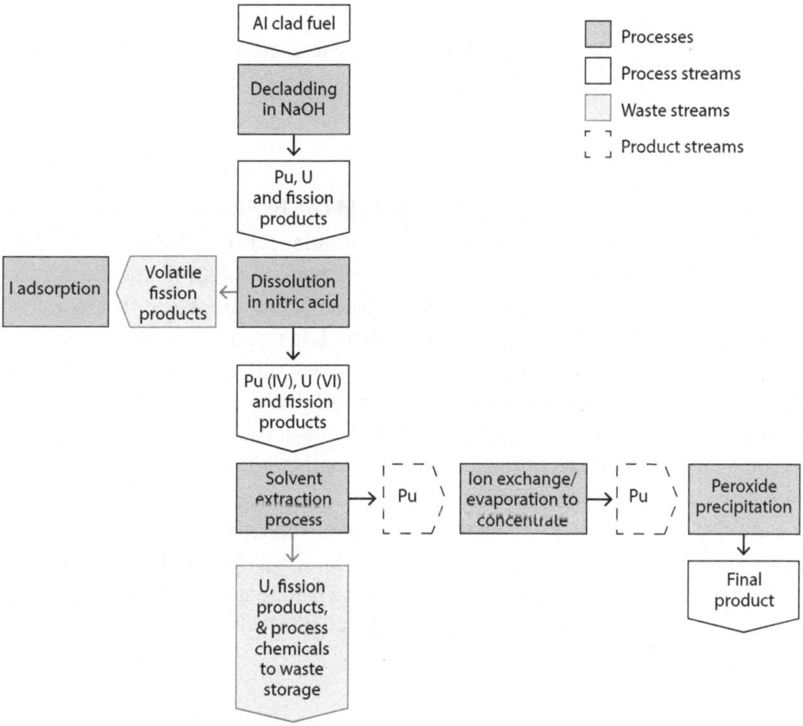

FIGURE 1.6 Savannah River Site recovery of plutonium.

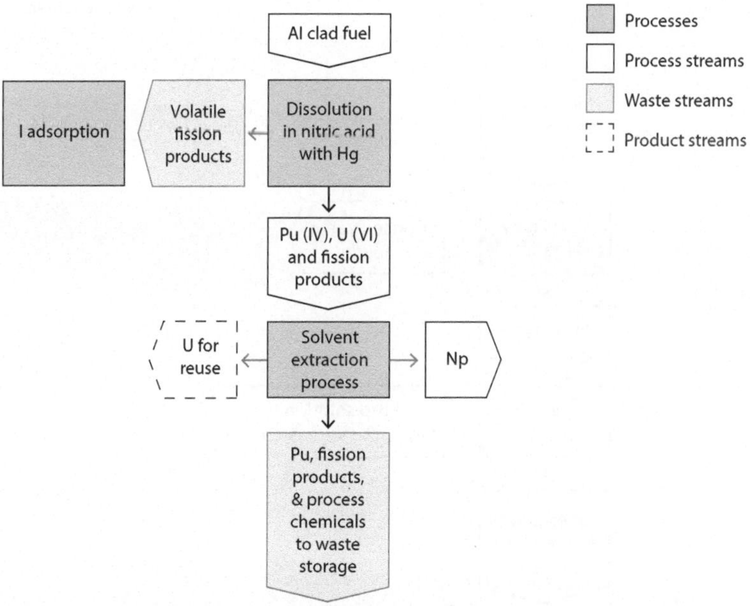

FIGURE 1.7 Savannah River Site recovery of uranium and neptunium.

plutonium. Other variations of the PUREX process have been employed, but in general, all rely on the same fundamental chemistry to recover the actinide species and then to separate plutonium from uranium.

1.1.5 Waste Processing

Waste from the processing of nuclear materials has been stored for up to 70 years in some cases. Concerns about the potential eventual release of that material to the environment have led to efforts to find more permanent solutions for these types of wastes. The primary disposition route is to immobilize the waste in a solid waste form, typically either a grout or glass waste form. It is frequently desirable to develop separate waste forms depending on the activity present in a particular waste stream. This has led to the development of process flowsheets for separating radionuclides from bulk waste streams so that a bulk-volume, low-activity waste form can be used for the majority of the waste volume, while most of the radionuclides are concentrated into a lower-volume, higher-activity waste form. Two primary examples of these processes are the West Valley Demonstration Project (see Chapter 6) and the Savannah River Site Modular Caustic Side Solvent Extraction Unit (MCU). Figure 1.8 provides a summary of the MCU process. Waste handling frequently involves more handling of solids, higher solids concentration, and solids with ill-defined physical properties.

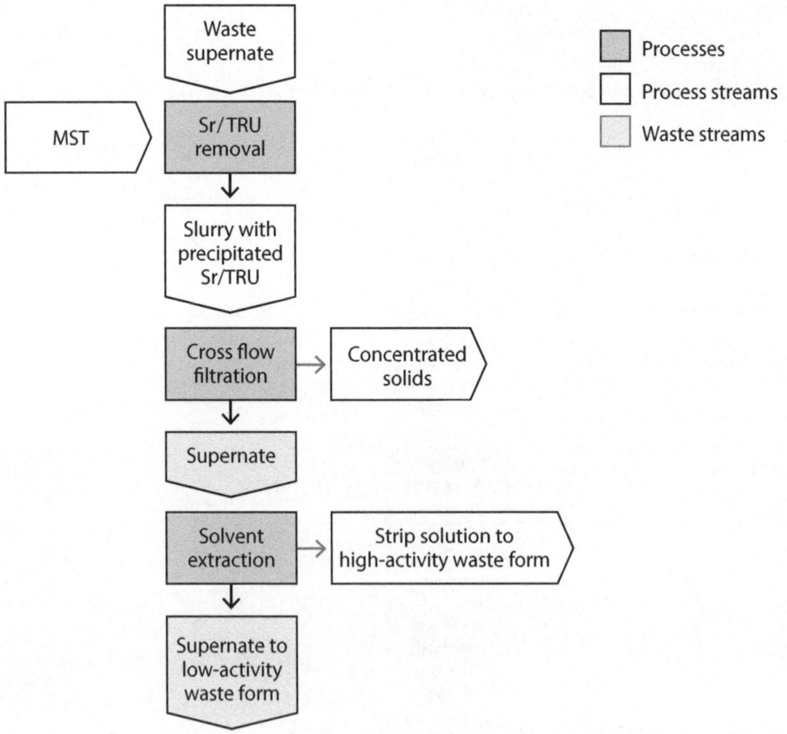

FIGURE 1.8 Savannah River Site waste separation process.

This poses unique challenges. Some of these challenges are discussed in Chapter 5 associated with filtration.

At the Hanford Site, a significant quantity of the soluble radionuclides have already been removed. However, there is still a significant quantity of soluble cesium that must be addressed to allow for production of a lower-activity waste form. At this time, that process is intended to be accomplished using an ion exchange process (see Chapter 6).

These process flowsheets form the basis for most of the existing flowsheets for processing of nuclear materials. There are new process flowsheets under development, some of which would utilize different unit operations. However, the focus of this book is on those processes that have seen extensive large-scale deployment for the treatment of nuclear materials.

2 Uranium Fuel Dissolution

Calvin H. Delegard

CONTENTS

2.1 INTRODUCTION

Industrial reprocessing of irradiated uranium-based metal or uranium dioxide (UO_2) fuels begins with chemical or mechanical removal of cladding or shearing of the cladding followed by oxidative dissolution of uranium-based fuel in hot and high concentration nitric acid (HNO_3). Reviews of UO_2, PuO_2, and mixed UO_2/PuO_2 dissolution in inorganic acids and mixtures were provided recently by Desigan et al. (2018).

Dissolution processing of other fuel types, such as highly alloyed uranium, uranium silicides, carbides, and nitrides, and thorium-based fuels; the use of dissolving agents other than those based on HNO_3 (e.g., other mineral acids, molten salts); and electrolytic dissolution are much more limited and thus not considered in this chapter aside from directing the reader to alternative information resources.[1]

2.2 SELECTION AND USE OF NITRIC ACID AND IMPACTS ON MATERIALS OF CONSTRUCTION

Dissolution of irradiated uranium metal and dioxide fuel in HNO_3 is used industrially for a variety of reasons: (1) the reaction is made safe by eliminating hydrogen gas release that would be inherent in use of other mineral acids, (2) the reaction dissolves all or nearly all of the fuel, (3) the dissolution is relatively rapid, and (4) the reaction is controllable despite the substantial reaction heat evolved. With proper regulation of the reagent parameters, the nitrate product solution has acid (~3 M) and uranium (~1 M) concentrations and the uranium and plutonium in oxidation states—U(VI) and Pu(IV), respectively—suitable for PUREX solvent extraction processing provided that all the uranium is dissolved.[2]

The HNO_3 that is not required to provide counterion to the dissolved U(VI) and Pu(IV) as their respective nitrate salts, $UO_2(NO_3)_2$ and $Pu(NO_3)_4$, and the supplemental HNO_3 required for salting strength in solvent extraction is recoverable as is nearly the entire complement of reduced oxides of nitrogen (NO and NO_2) arising from fuel dissolution. The HNO_3 reporting to the wastes following solvent extraction can also be recovered by distillation. These opportunities for chemical recycle further commend HNO_3 for the use in fuel dissolution.

Such favorable attributes are not the case for other mineral acids (e.g., hydrochloric acid, HCl, which generates flammable hydrogen gas during fuel dissolution),

---

[1] Investigations of other fuel compositions and non-HNO_3 acid fuel dissolution pathways are described elsewhere (e.g., Blanco and Watson (1961, 60–95); Slansky (1961); Long (1967, Chapter 4); Masson, Rovet, and Béziat (2008)). A review of alternative methods to dissolve uranium metal and its alloys and description of tests of some of these methods was prepared by Laue, Gates-Anderson, and Fitch (2004b, 2004a). The review was largely based on a separate review of methods used in analytical chemistry (Larsen 1959). Dissolution agents included HNO_3, aqua regia (HNO_3 + HCl), HNO_3 + hydrofluoric acid (HF), HCl + oxalic acid, H_2SO_4 + oxalic acid, bromine (Br_2) in ethyl acetate, HCl in ethyl acetate, hot phosphoric acid (H_3PO_4), and hydrogen peroxide in sodium hydroxide solution. More recently, dissolution of uranium metal and of irradiated UO_2 by alkaline peroxide/ammonium carbonate solution has been investigated and found to be effective (Soderquist, McNamara, and Oliver 2008; Soderquist et al. 2011). Dissolution of UO_2 in peroxide/sodium carbonate solution also has been studied (Peper et al. 2004).

[2] Note that if uranium metal is still present in the dissolver solution, some uranium is in the (IV) oxidation state and some plutonium in the (III) state. Once the uranium metal is dissolved, radiolysis and oxidation by nitric acid and residual nitrous acid, HNO_2, will convert the uranium to U(VI) and plutonium to Pu(IV).

which are difficult to handle in strong metallic vessels and are not amenable to subsequent uranium and plutonium separations processing.

Of the mineral acids required to dissolve uranium-based fuels, HNO_3 also is desirable because of its compatibility with the stainless-steel equipment and compatibility with separations processing. Low carbon or carbon-scavenged austenitic stainless steels, frequently 304L, are attractive in nuclear chemical processing applications because of strength, ease of construction, relatively low cost, and broad corrosion resistance under the oxidizing conditions imposed by HNO_3. The corrosion resistance is due to a passive layer of chromium oxide (Cr_2O_3) on the exposed metal surface (Decours et al. 1987; Fauvet et al. 2008). Carbon is always present to some extent in stainless steel. At above 0.03% carbon, chromium carbide ($Cr_{23}C_6$) forms during heat-treatment, such as welding, depleting the neighboring regions in the steel of chromium, preventing the formation of the protective Cr_2O_3 layer, and leading to enhanced corrosion in that region (Gupta 2017). Amounts of carbon in stainless steel are limited either by removal or by sequestration through carbon's reaction with scavengers such as niobium or tantalum. Thus, for example, the dissolvers in the first reprocessing plants at the Hanford Site used 25-12 S-Cb stainless steel (DuPont 1944, 408), akin to type 309 austenitic stainless steel, with columbium (Cb; i.e., niobium) added to scavenge carbon. The Magnox fuel dissolver at Sellafield, UK, operating from 1964 until 1978, used 18-13-Nb austenitic stainless steel (18% chromium (Cr), 13% nickel (Ni), with carbon scavenged by niobium (Nb) addition). Also at Sellafield, the Thermal Oxide Reprocessing Plant (THORP) dissolver, designed for reprocessing of commercial light water reactor UO_2 fuels, used nitric acid grade (NAG) low-carbon 310L stainless steel (25% Cr, 20.5% Ni, and 0.015% carbon) (Shaw 1990).

While stainless steel materials of construction generally are satisfactory for fuel dissolution (Blanco and Watson 1961, 60; but see Gnanamoorthy and Balasubramanian 1987), failures at heat-affected zones (e.g., welding lines) have been observed (Yamanouchi et al. 1987) and in exposed end grains formed by fabrication (Shaw 1990). In the latter case, failure occurred in the condensing region of the HNO_3 and not in the immersed region holding 3 M HNO_3 and 300 g uranium/L at 105°C. For these and other reasons, zirconium alloys have been selected for use in selected portions of reprocessing plant dissolvers in France that are highly difficult to replace or repair (Decours et al. 1987; Vaudano 2008a, 50).

The general performance of stainless steel in various reprocessing plant process systems is summarized by Fauvet et al. (2008).[3]

[3] Studies of stainless steel compatibility in HNO_3 have shown that particularly aggressive transpassive conditions exist in acid recovery concentrators (wherein, for example, dissolved chromium(III) ion Cr^{3+} can oxidize to chromium(VI) as dichromate, $Cr_2O_7^{2-}$), in plutonium product concentrators (where Pu(VI) likewise can provide aggressive conditions), and in concentrators from plutonium(IV) oxalate precipitation filtrate recovery (also likely containing manganese catalyst for oxalic acid destruction) causing early failures (Chauve et al. 1987; Decours et al. 1987; Gnanamoorthy and Balasubramanian 1987; Yamanouchi et al. 1987). Also, low but intolerable levels of stainless steel corrosion products contribute impurities to plutonium nitrate product solution undergoing evaporative concentration. In these process streams, equipment manufactured of tantalum and titanium and their alloys as well as zirconium have been employed with zirconium being most prominent (IAEA 1987). The ability of zirconium to resist corrosion and remain passive in hot HNO_3 containing Pu(VI) has been demonstrated (Nakahara, Sano, and Abe 2018).

2.3 CLADDING TREATMENT

Fuel cladding must be opened or removed to allow the dissolving agent (HNO_3) access to the core uranium-bearing fuel. Access to the core fuel is accomplished by (1) chemical dissolution of the cladding or (2) mechanical removal or mechanical breaching of the cladding. The first reprocessing plants used chemical cladding removal.

2.3.1 Chemical Dissolution of Cladding

Chemical removal of the cladding can be accomplished in two manners—using a chemical agent that dissolves both the cladding and the fuel or using a chemical agent that attacks the cladding only, leaving the fuel unaffected for subsequent processing.

For the Hanford Site, the initial steps in the reprocessing of uranium metal fuels were dissolution of the cladding followed by separate dissolution of the irradiated uranium metal fuel itself. Aluminum cladding was used for the fuel in the first eight of the Hanford reactors (named B, C, D, DR, F, H, KE, and KW) and Zircaloy-2 cladding in the ninth reactor (N). Chemical dissolution of the aluminum cladding was accomplished by treatment with sodium hydroxide (NaOH), which vigorously corrodes aluminum metal but has little effect on the underlying uranium metal core. Sodium nitrate, $NaNO_3$, was added to the NaOH to act as a hydrogen radical scavenger and suppress formation of flammable hydrogen gas. In this scavenging, the nitrate was chemically reduced to ammonia (NH_3) which remained in alkaline solution as NH_4OH, and sodium nitrite, $NaNO_2$. The cladding dissolution proceeded by the following reaction stoichiometry (Schneider 1951):

$$\begin{aligned}&Al_{(solid)} + 0.8125\,NaOH_{(solution)} + 0.9375\,NaNO_{3\,(solution)} + 1.6875\,H_2O_{(solution)} \rightarrow \\ &\quad NaAl(OH)_{4\,(solution)} + 0.75\,NaNO_{2\,(solution)} + 0.1875\,NH_4OH_{(solution)}\end{aligned} \tag{2.1}$$

Dissolution of Zircaloy-2 cladding from Hanford's irradiated N Reactor fuels was accomplished by the Zirflex Process (Swanson 1961). Zircaloy-2 is a zirconium alloy containing 1.2–1.7 wt% tin and smaller concentrations of iron, chromium, and nickel (Weakley 1979, II–1). Ammonium fluoride (NH_4F) was the cladding dissolution agent in the Zirflex Process. As was the case for the aluminum cladding removal step, nitrate—as ammonium nitrate (NH_4NO_3)—was used to scavenge hydrogen radicals and inhibit flammable H_2 gas evolution. The Zirflex Process reaction stoichiometry under PUREX Plant conditions using boiling 5.5 M NH_4F/0.5 M NH_4NO_3 solution was:

$$\begin{aligned}&Zr_{(solid)} + 6\,NH_4F_{(solution)} + 0.47\,NH_4NO_{3\,(solution)} + 4.94\,H_2O \rightarrow \\ &(NH_4)_2ZrF_{6\,(solution)} + 4.94\,NH_4OH_{(solution)} + 3.53\,H_2O_{(solution)} + 0.12\,H_{2\,(gas)}\end{aligned} \tag{2.2}$$

based on Moore, Walser, and Fritch (Section 4.2.1 of 1980). Although H_2 still evolved, its yield was about 6% of what it would have been in the absence of nitrate.

The cladding dissolver solution was made alkaline with the NaOH solution before discharge to the waste storage tanks. This step metathesized the dissolved $(NH_4)_2ZrF_6$ to form fine ZrO_2 precipitates, more NH_4OH, and dissolved sodium fluoride (NaF). About 1% of the uranium entering the Zirflex Process formed UF_4. Some of this UF_4 arose from the interdiffusion zone, about 12–14 μm thick, where gradated mixtures of Zircaloy cladding and uranium fuel were present. This zone, which transitioned in composition progressively from the cladding to the fuel, existed due to the swaging process in fabrication and likely irradiation recoil. The uranium present in this interdiffusion zone represented about 0.2% of the total uranium (Swanson et al. 1985). The UF_4, which also contained plutonium, likely as PuF_4, has low solubility and thus remained with the declad uranium metal fuel item within the dissolver. After cladding dissolver solution discharge, the NaOH solution was used to metathesize the UF_4 and PuF_4 to their respective hydrated oxides prior to treatment with HNO_3 in fuel dissolution (Moore, Walser, and Fritch 1980).

As at Hanford, aluminum-clad uranium metal fuel was used at Mayak in the USSR (Sokhina 2000). The first steps in the fuel reprocessing also were similar—the chemical removal of the aluminum cladding in strong NaOH solution containing $NaNO_3$ as a hydrogen radical scavenger—followed by dissolution of the uranium metal fuel cores in strong HNO_3. The Savannah River Site in the United States also processed aluminum-clad uranium metal fuel and used the same chemical decladding process, with $NaOH/NaNO_3$ solution, as was used at Hanford (Dupont 1957, 77).

The co-dissolution in HNO_3 of both the aluminum cladding and uranium core was used on a limited basis at Mayak. Mercuric nitrate $[Hg(NO_3)_2]$ was added to the HNO_3 solution to dissolve the aluminum cladding, which otherwise passivates in HNO_3. The dissolved mercuric ion (Hg^{2+}) is contact-reduced onto the aluminum, forming an aluminum-mercury amalgam. The amalgam itself dissolves readily in HNO_3 to form dissolved $Al(NO_3)_3$ and $Hg(NO_3)_2$, which can return catalytically to amalgamate with more aluminum. Mayak abandoned the mercury-catalyzed process because mercury damaged and deposited in process equipment and high plutonium losses were observed (Sokhina 2000, 23).

The co-dissolution in HNO_3 of both the aluminum cladding and uranium core was investigated for Hanford fuel elements but was never implemented. Higher throughput rates were projected for PUREX Plant operations by combining the cladding and core dissolutions compared with their as-practiced separate dissolutions, but maximum hydrogen generation rates reached about 35% of the lower flammability limit (Krieg and Van der Cook 1956). Mercury-catalyzed HNO_3 dissolution of irradiated aluminum-clad uranium-aluminum alloys occurred at the Savannah River Site with details on the effects of HNO_3, aluminum nitrate, and dissolved mercury concentrations and temperature on the rate provided by Hyder et al. (1979). Primary studies of mercury-catalyzed dissolution of aluminum cladding and aluminum-uranium alloy show that dissolution rates increased up to ~0.005 M mercuric nitrate but did not increase appreciably at greater concentrations (Wymer and Blanco 1957). Rates also initially increase with increasing total nitrate and HNO_3 concentrations up to about 3–8 M, depending on conditions, but decrease at higher concentrations.

This pattern was attributed to the opposing effects of increasing the metallic aluminum-mercury amalgam dissolution rate with increasing total nitrate and HNO_3 at their lower concentrations and the overtaking of amalgamation by dissolution of mercury itself, before it can amalgamate, at their higher concentrations (Wymer and Blanco 1957).

2.3.2 Mechanical Treatments

Magnox fuel, developed and used in the UK, has a cylindrical uranium metal core and a finned cladding of magnesium alloy (0.7%–0.9% aluminum with balance magnesium; Cronin and Collier (2012)). Both cladding and fuel can be dissolved in the HNO_3 solution. However, to avoid processing of the cladding with the uranium and the attendant increase in waste mass and volume, mechanical removal of cladding is performed. The cladding removal is accomplished by pressing the cylindrical uranium metal fuel through a circular die. The uranium metal, whose hardness is similar to steel, passes through the die, while the mechanically weaker magnesium alloy cladding readily peels off. Residual uranium metal can adhere to the cladding in the decladding process. Because the thin uranium residue is even more pyrophoric than the cladding, the peeled Magnox cladding generally is stored underwater (Fairhall and Palmer 1992). The uranium metal core, freed of most of the Magnox cladding, then may be dissolved in the HNO_3 solution.

Most commercial UO_2 and mixed oxide (MOX) fuels are fabricated as cylindrical ceramic pellets that are clad as stacks in stainless steel or zirconium alloy tubes (pins) with the tubes bundled in square or hexagonal arrays in fuel assemblies. Structural parts at the ends of the fuel assemblies, also constructed of stainless steel or zirconium alloy, serve to hold the tube arrays, incorporate lifting bales, and direct coolant flow, but themselves contain no nuclear material. Springs within the upper ends of the fuel pins press down on the fuel pellet stack while allowing radiolytic and thermal expansion of the pellets and providing space for krypton and xenon fission product gases to collect (Guillet and Guérin 2009, 9–12; Vaudano 2008a, 37–41).

Both stainless steel and zirconium resist corrosion in the HNO_3 solution. Thus, to allow the HNO_3 to access the irradiated fuel, the dissolution step is preceded by the removal of the structural parts at one end of the fuel assembly, down to the presence of the fuel, and then mechanical shearing of the fuel tubes into 1.25–5 centimeter-scale lengths by the so-called chop/leach or shear/leach process. The structural parts not bearing nuclear materials are examined to ensure their lack of nuclear material. The radiation-embrittled fuel tubes crack and splinter during shearing. The fuel pellets, already cracked by thermal and radiological stresses, are further fractured by the shearing (Vaudano 2008a, 46). This increases the available surface area for reaction with the leach HNO_3, increasing the dissolution rate. However, proper design and operation of the mechanical shear are important as pinching of the fuel tubes can block ready access of the HNO_3 to the contained irradiated fuels. The shearing mechanism also must operate in harsh and difficult-to-maintain environments.

A stepped tooth shear, with steps of width on the order of the fuel pin pitch, acting, with <0.08 mm clearance, perpendicularly against a V- or U-shaped notch

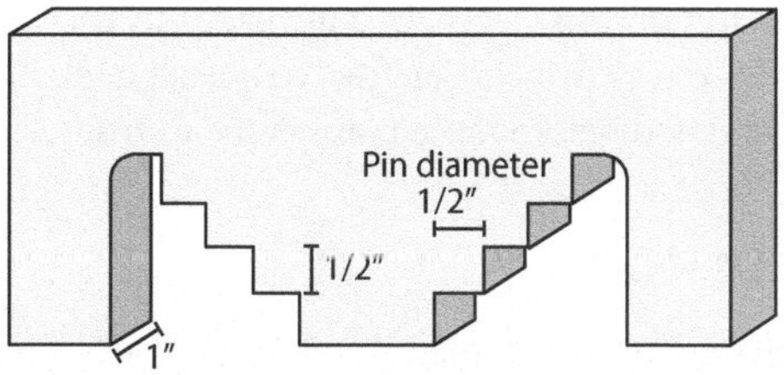

FIGURE 2.1 Stepped shear blade for chopping fuel rod bundles. (After Figure 5.40 of Long, J.T., *Engineering for Nuclear Fuel Reprocessing*, Gordon and Breach Science Publishers, New York, 1967.)

worked best of the designs tried, since it required less force with more breakage of the core fuel and showed little pinching of the cladding (Long 1967, 403–406; Watson et al. 1959). A stepped shear blade is illustrated in Figure 2.1. A similar stepped shear, though of a different geometry, currently is used at the Cap La Hague reprocessing plants UP3 and UP2-800 in France to chop oxide fuel pins from fuel bundles into ~3-cm lengths (Vaudano 2008a, Figure 29). The shearing is conducted under a nitrogen gas sweep to eliminate potential fires from ignition of finely divided pyrophoric zirconium metal.

Shear/leach processing of Hanford N Reactor Zircaloy-2 clad uranium metal fuel occurred from 1966 until 1971 at the Nuclear Fuel Services reprocessing plant at West Valley, New York (Schulz 1972). The N Reactor fuel consisted of two concentric tubes, each 53 cm or 66 cm long, with the outer tube being ~6.16 cm outer diameter (OD) and 4.32 cm inner diameter (ID) and the inner tube ~3.25 cm OD and 1.22 cm ID. The inner and outer tubes were separated by a metallic locking spacer. In fabrication, the uranium metal and Zircaloy-2 cladding were co-extruded, cut to length, and milled; end caps welded before being annealed (to randomize the uranium metal grains aligned by the extrusion process); and the Zircaloy-2 passivated and leak-tested in steam autoclaves (Schulz 1972; Taylor 2000). In chop/leach processing at the Nuclear Fuel Services plant, the fuel was sheared into ~1-inch (~2.5 cm) lengths under an argon cover gas. In shearing operations, it was found that the outer tube fractured by crushing whereas the inner tube broke cleanly. No details on the design of the shear used at the Nuclear Fuel Services plant were found, but research into the shear design was provided by Watson et al. (1961).

Despite the inert cover gas, several fires occurred in the Nuclear Fuel Services plant during periods of improper shearing operations (Schulz 1972). Early lab tests into the cause of the fires centered on ignition of the thin Zircaloy-2 cladding or of sensitized weld beads in the cladding as the starting point (Schulz 1972). However, later lab investigations instead suggested that ignition of the likewise pyrophoric and highly fractured uranium metal, exacerbated by formation of uranium hydride (UH_3) was a more likely source of the fires (Swanson et al. 1985).

Crush/leach processing had been examined, though not implemented, for N Reactor fuel at the Hanford Site PUREX Plant (Schulz 1972). Later consideration was given to the shear/leach processing of N Reactor fuel at the PUREX Plant (Swanson et al. 1985), and a shear concept with opposing semicircular blades to encompass the fuel almost radially and cut to ~3.8 cm (1.5-inch) lengths was designed. However, the Hanford PUREX Plant was shut down before these plans came to fruition.

2.4 DISSOLVER DESIGN

Dissolutions of irradiated uranium metal and oxide fuels, and of their cladding where applicable, historically have been performed batch-wise. However, a continuous dissolver applied to sheared UO_2 and MOX fuel pins has been developed and implemented at Cap La Hague plant UP3 and UP2-800 in France. Dissolver designs must account for several important process factors: (1) corrosion, (2) dense fuel materials (uranium metal density is ~19 g/cm^3; UO_2 and MOX is ~11 g/cm^3), (3) heating/cooling, heat generation, and handling of water vapors, (4) chemical off-gas, primarily oxides of nitrogen (NO_x) from the reaction of HNO_3 with the fuels, (5) gaseous radioactive fission products (i.e., krypton, xenon, and iodine), (6) criticality safety for enriched uranium or MOX fuels, and (7) handling of solid (undissolved or precipitating) dissolver heels.

The first two considerations drive robust designs using corrosion-resistant materials of construction. The next three considerations require design of gas and vapor handling systems to oxidize and return NO_x vapors to the dissolver and to the plant in the form of HNO_3 (to be addressed in a subsequent section of this chapter), the capture of radioiodine, and the capture or safe release of krypton and xenon. Criticality safety for enriched uranium or MOX fuel is addressed by designing critically dimensioned dissolver vessels, usually in annular or slab geometries, and by the use of neutron-absorbing wall materials. Finally, provision for removal of dissolver heels, as well as for clarification of the product dissolver solution, must be made. The means to address these design issues are illustrated by the following examples taken from the beginning of reprocessing to the present day.[4]

2.4.1 Bismuth Phosphate Process Batch Fuel Dissolver

The world's first reprocessing plant dissolvers were at Hanford's T and B Plants. The dissolver design (Figure 2.2) was of an austenitic 25-12 S-Cb stainless steel (type 309) construction with a heavily sloped bottom to support the 4-metric ton loading of irradiated aluminum-clad natural uranium metal slugs and ~2.4 tons of 60%–62% (12.9–13.5 M) HNO_3. The dissolver outer diameter was ~2.13 m and ~2.09 m tall with ~6,400-L working volume. Because the uranium metal fuel being charged to the dissolver was of natural enrichment, no dimensional provisions for criticality safety were required. Each slug contained about 3.44 kg of uranium metal and 0.12 kg of aluminum cladding and aluminum-silicon bonding agent. The slugs were charged by the dumping of buckets containing up to 105 slugs into the drained tank, thus delivering heavy impact to the tank structures. A cover was placed over the bucket aperture, once the fuel slug charging was complete, to contain gases and

[4] Further general information on dissolver design and operation is presented in Sections 4.4 through 4.6 of Benedict, Pigford, and Levi (1981).

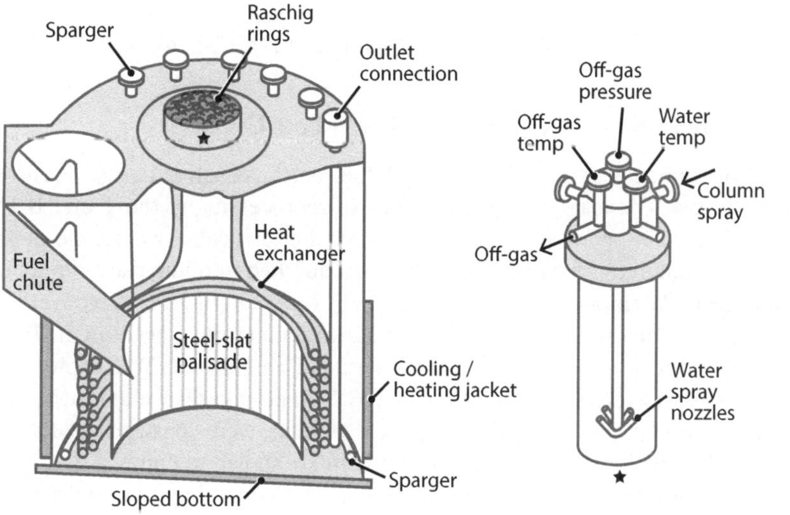

FIGURE 2.2 Hanford Bismuth Phosphate Process B and T Plant dissolver (left) and condenser top (right). (After Figure 2 of DuPont, *Hanford Engineer Works Technical Manual*, General Electric, Hanford Atomic Products Operation, Richland, WA, 1944, HW-10475 ABC, https://pdw.hanford.gov/arpir/pdf.cfm?accession=0076415H.)

vapors and maintain sealing at the imposed ~0.9 atmospheres internal pressure. A 1.60-m diameter circular interior palisade of stainless steel slats kept the slugs in the middle of the tank and away, outside the slats, from the heat exchanger pipe coils used to raise the tank temperature, with steam, to initiate boiling or for cooling water. Once the uranium metal fuel dissolution started, reaction heat maintained boiling conditions. The sloped bottom allowed solution drainage to the outlet collection pipe at the lowest region of the tank. A sparger ring at the tank bottom periphery was used for solution mixing, and an external jacket was used for further heating or cooling.

An instrumented water-cooled condenser, ~61 cm diameter and 3.66 m tall, was present at the top of the dissolver vessel to return vapors and register the off-gas pressure and temperature and the cooling water temperature. The lower part was packed with Raschig rings to collect droplets and condensate. A water spray scrubbed some of the NO_x fumes back into the vessel in the upper water-cooled section of the condenser.[5] Even with the scrubbing, about 60% of the nitrogen in the starting HNO_3 was discharged to the off-gas. The NO_x off-gas, with krypton and xenon fission product gases, were removed by steam jet eduction and air dilution (dilution factor of ~1850) to the ~61-m tall T or B Plant stacks. Radioiodine also was present in the off-gas, more so during the initial dissolution before reflux, little during reflux in the initially more concentrated acid, and increasing as the acid concentration decreased over the course of the batch dissolutions. For personnel safety, dissolver operations

[5] NO_x reactions and equilibria are addressed in a later part of this chapter.

did not occur during atmospheric inversions, fog, or calm air conditions. No clarification of the dissolver solution was noted (DuPont 1944, 154, 403–419).

2.4.2 Late Hanford PUREX Plant Batch Fuel Dissolver

Significant refinements in uranium metal fuel batch dissolver design occurred at Hanford in the ~35 years between the start of reprocessing at the T and B Plants and PUREX Plant operations in the late 1980s. For criticality safety, the dissolver vessel itself required a redesign to handle low-enriched uranium fuels (up to 1.25% uranium-235). Additional refinements were in off-gas handling: (1) to recover HNO_3, which served to dissolve fuel and as a solvent extraction salting agent in the PUREX Process, (2) to recover ammonia from the Zirflex Process, and (3) to capture radioiodine. The dissolvers in the Hanford PUREX Plant, and all wetted pieces in the off-gas train except for the fiberglass particulate filter, were constructed of stainless steel. The dissolver and off-gas handling system are shown in Figure 2.3.

The Bismuth Phosphate Process was based on coprecipitation and contained many centrifugation steps for solution clarification. Therefore, no solution clarification step was necessary following the irradiated fuel dissolution. In contrast, the product solution from the Hanford PUREX Plant dissolver required clarification to remove solids, which could interfere with the subsequent solvent extraction process. Centrifuges were used in the PUREX Plant to remove UF_4 formed in the

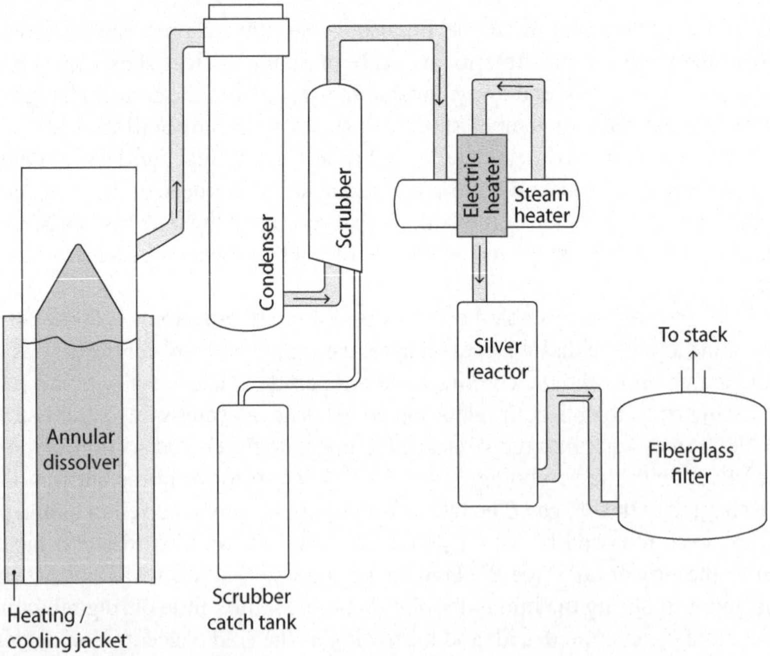

FIGURE 2.3 Hanford PUREX Plant uranium metal fuel dissolver and dissolver off-gas system. (After Figure 3-5 of Moore, J.D. et al., *PUREX Technical Manual*, Rockwell Hanford Operations, Richland, WA, RHO-MA-116, 1980.)

Zirflex Process cladding removal. The UF_4 was metathesized using a potassium hydroxide solution (KOH) to form hydrated UO_2, treated with aluminum nitrate to complex any remaining fluoride, then dissolved in HNO_3, and added to dissolved fuel to feed the solvent extraction process (Sections 3.5 and 4.2 of Moore, Walser, and Fritch 1980).

For criticality safety to allow large process volumes, the Hanford PUREX Plant dissolver vessel was a three-region design having a stainless steel sheathed inner concrete cylinder with a neutron-absorbing cadmium liner as the innermost region, an annular space holding the fuel elements as the second region, and an outer annular region housing heat exchanger coils. The inner 1.6-m diameter cylinder had a conical top to distribute the fuel elements radially when they were dumped from fuel baskets during dissolver loading. An open barrier, of vertical stainless steel bar construction, surrounded the fuel-bearing dissolver annular space (the second dissolver region) at a 30-cm distance. As in the Bismuth Phosphate Process dissolver, this palisade prevented damage to heat exchanger coils lying outside of the fuel dissolution region. The annular third outer region, at another 30-cm spacing, thus lay between the palisade and the outer dissolver wall and contained the heat exchanger coils. The second region, i.e., the annulus holding the fuel elements, had a perforated floor about 23–26 cm above the dissolver bottom. This space housed a (air or steam) sparge ring to provide solution mixing and eliminate the quiescent zones around the fuel, which would otherwise occur at the vessel bottom. The dissolver was topped with a water-sealed flat lid. The water sealing allowed gas release in the case of over pressurization and could hold about 0.02 atmospheres vacuum. The separate cladding and fuel dissolutions occurred under slight vacuum to prevent vapors from entering the process canyon (Section 3.3 of Moore, Walser, and Fritch 1980).

As shown in Figure 2.3, the off-gas from the dissolver vessel entered the dissolver tower, a condenser where the off-gas was directed downward over 511 water-cooled tubes, helically finned to improve heat transfer and prevent channeling. The dissolver tower, about 1.3 m in diameter and nearly 6 m tall, contained a water spray to enhance fume absorption, including radioiodine. The condensate contained HNO_3 from uranium dissolution as well as HNO_3 formed by disproportionation of nitrogen dioxide gas (NO_2) in water. Nitric oxide gas (NO) also formed in uranium dissolution and by NO_2 disproportionation. The NO was oxidized by oxygen provided by controlled air in-leakage and air sparging in the dissolver to form more NO_2. The reactions, described more fully later in this chapter, are summarized below:

$$\begin{gathered} HNO_{3\,\text{vapor}} + H_2O_{\text{vapor}} \rightarrow HNO_{3\,\text{solution condensate}} \\ 3\,NO_{2\,\text{gas}} + H_2O_{\text{liquid}} \rightarrow 2\,HNO_{3\,\text{solution}} + NO_{\text{gas}} \\ 2\,NO_{\text{gas}} + O_{2\,\text{gas}} \rightarrow 2\,NO_{2\,\text{gas}} \end{gathered} \tag{2.3}$$

The HNO_3 condensates with captured radioiodine were returned to the dissolver vessel. The downdraft condenser and provision of atmospheric oxygen greatly enhanced HNO_3 utilization and effectively added atmospheric oxygen as a uranium metal dissolution reagent while decreasing NO_x discharge. Water and ammonia from cladding dissolution likewise were captured, but for disposal as waste, while the off-gas cooling allowed better performance for the ammonia scrubbers located immediately downstream (Section 3.4.1 of Moore, Walser, and Fritch 1980).

The ammonia scrubber was about 76 cm in diameter and 3.66 m tall. The fumes passed upward through a tortuous path afforded by an ~1-m thick bed of stainless steel Pall™ rings and were directed, at the top, through a series of three 7-cm deep weirs containing the fresh scrub water, which percolated downward through the Pall™ ring bed. The ammonia-bearing scrubber solution was concentrated and discharged to the Hanford underground waste tanks, while the scrubbed vapors proceeded to a steam and electrical heater (Sections 3.4.2 and 3.7.2 of Moore, Walser, and Fritch 1980).

The fumes leaving the ammonia scrubber entered steam and electrically-augmented heaters to raise the off-gas temperature to 193°C in preparation for the radioiodine-bearing off-gas encounter with the silver reactor. The silver reactor consisted of two identical sets of unglazed ceramic Berl saddle beds coated in silver nitrate ($AgNO_3$), administered as solution and dried. Iodine reacted with the $AgNO_3$ and was fixed on the Berl saddles. Although the captured radioiodine remained with the Berl saddles, the saddles could be, and were, retreated with the $AgNO_3$ solution to restore their efficacy. The silver reactor was about 1.4 m in diameter and 4.1 m tall. Each bed consisted of an ~86-cm deep layer of ~1.3-cm Berl saddles, an ~60-cm space, and an ~30-cm deep layer of ~2.5-cm Berl saddles (Sections 3.4.3 and 3.4.4 of Moore, Walser, and Fritch 1980).

The treated off-gas then passed through a dual set of fiberglass particle filters, one of which is shown in Figure 2.3 (Section 3.4.5 of Moore, Walser, and Fritch 1980). In a final step before discharge, the filtered off-gas was passed to the so-called backup facility where further scrubbing of NO_x fumes occurred to effect HNO_3 recovery via two nearly identical bubble-cap absorbers, each with eight trays and about 6.4 m tall (Section 3.13.2 of Moore, Walser, and Fritch 1980).

2.4.3 Basket Dissolvers for Fuel Dissolution in Chop/Leach Processing

Both the Hanford Bismuth Phosphate and PUREX Plants accessed the irradiated uranium fuel by dissolving the cladding itself. In modern reprocessing plants, however, cladding is not dissolved but rather broken or sheared in a chop/leach process so that the uranium-bearing fuel can be dissolved in HNO_3 and the leached cladding hulls discarded. Mechanical means to handle the chopped fuel pieces and the resulting cladding hulls often rely on perforated basket designs that allow access of the dissolver solution to the chopped fuel pins but with apertures small enough to retain the cladding pieces after irradiated fuel dissolution. Alternative leach vessel designs use perforated helical screws to advance the sheared fuel pins through cylindrical dissolver vessels, which themselves can be perforated to allow drainage of the dissolver solution and hull rinses. Basket perforations could be circular or slotted with apertures ranging from ~0.38 to 1.6 mm or grid (screen cloths) with ~0.06 mm pores. Groenier (1971) prepared an excellent overview of alternative fuel dissolver designs based on the shear/leach method as well as designs featuring cladding dissolution.

Because the modern reprocessing plants use the PUREX process, the fuel dissolver designs using baskets to hold the chopped fuel pins still must address the other factors that the Hanford PUREX Plant addressed—HNO_3 recovery and economy, radioiodine capture, and removal of solids from the dissolver solution before solvent extraction—as well as have geometries favorable to criticality safety. Aspects of the dissolver design for the Nuclear Fuel Services reprocessing plant are shown in Figure 2.4. The fuel baskets were up to ~20 cm in diameter, ~2.1 m tall, and were placed in vertical circular shafts about 25 cm in diameter. Six such vertical shafts were arrayed radially at the periphery of the ~2.44-m ID and 2.59-m OD dissolver vessel leaving an annulus of ~7.6 cm thick. The bottom of the annular space had an air-driven sparger ring to maintain mixing of the fuel particle solids and thus enhance their dissolution. For further criticality safety, the inner space in the dissolver was filled with borated concrete. The baskets were perforated but had an acid-consumable liner of mild steel to retain fines until loading into the dissolver. The baskets and the contained fuel hulls were rinsed by a spray ring after fuel dissolution and removal of the dissolver solution.

Another batch dissolver design featuring sheared fuel baskets is depicted on the left side of Figure 2.5. This design was adopted by the UP2-400 plant at Cap La Hague, France (Figure 34 of Vaudano 2008a). The sheared fuel pins fall into the perforated basket(s) arrayed around a central annular dissolver solution collection vessel. The annular dissolver solution tank is vented for off-gas treatment; i.e., recovery of HNO_3 by oxidation and scrubbing of NO_x fumes and radioiodine removal. A simultaneous thermosiphon and gas-lift loop is imposed by the irradiated UO_2 dissolution reaction, augmented by an external superheated water jacket. At completion of the dissolution reaction, the basket containing the leached cladding hulls is retrieved and the hulls dumped and rinsed before discard. Similar concepts but featuring circular leaching vessels and slab dissolver solution collection tanks also have been designed (bottom of Figure 2.5, after Groenier 1971).

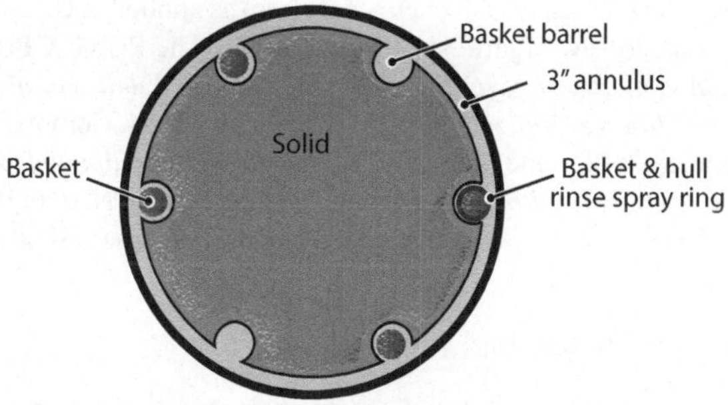

FIGURE 2.4 Nuclear Fuel Services, West Valley, NY, dissolver cross section. (After Figure 15 of Groenier, W.S., *Equipment for the Dissolution of Core Materials from Sheared Power Reactor Fuels*, Oak Ridge National Laboratory, Oak Ridge, TN, ORNL-TM-3194, 1971, https://www.osti.gov/servlets/purl/4008629.)

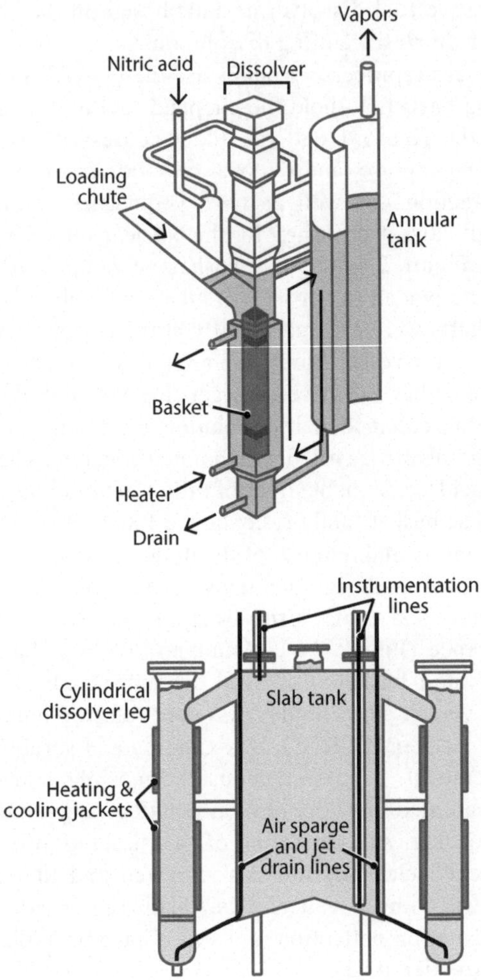

FIGURE 2.5 Batch shear/leach dissolver concepts (top) as applied at UP2-400 and (bottom) by ORNL. (After, respectively, Figure 34 of Vaudano, A., The PUREX Process: Head-End Operations, in *Treatment and Recycling of Spent Nuclear Fuel: Actinide Partitioning–Application to Waste Management*, edited by M. Lecomte, 45–54, Commissariat à l'Énergie Atomique, Paris, France, 2008a; and Figure 3 of Groenier, W.S., *Equipment for the Dissolution of Core Materials from Sheared Power Reactor Fuels*, Oak Ridge National Laboratory, Oak Ridge, TN, ORNL-TM-3194, 1971, https://www.osti.gov/servlets/purl/4008629.)

2.4.4 Continuous Dissolvers

Continuous dissolvers represent the highest evolution of dissolver design and feature continuous or near-continuous feed of sheared fuel, simultaneous HNO_3 addition, and constant overflow removal of dissolver solution. A continuous dissolver design eliminates off-gas surging, as occurs in batch dissolvers, thus simplifying off-gas handling design and diminishing off-gas handling rate requirements. The major

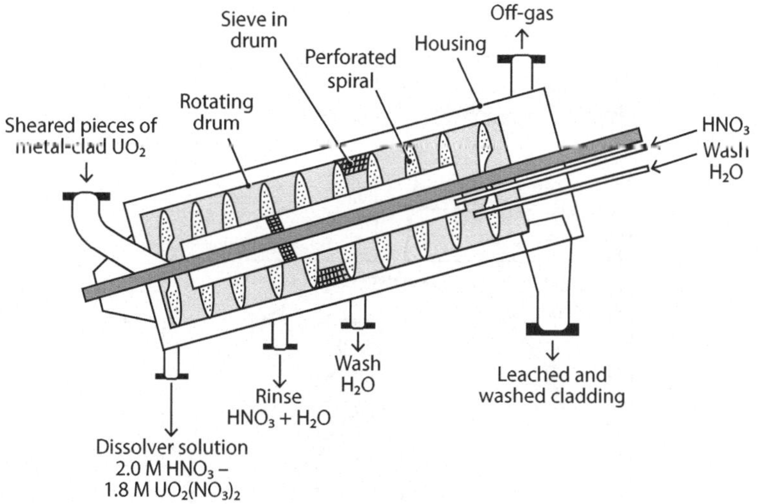

FIGURE 2.6 Continuous helical dissolver concept. (After Figure 13 of Groenier, W.S., *Equipment for the Dissolution of Core Materials from Sheared Power Reactor Fuels*, Oak Ridge National Laboratory, Oak Ridge, TN, ORNL-TM-3194, 1971, https://www.osti.gov/servlets/purl/4008629.)

issue in continuous dissolver design lies in devising ways to remove the cladding hulls and other insoluble debris.

The inclined helical rotary spiral continuous dissolver design shown in Figure 2.6 is one such concept. The sheared fuel is fed to the bottom of the dissolver and HNO_3 reagent fed to the middle. The helical screw, within a rotating drum, is sieved and advances the fuel pieces up the dissolver vessel, countercurrent to the descending HNO_3 solution. This means that the last vestiges of irradiated UO_2 remaining in the cladding encounters the solution highest in acid strength and, importantly, nitrate concentration, thus giving the highest dissolution rate. Further up the rotating drum, the leached hulls encounter water rinses as they tumble, washing away the remaining uranium-bearing solution. The outer drum is dammed at two locations, directing the leaches and water rinses. The leached and rinsed hulls are advanced to the top of the dissolver where they drop out through a bottom chute. The NO_x off-gas with the other fumes, including radioiodine, exits the top of the dissolver, while the dissolver overflow solution drains from the dissolver bottom. This design underwent extensive testing, which revealed two major design weaknesses: inadequate heat transfer to the inner drum where the fuel resides and inadequate space for gas disengagement that could lead to foaming and thus solution transfer to the off-gas system.

Continuous irradiated fuel dissolution has been implemented commercially in the UP3 and UP2-800 reprocessing plants at Cap La Hague in France (Vaudano 2008a, 49–50). The rotating wheel dissolver design was examined as early as the 1950s (Figure 8.9 of Culler Jr. et al. 1957) and has undergone dedicated development and refinement in France. The dissolver as currently implemented at Cap La Hague consists of a 12-compartment wheel rotating half-immersed in the

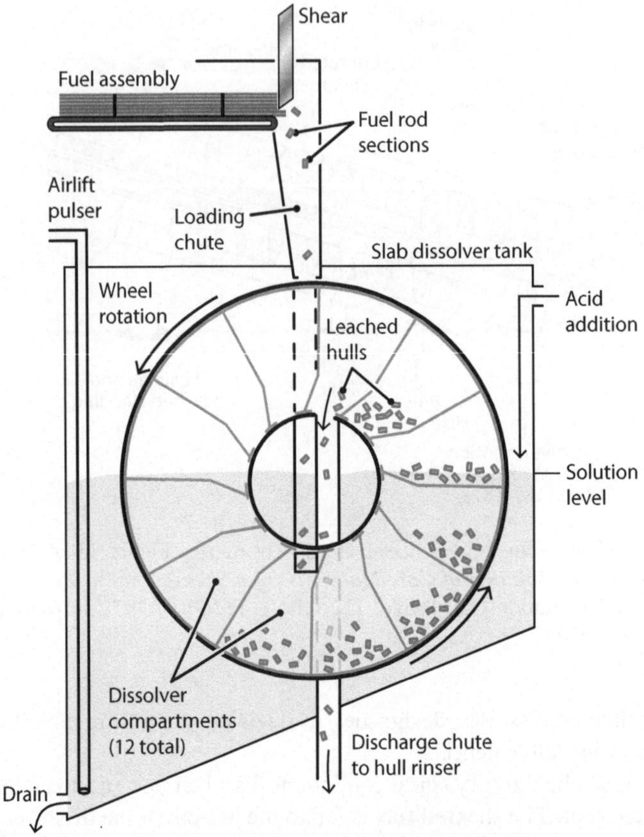

FIGURE 2.7 Continuous rotating wheel dissolver. (After Figure 35 of Vaudano, A., The PUREX Process: Head-End Operations, in *Treatment and Recycling of Spent Nuclear Fuel: Actinide Partitioning–Application to Waste Management*, edited by M. Lecomte, 45–54, Commissariat à l'Énergie Atomique, Paris, France, 2008a.)

dissolvent (Figure 2.7) within a critically safe heated slab tank. Fresh HNO_3 solution is fed continuously as dissolver solution overflows for feed adjustment and solvent extraction. The wheel is rotated in a stepwise fashion, at one-twelfth of a rotation between paused steps. At each step, sheared fuel pieces enter one compartment while leached cladding hulls are discharged into a chute from another compartment. The drum and compartments are constructed of stainless steel, and the heated slab dissolver tank constructed of zirconium. The rotating wheel is supported by roller bearings and driven by an external toothed cog (see also Vaudano 2008a, Figure 35) with speed adjusted to assure complete dissolution of the irradiated uranium.

The leached hulls, once discharged from the rotating wheel dissolver, are sent to a vessel of helical design, perhaps similar to that shown in Figure 2.6, which tumbles the cladding hulls under the rinse solution and allows complete removal of the dissolver solutions.

The modern French reprocessing plants implement NO_x capture and oxidation to minimize undue loss of HNO_3 reagent (Vaudano 2008a, 52).

2.5 MODERN RADIOIODINE REMOVAL

Radioiodine can manifest in multiple chemical forms—as hydriodic acid (HI), hypoiodous acid (HIO), iodic acid (HIO_3), or elemental iodine. Iodine carried over to the subsequent solvent extraction processes can react with the organic phases and compromise their performance. Radioiodine present in the irradiated fuel is desorbed from the dissolver solution effectively by steam stripping through partial distillation of the dissolver solution (Section 4.6 of Benedict, Pigford, and Levi 1981). The Cap La Hague plants remove >97% of the radioiodine by this method. The radioiodine then is scrubbed from the off-gas, downstream of the NO_x recycle, by absorption into sodium hydroxide (NaOH) solution and discharge to the sea. Absorption of radioiodine onto $AgNO_3$-impregated ceramics provides further iodine removal both in the dissolver off-gas and in other plant ventilations, leaving less than 0.5% of the original radioiodine lost to the atmosphere (Vaudano 2008a, 52).

2.6 UNDISSOLVED RESIDUES

Undissolved and precipitated residues remain after the dissolution step, particularly for high-burnup fuels containing high fission product concentrations. These residues include cladding shards that are too fine to be retained in the baskets, zirconium/molybdenum (Zr/Mo) fission product compounds, and metallic fission product inclusions (the epsilon phase). Low solubility plutonium dioxide, PuO_2, residues also may be present from MOX fuels as are impurities from the fuel itself (Vaudano 2008a, 47). The Zr/Mo fission product compound nominal formula is $Mo_2Zr_{(1-\alpha)}Pu_\alpha O_7(OH)_2 \cdot 2H_2O$ and can deposit on equipment surfaces as cruds (Vaudano 2008a, 50–52). The fission product epsilon phase (or ε-metal), also known as "white inclusions," "the five metal phase," or the "noble metal phase" of molybdenum, palladium, rhodium, ruthenium, and technetium, collect within the irradiated fuel matrix with sub-μm dimension (McNamara et al. 2014). From 3 to 5 kg of insoluble residue per metric ton of light water reactor fuel irradiated to 33,000 megawatt-day per metric ton of heavy metal (MTHM) ordinarily is collected during feed clarification (Section 3.2.1.2 of OECD 2005). The feed can be clarified by filtration but, for process reliability, is done by centrifugation (Vaudano 2008a, 53–54).

2.7 URANIUM METAL AND URANIUM DIOXIDE DISSOLUTION REACTION MECHANISMS AND STOICHIOMETRIES

The dissolutions of uranium metal and UO_2 in HNO_3 are oxidation/reduction reactions in which the uranium metal and UO_2 are oxidized to uranyl nitrate [$UO_2(NO_3)_2$] and the HNO_3 is reduced to gaseous products, primarily nitric oxide (NO) and nitrogen dioxide (NO_2). These two gases, collectively dubbed NO_x, constitute, in varying ratios, the primary HNO_3 reduction products. In one study for uranium metal, about 0.03 moles of elemental nitrogen (N_2) per mole of U also arise from reduction of the HNO_3 with trace quantities of nitrous oxide (N_2O; about 0.0004 moles per mole of U) and hydrogen (H_2; about 0.0002 moles per mole U) (Blanco and Watson 1961, 60–64). Small quantities of carbon monoxide (CO) and carbon dioxide (CO_2) evolve from the reaction of carbon present largely as uranium carbide (UC) inclusions in metallic fuel.[6]

Recent reviews of UO_2 (Marc et al. 2017) and actinide dioxide (UO_2, PuO_2, MOX, and thorium dioxide, ThO_2) (Desigan et al. 2018) dissolutions in HNO_3 have been published. These reviews may be consulted to provide greater levels of detail and additional references.

The plutonium and fission products that are produced by irradiation constitute less than 1% of the irradiated uranium for fuel used in weapons plutonium production. Even light-water reactor fuel with 33,000 megawatt-days per metric ton irradiation contains only ~3 wt% fission products (discounted for volatile krypton, xenon, and iodine) and ~1 wt% reactor-produced non-uranium actinide isotopes of which ~0.9 wt% is plutonium (Table 8.7 of Benedict, Pigford, and Levi 1981). For simplicity, the fission product and non-uranium reactor-produced actinide isotope corrosion reactions are neglected in the following discussion on reaction stoichiometry.

Oxidative reactions involving HNO_3 are notorious for their induction times and sigmoidal reaction rates (negligible initial rates that then increase with time before tapering off as the reactants are consumed). The induction times and sigmoidal reaction rates are characteristic of autocatalytic reactions in which a reaction product also participates in an earlier reaction step. Both uranium metal and UO_2 react initially with HNO_3 to produce nitrous acid, HNO_2. Nitrous acid then acts in an autocatalytic manner to increase reaction rate as its concentration in solution increases. The role of autocatalysis is attested by the sigmoidal shape of the reaction kinetics curve, particularly at lower temperatures and discernably moderate rates, and the presence of induction times wherein initially low and then increasing rates are observed (see, for example, Lacher et al. 1961 for uranium metal; Shabbir and Robins 1969 and Figure 8 of Herrmann et al. 1984 for UO_2). The oxidative dissolution reaction of UO_2 with HNO_3 to form UO_2^{2+} occurs at the solid/liquid interface during the induction time (Cordara et al. 2017). The kinetics has been investigated at the necessary mild conditions of 0.1–4 M HNO_3 and 22°C–60°C with the transition to the autocatalytic regime found at $(1–1.6) \times 10^{-4}$ M U(VI) or, equivalently by stoichiometry, $(1–1.6) \times 10^{-4}$ M HNO_2.

The participation of HNO_2 in an autocatalytic mechanism is shown by the decreased, and even nullified, uranium metal reaction rate when HNO_2 is scavenged by urea (Lacher et al. 1961) or by hydrazine (Section 3.2.6 of Swanson et al. 1985) and by an increased rate when nitrite salt is added (Lacher et al. 1961). Uranium metal dissolution rates also decrease with agitation and as solutions reach boiling

[6] Evolution of carbon as CO_2 but with up to 50% of the carbon remaining in solution as organic acids (e.g., formic, acetic, oxalic) and aromatic compounds with phenolic and carboxylic acid functional groups (e.g., mellitic acid; Glatz, Bokelund, and Zierfuss (1990)) is reported for uranium and plutonium carbide fuels (page 99 of Masson et al. 2008).

temperatures. Both factors lower HNO_2 concentration at the metal surface by dispersal while boiling also decreases the rate by stripping HNO_2 to the vapor phase.

The role of HNO_2 in UO_2 dissolution is similar to that in uranium metal dissolution. Thus, the addition of sodium azide (NaN_3, a potent HNO_2 scavenger) to the HNO_3 solution lengthens the induction period for UO_2 dissolution by over an order of magnitude, from 1 or 2 minutes to 30 minutes (Herrmann et al. 1984). The hydrazine addition likewise stunts the UO_2 dissolution reaction by scavenging HNO_2 (Figures 2.6 and 2.7 of Desigan et al. 2017). As was observed for uranium metal dissolution (Lacher et al. 1961), the addition of (sodium) nitrite salt ($NaNO_2$) accelerates the UO_2 dissolution reaction (Shabbir and Robins 1969). Confirmatory evidence of the role of HNO_2 is given by the presence of a green color adjacent to the dissolving UO_2 because of the mixed color of the yellow hexavalent uranium, U(VI), and blue dissolved N_2O_3 (arising from condensation of $2HNO_2 \rightarrow H_2O + N_2O_3$) and not U(IV), which could not be detected in solution (Herrmann et al. 1984). Direct measurements of HNO_2 concentrations during UO_2 dissolution in HNO_3 have been made by solution sampling, rapid quenching in excess NaOH solution, and analysis of nitrite and nitrate ion concentrations (Desigan et al. 2017). The bulk solution HNO_2 concentrations reached as high as 0.35 M in tests starting with 8 M HNO_3 and run at 80°C. Concentrations near the dissolving UO_2 would be expected to be greater.

In this autocatalytic cycle, the HNO_2 formed by the reduction of HNO_3 reacts with more UO_2 to produce NO (Desigan et al. 2017):

$$UO_2 + 2\,HNO_2 + 2\,H^+ \rightarrow UO_2^{\,2+} + 2\,NO + 2\,H_2O \tag{2.4}$$

Presumably, uranium metal follows a similar cycle. The HNO_2 is unstable in acid and disproportionates to form NO and NO_2 (Desigan et al. 2017):

$$2\,HNO_2 \rightarrow NO + NO_2 + H_2O \tag{2.5}$$

The suitability of autocatalysis to describe the reactions of HNO_2 in uranium metal and UO_2 dissolution has been questioned, largely on the grounds that mechanisms for HNO_2 production from reaction products have not been identified or demonstrated (Sections 4.2 and 4.3 of Marc et al. 2017). Nevertheless, it is clear that rapid uranium metal or UO_2 dissolution rates are not achieved until HNO_2 reaches some critical concentration.

The relative concentrations of NO and NO_2 reporting to the off-gas, however, are not established by the uranium metal or UO_2 but rather by their interactions with HNO_3 as functions of the HNO_3 concentration and temperature. This has been demonstrated by careful sampling of off-gas just above UO_2 dissolution test solutions, whereby the relative NO_2 and NO partial pressures were shown to agree with results obtained for "pure" NO_2, NO, and HNO_3 aqueous solution systems containing no other components (Herrmann 1984).

$$3\,NO_{2\,gas} + H_2O_{liquid} \leftrightarrow 2\,HNO_{3\,solution} + NO_{gas};\ \Delta H = -73\ kJ/mole \tag{2.6}$$

As the above reaction shows, the dissolver solution essentially acts as the primary fume scrubber with the equilibrium distribution of NO and NO_2 gases determined by the system's HNO_3 aqueous solution concentration and temperature. The above equilibrium shows that as the NO and NO_2 gas pressures decrease (e.g., by dilution with air at the completion of the dissolution reaction), the fraction of NO_2 within the NO_x increases steeply. The equilibrium also shows that NO_2 is favored at higher HNO_3 concentration and at higher temperature. Because the equilibrium in NO, NO_2, H_2O, and HNO_3 is independent of the source of the NO and NO_2, the same conclusions should apply to uranium metal dissolution.

Nitrous oxide, N_2O, also is observed in dissolutions of uranium metal and UO_2. For uranium metal dissolution, the fraction of N_2O in the off-gas is low, ~0.0004 or 0.0005 moles per mole uranium (Blanco and Watson 1961, 60–64; Wymer and Foster 1956, respectively). In contrast, significant N_2O quantities were found in the off-gas from UO_2 dissolution (Herrmann et al. 1984). In this work, the amount of UO_2 dissolution is accounted by N_2O production by way of the reaction

$$UO_2 + 3\,HNO_3 \rightarrow UO_2(NO_3)_2 + 0.5\,N_2O + 1.5\,H_2O \tag{2.7}$$

and is increased from ~0% to ~16% as the HNO_3 concentration increases from 4.1 M to 12.4 M at 90°C; higher temperatures also increased N_2O. Later tests at the same institute failed to find N_2O in the off-gas from UO_2 dissolution even though N_2O was found in dissolving uranium carbides, (U, Pu) carbide, and (U, Pu)O_2 (Glatz, Bokelund, and Zierfuss 1990, Table 2.1). A recent review of the technical literature

TABLE 2.1
Enthalpies of Uranium Metal and UO_2 Dissolutions in Nitric Acid

Reactions[a]			ΔH_{rxn}, kJ/mole U[b]
Uranium Metal			
$U + 4\,HNO_3$	→	$UO_2(NO_3)_2 + 2\,H_2O + 2\,NO$	−1069.9
$U + 8\,HNO_3$	→	$UO_2(NO_3)_2 + 4\,H_2O + 6\,NO_2$	−872.5
$U + 5.5\,HNO_3$	→	$UO_2(NO_3)_2 + 2.75\,H_2O + 1.25\,NO + 2.25\,NO_2$	−995.9
$U + 4\,HNO_3 + O_2$	→	$UO_2(NO_3)_2 + 2\,H_2O + 2\,NO_2$	−1184.1
$U + 2\,HNO_3 + 1.5O_2$	→	$UO_2(NO_3)_2 + H_2O$	−1339.9
UO_2			
$UO_2 + 2.667\,HNO_3$	→	$UO_2(NO_3)_2 + 1.333\,H_2O + 0.667\,NO$	−165.0
$UO_2 + 4\,HNO_3$	→	$UO_2(NO_3)_2 + 2\,H_2O + 2\,NO_2$	−99.2
$UO_2 + 3.1667\,HNO_3$	→	$UO_2(NO_3)_2 + 1.5833\,H_2O + 0.4127\,NO + 0.75\,NO_2$	−140.7
$UO_2 + 2.667\,HNO_3 + 0.333\,O_2$	→	$UO_2(NO_3)_2 + 1.333\,H_2O + 0.667\,NO_2$	−203.0
$UO_2 + 2\,HNO_3 + 0.5\,O_2$	→	$UO_2(NO_3)_2 + H_2O$	−255.0

[a] The U and UO_2 are present as solids. The HNO_3 is present as 50 wt% solution (~10.3 M). The $UO_2(NO_3)_2$ is present as 1.0 moles of $UO_2(NO_3)_2$ with 1.6 moles of HNO_3 and 50 moles of H_2O or ~1 M $UO_2(NO_3)_2$ in ~1.6 M HNO_3. The H_2O is present as liquid. The O_2, NO, and NO_2 are present as gases at 1 atmosphere.

[b] Calculated based on the thermodynamic value compilation of Wagman et al. (1982).

regarding UO_2 dissolution likewise failed to find any other studies identifying N_2O even though a similar method, infrared spectrometry, was used in the separate studies (Marc et al. 2017). Based on these other studies, and the very low N_2O quantities found in uranium metal dissolution in HNO_3, it is likely that N_2O formation from UO_2 dissolution is, in fact, negligible.

Besides NO_2 (nitrogen oxidation state 4), HNO_2 (oxidation state 3), NO (oxidation state 2), and N_2O (oxidation state 1), other reduced compounds of HNO_3 (oxidation state 5) can be postulated. These additional chemically reduced nitrogen compounds in the off-gas or in the solution might include diatomic nitrogen gas, N_2 (oxidation state 0); hydroxylamine, NH_2OH (oxidation state −1); hydrazine, N_2H_4 (oxidation state −2); and ammonia NH_3 (the most reduced oxidation state for nitrogen, oxidation state −3). However, hydroxylamine, hydrazine, and ammonia are not stable in the presence of HNO_2 under the given reaction conditions. Despite careful search with sensitive analytical methods, neither N_2 nor NH_3 could be detected (Herrmann et al. 1984). Some nitrogen (0.043 moles/mole of uranium) has been observed in dissolution of uranium metal in HNO_3 (Wymer and Foster 1956).

Therefore, NO and NO_2 may be considered the overwhelmingly dominant products of the dissolution of uranium metal and UO_2 in HNO_3 solution.

2.8 ACID CONSUMPTION IN THE URANIUM METAL AND URANIUM DIOXIDE DISSOLUTIONS

Acid consumption varies broadly according to the oxidation/reduction reactions to produce NO and NO_2:

$$U + 4\,HNO_3 \rightarrow UO_2(NO_3)_2 + 2\,H_2O + 2\,NO \tag{2.8}$$

$$UO_2 + 2.667\,HNO_3 \rightarrow UO_2(NO_3)_2 + 1.333\,H_2O + 0.667\,NO \tag{2.9}$$

$$U + 8\,HNO_3 \rightarrow UO_2(NO_3)_2 + 4\,H_2O + 6\,NO_2 \tag{2.10}$$

$$UO_2 + 4\,HNO_3 \rightarrow UO_2(NO_3)_2 + 2\,H_2O + 2\,NO_2 \tag{2.11}$$

It has been observed that higher total nitrate concentrations, higher temperatures, and dilution of the NO_2 and NO off-gas favor the second, less efficient, reactions to produce NO_2. In general, acid concentrations above ~8 M HNO_3 highly favor NO_2; acid concentrations below 8 M HNO_3 favor NO. As such, NO_2 is found in high concentrations earlier in batch dissolver runs when acid concentrations are high.

According to Wymer and Foster (1956), the stoichiometry of uranium metal batch dissolution in 55% (11.7 M) HNO_3 is:

$$\begin{gathered} U + 4.5\,HNO_3 \rightarrow \\ UO_2(NO_3)_2 + 2.25\,H_2O + 1.57\,NO + 0.84\,NO_2 + 0.0005\,N_2O + 0.043\,N_2 \end{gathered} \tag{2.12}$$

In studies using an updraft condenser similar to that used in the Bismuth Phosphate Process T and B Plants at Hanford, a near average of the individual reactions to produce NO and NO_2 occurs (Blanco and Watson 1961). The reaction has the following much less efficient stoichiometry:

$$U + 5.5\,HNO_3 \rightarrow UO_2(NO_3)_2 + 2.75\,H_2O + 1.25\,NO + 2.25\,NO_2 \tag{2.13}$$

The reaction stoichiometry also can be influenced by the oxidation of the NO vapors by atmospheric oxygen (O_2) in the condensing regions of the dissolver and by the capture and disproportionation of NO_2 in condensing water to produce HNO_3 and NO. The relevant reactions are:

$$2\,NO + O_2 \rightarrow 2\,NO_2 \tag{2.14}$$

$$3\,NO_2 + H_2O \leftrightarrow 2\,HNO_3 + NO \tag{2.15}$$

These equilibria are exploited in later reprocessing plants to conserve HNO_3. In the extreme, the so-called "fumeless dissolution" can take place in which virtually all of the NO_x off-gas is captured and converted to HNO_3. The NO fumes first react with oxygen in an air downdraft to oxidize NO to NO_2. The water-rich condensate scrubs the NO_2 rising from both the oxidation of uranium metal and the oxidation of NO to form HNO_3 and NO, as shown in the equilibration reactions 2.14 and 2.15. The product NO from these equilibria, in turn, is oxidized by atmospheric O_2. The condensate is returned to the dissolver vessel, and the residual fugitive vapors are removed by downstream scrubbers. Overall, in fumeless dissolution, the uranium metal or UO_2 can be oxidized entirely by atmospheric O_2, as mediated by nitrate, with HNO_3 consumption limited to that required to form $UO_2(NO_3)_2$ according to the following net reactions:

$$U + 2\,HNO_3 + 1.5\,O_2 \rightarrow UO_2(NO_3)_2 + H_2O \tag{2.16}$$

$$UO_2 + 2\,HNO_3 + 0.5\,O_2 \rightarrow UO_2(NO_3)_2 + H_2O \tag{2.17}$$

Such efficiencies may not be achieved in practice for the dissolver alone. For example, in the Hanford Site PUREX Plant, the observed HNO_3 consumption is 3.71–3.75 moles per mole of uranium metal after the downdraft condenser but before the follow-on acid recovery steps, which cycle the acid to other areas of the plant (Allen, Jacobs, and Reberger 1982; Blanco and Watson 1961). With the downdraft scrubber functioning in Hanford's PUREX Plant, the NO:NO_2 ratio in the off-gas before the subsequent acid recovery steps is ~9:1. Therefore, the net reaction in the Hanford PUREX Plant dissolver is:

$$U + 3.72\,HNO_3 + 0.295\,O_2 \rightarrow UO_2(NO_3)_2 + 1.86\,H_2O + 1.55\,NO + 0.17\,NO_2 \tag{2.18}$$

wherein atmospheric O_2 provides just under 20% of the required oxidant. Fumeless dissolution evidently is employed in the dissolution of metallic uranium Magnox fuel at Sellafield (Shaw 1990). The dissolution is conducted continuously (overflow) at nominally 3 M HNO_3 and 300 g U/L (about 1.3 M) uranyl nitrate to give about 5.6 M total nitrate.

2.9 HEAT GENERATION BY URANIUM METAL AND URANIUM DIOXIDE DISSOLUTIONS IN NITRIC ACID

The reaction of uranium metal with HNO_3 is highly exothermic (heat-producing) with the enthalpy (heat) of the dissolution reaction ($\Delta H_{rxn.}$) depending on the reaction stoichiometry and products (Table 2.1). Once the dissolver charge is heated, the ensuing evolved reaction heat often allows reprocessing plant dissolvers to operate at boiling, provided sufficient initial acid concentrations and uranium metal surface areas are exposed for reaction. The enthalpies of the corresponding UO_2 dissolution reactions are much lower than those for uranium metal, reflecting the fact that UO_2 is more oxidized (oxidation state four) than the metallic uranium (oxidation state zero). It is seen that the respective uranium metal and UO_2 reactions involving O_2 as an oxidant provide greater uranium-specific heat compared with those using HNO_3 exclusively or partially as oxidant. This supplemental heat, however, is not deposited in the dissolver itself but rather in the off-gas handling system where the O_2 encounters the NO fumes. Additionally, the uranium-specific enthalpies for uranium metal compared with UO_2 are from ~5 times greater in the case of "fumeless" dissolution, in which uranium oxidation occurs by atmospheric oxygen, to ~9 times greater for the metal dissolution than for the corresponding UO_2 dissolution to form the NO_2 HNO_3 reduction product.

The higher uranium-specific enthalpies provided by metal dissolution compared with UO_2 dissolution are at least partially overcome by the higher specific surface area of the irradiated UO_2 fuel pellets compared with the irradiated uranium metal fuel slugs to provide sufficient heating rates to maintain boiling.

2.10 URANIUM METAL AND URANIUM DIOXIDE DISSOLUTION RATES

Factors that may affect the uranium metal and UO_2 dissolution reaction rates are temperature, acid concentration, total nitrate (NO_3^-) concentration, which arises from HNO_3, the product uranyl nitrate ($UO_2(NO_3)_2$) and other nitrate salts present in the dissolver, stirring or agitation rate, catalysts or additives, fuel grain structure/orientation, surface roughness, and fuel irradiation. The effects of these factors on the dissolution rates are examined.

2.10.1 Temperature Effects

The rate of uranium metal dissolution in HNO_3 generally increases with temperature (Lacher, Salzman, and Park 1961; Swanson et al. 1985). The activation energy for uranium dissolution is 51.0 kJ/mole in 8.1 M HNO_3 and 66.5 kJ/mole in 15.6 M HNO_3 in tests run between 0°C and 50°C (Lacher, Salzman, and Park 1961). In tests conducted in 7.8 M HNO_3 between about 71°C and 105°C, the activation energy was found to be about 46.0 kJ/mole per mole (Swanson et al. 1985). Based on the Arrhenius equation, this means that the rate increases about a factor of 2 for each 15°C increase in temperature between 71°C and 105°C. However, as temperatures approach boiling, the rate decreases. As already noted, the decrease likely is because the HNO_2 required to sustain the reaction is swept from the solution by the water and HNO_3 evaporating during boiling. As a result, the uranium metal dissolution rate in boiling (112°C) 7.8 M HNO_3 is the same as that at 100°C (Swanson et al. 1985).

As noted in a recent review of studies of UO_2 dissolution in HNO_3 (Marc et al. 2017), temperature increase leads to increased rates, up to about 90°C–95°C, but that the rates decrease as temperature approaches boiling. These findings, therefore, are similar to those observed for uranium metal. The same review summarized nine studies of UO_2 dissolution in HNO_3 that showed activation energies ranging from 8.3–97.5 kJ/mole. The review authors attributed the wide range to the variety of test conditions (acid concentrations, temperatures) examined and changes over those ranges in the reaction mechanisms and contributing factors such as gas solubilities and boiling. In general, higher activation energies corresponded to greater HNO_3 concentrations.

Because of the high heat production of the dissolution reactions, the desire to decrease process times, and the difficulty of maintaining temperatures just below boiling (to obtain the higher rates), most if not all plant-scale uranium metal and even many UO_2 dissolvers operate at boiling or near-boiling (>90°C; Vaudano 2008a, 47). Therefore, for the purposes of the present review, further considerations of uranium metal and UO_2 dissolution in HNO_3 largely are confined to boiling temperatures.

2.10.2 Nitric Acid and Nitrate Concentration Effects

Uranium metal and UO_2 dissolution rates in HNO_3 may be denominated in various ways according to the experimental approach. For example, the weight loss of a uranium metal coupon after removal of the corrosion layer or the change in metal coupon size might be determined as functions of time to derive dissolution rates. Because uranium metal and UO_2 corrosion in HNO_3 generally are found to be uniform (i.e., occur uniformly over the entire exposed surface and, while surface roughening occurs, show relatively low pitting or lamellar corrosion that would drastically increase surface area), corrosion rates are expressed in units dependent on the geometric surface area (i.e., weight loss per unit area per unit time). For example, dissolution rates may be expressed as (weight loss, mg)/(cm^2-hour). The generally isotropic corrosion means that the dissolution rate also may be expressed as a linear penetration. Ignoring microscopic surface roughness, this is done by dividing the geometric surface area-dependent rates by the uranium metal fuel density (18.7–19.1 g/cm^3) or UO_2 density (10.97 g/cm^3 particle density; powder compacts and sintered fuel pellets generally attain ~94% theoretical density, or ~10.3 g/cm^3; Uriarte and Rainey (1965)).

For example, for uranium metal, a mass loss per unit area rate of 1,000 mg/cm²-hour may be converted to linear penetration rate, in µm/hour, by the following calculation:

$$\frac{1{,}000 \text{ mg U}}{\text{cm}^2\text{-hr}} \times \frac{\text{g U}}{10^3 \text{ mg U}} \times \frac{\text{cm}^3}{19.1 \text{ g U}} \times \frac{10^4 \mu\text{m}}{\text{cm}} = \frac{525 \mu\text{m U}}{\text{hour}} \tag{2.19}$$

Expressions in terms of linear penetration rates allow times to complete dissolution of uranium metal or UO_2 articles to be determined readily. For example, a 3.39-cm diameter Hanford Site uranium metal slug, as used to fuel the single-pass plutonium production reactors (B, C, D, DR, F, H, KE, and KW) corroding at 525 µm/hour, would be completely dissolved after (16,900 µm radius/525 µm/hour ≅) 32 hours.

Tests have shown that, at non-boiling temperatures, uranium metal dissolution rate in HNO_3 increases with the square of the HNO_3 concentration [HNO_3] (Swanson et al. 1985). Review of the technical literature for UO_2 reaction with HNO_3 shows the dissolution rate increases with $[HNO_3]^n$, where n ranges from 1 to 3.3 with most values around 2–2.3 (Marc et al. 2017). However, numerous studies have shown the more useful relation that, provided the solution remains acidic, the dissolution rates for both uranium metal and UO_2 depend on the total nitrate concentration. The total nitrate concentration includes the nitrate ion contributions from nitric acid ($UO_2(NO_3)_2$) produced during dissolution of the uranium metal or UO_2, and other nitrate salts, e.g., $Al(NO_3)_3$, which may be present in the dissolver solution.

Uranium metal dissolution rate data (138 points) from thirteen different studies for boiling HNO_3 solution systems, some of which also contained $UO_2(NO_3)_2$ and/or $Al(NO_3)_3$ including three tests having $Al(NO_3)_3$ only and no "free" HNO_3 (Hyde et al. 1944), are presented in Figure 2.8. The studied uranium metals were low alloys used or considered for production reactors. Nearly all of the data were for unirradiated uranium, but five tests for irradiated specimens (from Swanson 1988) dissolved at rates consistent with their unirradiated counterparts.

The data from the various studies for uranium metal are seen to be remarkably consistent considering the variety of uranium materials, solution compositions, and experimental techniques used. The observed rates generally range over a factor of two or less for a given nitrate concentration. As seen in Figure 2.8, the rates were mathematically fit and found to be proportional to the ~2.5 power of the nitrate concentration:

$$\text{Uranium metal dissolution rate, } \mu\text{m/hour at boiling} = 1.702 \left[\text{Nitrate, in moles per liter}\right]^{2.457} \tag{2.20}$$

Part of the cause for the increased rate with increasing nitrate concentration can be attributed to the higher boiling temperatures of the more concentrated solutions. The boiling points of the solutions, whose behaviors are shown in Figure 2.8, range from 100°C for 1 M nitrate to 122°C for 15.7 M nitrate. Plots of boiling point data (not shown here) for 2–15.6 M HNO_3 containing 0–1.5 M $UO_2(NO_3)_2$ or 0–0.3 M

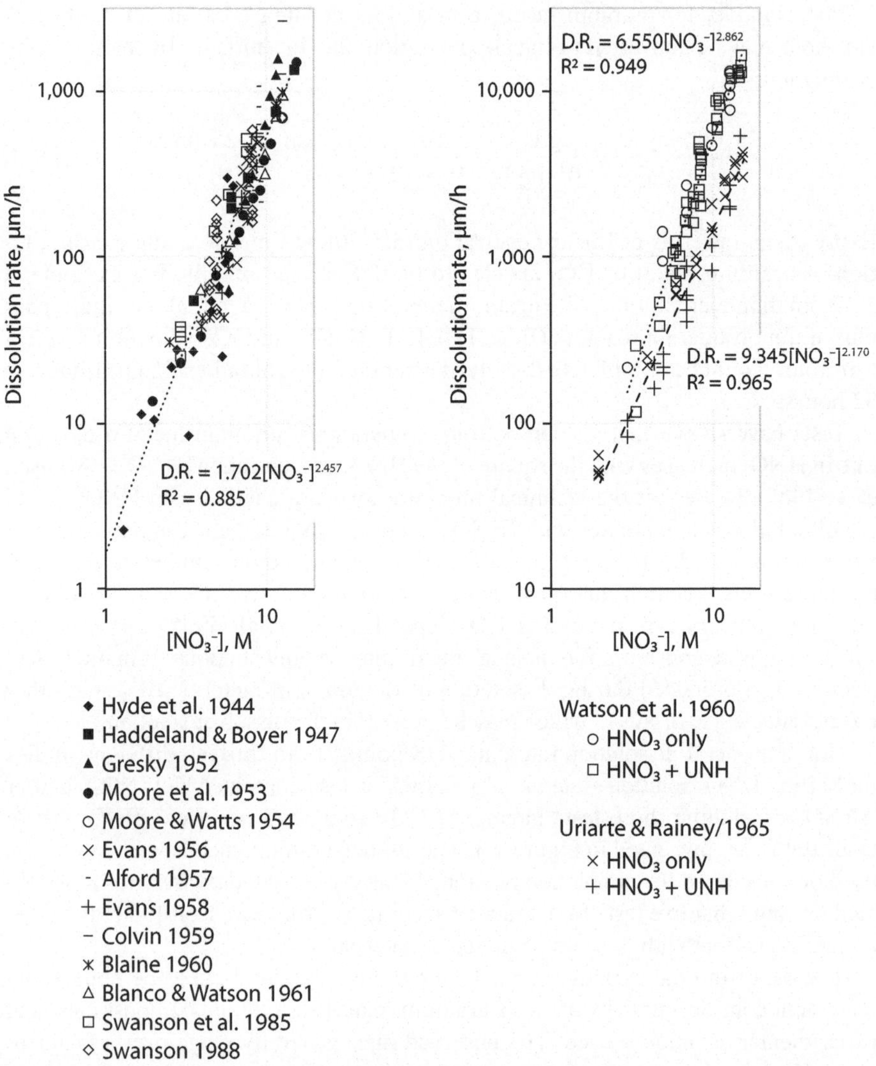

FIGURE 2.8 Dependencies of uranium metal (left) and uranium dioxide (right) dissolution rates on total nitrate concentration at boiling.

$Al(NO_3)_3$ show the boiling points vary linearly with total nitrate concentration (based on data given in Tables A-1, A-4, and A-6 of Uriarte and Rainey 1965).

The rates of UO_2 dissolution in HNO_3 and $HNO_3/UO_2(NO_3)_2$ solutions at boiling were abstracted from two studies (Uriarte and Rainey 1965; right side of Figure 2.8; Watson et al. 1959). Results from both studies showed steep dependencies on the total nitrate concentration similar to those found for uranium metal, and both were independent of the nitrate source; i.e., as HNO_3 or as $UO_2(NO_3)_2$. However, the slope of the total nitrate dependence for the latter tests is lower (Uriarte and Rainey 1965). This is likely because, in these tests, the UO_2 (as pellets) were only immersed for two minutes. Such short times would include significant periods under the lower-rate reaction induction time during which HNO_2 concentrations were still building. Thus, the results for the earlier tests (Watson et al. 1959), which show the dissolution rate proportional to the ~2.9 power of the total nitrate concentration, likely are more representative of UO_2 dissolutions at extended times.

$$UO_2 \text{ dissolution rate, } \mu\text{m/hour at boiling } (\text{Watson et al. 1959}) = 6.550 \left[\text{Nitrate, in moles per liter}\right]^{2.862} \quad (2.21)$$

$$UO_2 \text{ dissolution rate, } \mu\text{m/hour at boiling } (\text{Uriarte and Rainey 1965}) = 9.345 \left[\text{Nitrate, in moles per liter}\right]^{2.170} \quad (2.22)$$

It is also seen that the linear penetration rates for UO_2 dissolution are about an order of magnitude greater than those for uranium metal. Taking into account the relative densities of the uranium materials (~19 g/cm^3 for metal and ~10.3 g/cm^3 for compacted and sintered UO_2), the rate of uranium entering the solution at boiling under similar total nitrate concentration and solid surface area is about two times greater for UO_2 compared with uranium metal.

At Hanford, the strong dependence of dissolution rate on the nitrate concentration and the desire for high dissolved uranium (~2 M) and low (<2 M) HNO_3 concentrations in the initial solvent extraction stage led to longer dissolution times. To address these process exigencies, three approaches to decreasing the dissolution time were adopted in Hanford's PUREX Plant:

- Adding aluminum nitrate, $Al(NO_3)_3$, to the dissolver solution to increase nitrate ion concentration;
- Carrying over of an undissolved uranium metal "heel," of higher specific surface area, from one dissolution cycle to the next;
- Using "cored" and internally and externally cooled (I&E) fuel elements of hollow cylindrical construction.

Although adding the $Al(NO_3)_3$ to the dissolver solution enhances dissolution rates, and even enhances uranium and plutonium extraction in the PUREX Process, the aluminum is not recyclable and increases the waste. For this reason, $Al(NO_3)_3$ addition to the dissolver solution was used at Hanford more as a sequestering agent for fluoride from the Zirflex Process rather than to enhance nitrate concentration.

In the second approach, a metal heel is left by not dissolving the prior dissolver charge to completion. The undissolved metal is left for the succeeding dissolution "cuts" or acid strikes. For example, the first fuel charge could be dissolved until about half of the slug diameter remains. The solution would be discharged, and a fresh charge of irradiated fuel slugs and acid then would be added to the dissolver containing the partially dissolved slugs from the prior cut. This cycling could continue indefinitely. The net effect, from the second cut onwards, would be to increase the specific surface area of uranium metal exposed to the HNO_3, thus decreasing the time needed to reach the target dissolver solution concentration. Because the plutonium concentration is greater near the slug surface, the practice of leaving partially dissolved irradiated slugs also would flatten the concentration of plutonium relative to uranium in the dissolver solution to be fed to the solvent extraction process (Blanco and Watson 1961).

Fuel dissolution rate improvements, though important, were secondary to the Hanford Site's adoption, for single pass reactors, of cored and I&E fuel designs. Both fuel types were more difficult to fabricate than the original solid cylindrical fuel design. However, their greater mechanical flexibility and heat rejection capability allowed for a significant increase in reactor power and consequent plutonium production rate. The cored fuel (used 1954–1957) was cylindrical but with a lengthwise axial hole machined to allow inward expansion of the uranium by plastic deformation during irradiation and heat. The holes at either end were plugged by aluminum, and the entire element clad in aluminum. The aluminum-clad I&E tubular fuel design was adopted in 1957. With the I&E design, the cooling water could pass both around the element and through the aluminum-clad axial inner passage, improving fuel heat dissipation to allow higher reactor power and correspondingly shorter irradiation times (Gerber 1982; DOE 2002). The I&E design also would have improved the neutron absorption profile to achieve a more uniform plutonium distribution through the fuel. Significantly, the later Hanford N Reactor fuel design adopted the I&E principles, using two concentric and individually clad fuel tubes.

Calculation analyses may be run to evaluate and compare the following alternative dissolution strategies:

- A nominal 3.39-cm ∅ × 21.15-cm long solid cylindrical Hanford fuel slug (~3,653 g uranium metal) in 7.67 L of solution to give 2 M uranium at total slug dissolution.
- A 3.39-cm ∅ × 21.15-cm long fuel slug present with a heel slug carried over from the prior dissolution cut. In this analysis, the heel slug is already dissolved to half its initial diameter and, thus, has the dimensions 1.02-cm ∅ × 17.75 cm. The end of the dissolution step, in the latter case, results in complete dissolution of the initial heel slug leaving the starting whole slug dissolved to the initial size of the heel slug, thus becoming feed for the following dissolution cut.
- An I&E slug of the same (3.39-cm ∅ × 21.15-cm long) outer dimensions as the whole slug but with a 1.27-cm ∅ axial hole, again dissolved to the target 2 M uranium concentration. The solution volume becomes 6.60 L because of the lower (3,141 g) slug mass.

In all cases, the starting dissolvent is 8.45 M HNO_3/0.3 M $Al(NO_3)_3$, and the product solution is 2 M $UO_2(NO_3)_2$/1 M HNO_3/0.3 M $Al(NO_3)_3$. The dissolution was assumed to occur at boiling following the stoichiometry described previously for the Hanford PUREX Plant and a slightly steeper rate law compared with that shown in Figure 2.8 derived for solutions containing >5 M total nitrate:

$$U + 3.72\,HNO_3 + 0.295\,O_2 \rightarrow UO_2(NO_3)_2 + 1.86\,H_2O + 1.55\,NO + 0.17\,NO_2 \qquad (2.23)$$

Dissolution rate, μm/hour at boiling = 0.3403 [Nitrate, in moles per liter]$^{3.177}$

The calculation showed that the time to reach the target 2 M uranium concentration would be ~96 hours for the whole slug (to complete dissolution), ~50 hours for the whole-plus-heel slug (leaving a heel slug), and ~28 hours for the I&E slug (to complete dissolution). The I&E slug design proves to be particularly effective for dissolution, maintaining a high relative surface area over the entire dissolution cycle because, although the outer surface area of the I&E slug decreases with time as it corrodes inward, the surface area of the axial inner channel increases as it corrodes outward.

2.10.3 Agitation Effects

The dissolution rates of uranium metal and UO_2 decrease with agitation. This is because the dissolution reaction is autocatalytic with HNO_2 serving as both the primary HNO_3 reduction product and the active agent in attacking the uranium metal (Lacher, Salzman, and Park 1961) and UO_2 (Herrmann 1984; Herrmann et al. 1984). Agitation serves to move the HNO_2 reactant, which forms on uranium metal and UO_2 surfaces by HNO_3 reduction, away from the dissolving solids surface. In practice, however, the reaction of uranium metal or UO_2 with HNO_3 solution normally takes place at boiling, producing abundant H_2O and HNO_3 vapors and NO_x product gases. Thus, agitation is unavoidable.

2.10.4 Catalysts or Additives Effects

The catalytic effect of HNO_2 for both uranium metal and UO_2 dissolution has been noted both by the quenching effects that nitrite scavengers, such as hydrazine, urea, and sodium azide, have on the dissolution rates and by the positive effect nitrite salt addition has on the rate.

Catalysts or other additives to the HNO_3 (e.g., fluoride, phosphate, fluosilicate, and sulfate) have been tested to accelerate the uranium metal dissolution reaction for Hanford processing (Moore, Schulz, and Walter 1953; Moore and Watts Jr 1954). Although substantial initial rate increases could be gained by the use of sodium sulfate (Na_2SO_4) and particularly phosphoric acid (H_3PO_4), ammonium fluosilicate [$(NH_4)_2SiF_6$], and hydrofluoric acid (HF) additives if dissolutions were terminated before acid depletion, no overall time savings could be attained when the entire dissolution cycle (to low free acid) was considered. It was postulated that the catalysis is inhibited by the scavenging of the dissolved uranium occurring due to uranium sulfate and fluoride complexes and, ultimately, precipitation of the corresponding solid phases.

Evans (1958) conducted parametric studies of the effects of H_2SO_4 (at 0 M and 0.1 M), $NaNO_2$ (to form HNO_2 at 0 M and 0.2 M), HNO_3 (at 1 M and 3 M), and total nitrate (using $UO_2(NO_3)_2$ to attain goal 5- and 9-M total nitrate concentrations) on the corrosion rate of uranium metal. Increases in either HNO_3 or total nitrate were found to increase the dissolution rate strongly. However, 8% and 6% respective increases in rate associated with increases in H_2SO_4 and $NaNO_2$ concentration were found to be within the statistical variability of the experiments.

As a result of studies such as those of Evans (1956); Moore, Schulz, and Walter (1953); and Moore and Watts Jr (1954), dissolution additives were not used at the Hanford Site for uranium metal dissolution.

As noted earlier in this chapter, in an attempt to accelerate processing, co-dissolution in HNO_3 of both the aluminum cladding and uranium core of Hanford fuel elements also was investigated in which mercuric nitrate was added to catalyze aluminum cladding dissolution in HNO_3 (e.g., Krieg and Van der Cook 1956). However, the co-dissolution of aluminum cladding and fuel never was practiced at Hanford, though it was used at the SRS and, briefly, at the Mayak plant in the USSR.

2.10.5 Grain Size and Orientation Effects

The fabrication of uranium metal fuel occurs by metallurgical processes (swaging, rolling, heating, and the intentional or inadvertent introduction of other elements) that can affect the presence, size, and orientation of metal grains and inclusions. The effects of carbon content on grain size and uranium metal dissolution in HNO_3 studied at Hanford showed increasing carbon concentration (40–1,570 ppm) corresponded to decreased metallurgical grain size (0.27–0.06 mm). Decrease in grain size from 0.45 to ~0.06 mm, in turn, increased dissolution rate in boiling 8 M HNO_3 about a factor of 20 (from ~80 to ~1,200 μm/hour). For some reason, further grain size decreases to 0.03–0.04 mm and abruptly decreased dissolution rate to ~60–80 μm/hour (Bement and Swanson 1957; Moore, Schulz, and Walter 1953).

Corrosion of uranium metal fuel slug ends also was observed at Hanford to proceed more rapidly than corrosion through the side of the fuel slug. For Hanford single-pass reactor fuel, the end-on ("end grain") corrosion rate was about 3.6 times that of the side-on penetration rate (Blaine 1960). The end-to-side corrosion rate ratio of Hanford slugs was about 3.5 in tests by Alford (1957) and about 2.5 in tests by Evans (1956). For N Reactor fuel, about 30% higher dissolution rates were observed for end grain corrosion than for side grain corrosion. This anisotropy in dissolution rate was attributed to metal grain alignment caused by the extrusion process (Swanson 1988). However, earlier research on N Reactor fuel dissolution by the same laboratory showed no measurable difference between end and side grain corrosion rates (Swanson et al. 1985).

2.10.6 Surface Roughness Effects

Most of the uranium metal dissolution tests began with coupons that were machine-surfaced and thus relatively smooth. Many of the studies, however, measured dissolution rates over several subsequent intervals with the same corroding coupon to provide information on the effect of the surface condition on the dissolution rate. Though the dissolving uranium metal surfaces generally did not show severe pitting or lamellar corrosion that would dramatically increase surface area, development of surface roughness and higher effective surface area was noted.

The effect of surface roughness on dissolution rate was studied by Swanson and colleagues (1985). Dissolution rates of fresh as-cut specimens were found to be 3–4 times lower than the rates for specimens that had already undergone dissolution to achieve a 0.16-mm or more penetration.

Evans' (1958) parametric studies of the effects of H_2SO_4, $NaNO_2$, HNO_3, and total nitrate (as HNO_3 and $UO_2(NO_3)_2$) on the corrosion rates were conducted for each test condition over two successive 30-minute immersion intervals and began with machined 0.32 × 1.27 × 2.54-cm (~19-gram) coupons. The dissolution rates for the second immersions were about 1.6 times higher than the first immersion rates for the tests conducted with 9 M total nitrate (which suffered from 0.12 to 0.3 mm corrosion penetration) but were statistically indistinguishable from the first immersion for tests at 5 M total nitrate (which had only 0.005–0.02 mm penetration).

In a second parametric study, Evans (1958) examined the corrosion of the same coupon dimensions but allowed the corrosion to occur in each of the three successive immersions until ~1 gram, equivalent to ~0.06-mm penetration, had been dissolved. The 18 test solutions ranged from 0–3.6 M HNO_3 and 5.4–9.0 M total nitrate with $Al(NO_3)_3$ and $UO_2(NO_3)_2$, at 0–0.3 mole ratios, providing the balance of the nitrate. The average corrosion rate increased by a factor of ~1.4 from the first to the second immersion, increased by a factor of ~1.5 from the second to the third immersion (both ~0.06-mm penetration), and increased by a factor of ~2.0 from the first to the third immersion (~0.12-mm penetration).

Surface roughening of UO_2 pellets also is observed from dissolution in HNO_3 (Marc et al. 2017). In this review, it was noted based on the findings from two studies that the UO_2 specific surface area increases by a factor of four over the course of pellet dissolution. The review also notes that certain crystallographic planes of the UO_2 and grain boundaries are more susceptible to dissolution attack.

2.10.7 Irradiation Effects

Irradiation decreases uranium metal corrosion rates. The dissolution rates of unirradiated and irradiated uranium metal fuel in boiling 5 M HNO_3 were measured in both end-grain and side-grain directions (Table 4.14 in Swanson 1988). The testing showed that irradiation decreased the end-grain dissolution rate by about a factor of three and decreased the side-grain dissolution rate by a factor of ~one-third. In a single limited test, Colvin (1959) found that irradiated uranium dissolved about half as fast as unirradiated uranium in ~12.9 M HNO_3. No explanations were offered for these trends.

Irradiation of UO_2 pellets leads to radial cracking (Noirot 2009, 18) that greatly enhances the surface area available for attack by the HNO_3. Because of this high degree of fracturing, shearing of the UO_2-bearing fuel pins prior to dissolution reduces to powder much of the pelletized contained fuel (Vaudano 2008a, 46).

2.11 PLUTONIUM AND FISSION PRODUCT CONCENTRATION VARIATIONS WITHIN URANIUM METAL FUEL SLUGS AND URANIUM DIOXIDE PELLETS AND THEIR RELEASE RATES FROM SLUGS DURING DISSOLUTION

Plutonium and the fission products are not uniformly distributed within irradiated uranium metal fuel slugs but are more concentrated near the outer surfaces where neutron moderation by the cooling water is greater, thus increasing the likelihood of neutron capture by ^{238}U to form ^{239}Pu or be captured by ^{235}U and ^{239}Pu to form fission products.

Thus, the concentrations of fission product radionuclides within the slugs should also vary both in the axial and radial dimensions. As a result, it would be expected that plutonium and fission product release, such as release of the inert krypton and xenon fission product gases, would be greater during the initial times of dissolution when the outer slug metal is being dissolved because of the higher plutonium and fission product concentrations (per unit of uranium) near the fuel surface. This phenomenon was noted in the Manhattan Project pilot development of the Bismuth Phosphate Process, wherein 2.54-cm ∅ × 10.2-cm long irradiated slugs' outer, middle, and core materials had relative plutonium concentrations of 1.00/0.87/0.73 (DuPont 1944, 406).

The radionuclide distributions in N Reactor fuel were evaluated to determine the magnitudes of the irradiation effects. The N Reactor fuel is of a tube-in-tube design in which a hollow inner cylindrical tube is fit, with spacers, into an outer hollow cylindrical tube. Reactor physics calculations were performed for N Reactor fuel to predict the peak power, which corresponds to peak fissioning, for the ends of the outer elements relative to the entire assembly (normalized inner plus outer element) to estimate the maximum enhancement of plutonium and fission product concentration across the fuel. The calculations that showed a peaking factor of 1.19 could be applied to the (axial) end of the outer element and a factor of 1.27 applied to the radial peak at outer edge of the element (Schwinkendorf 2002). Therefore, the greatest power occurs at the outer surface adjacent to the end cap with a peaking power of $1.19 \times 1.27 = 1.52$; i.e., the irradiation at these areas is about 50% higher than it is for the fuel element overall.

Based on these findings, an analysis was performed to estimate the radionuclide compositions at the ends and outer surfaces of the fuel elements. As shown in Table 2.2, the maximum enhancement (for $^{239,240}Pu$ and ^{241}Am) is about 1.4 times that in the fuel overall (Schwinkendorf 2002). The fission product concentrations at the fuel peripheries (^{85}Kr, ^{90}Sr, ^{99}Tc, ^{137}Cs, and ^{154}Eu) are about 1–1.2 times higher than in the overall fuel.

TABLE 2.2
Specific Isotope Concentrations for Outer and End Regions of N Reactor Fuel

Isotope	Specific Concentration, Ci/metric ton U: Overall[a]	Outer & End[b]	(Outer & End)/Predicted
^{60}Co	1.63E+0	1.09E+0	0.67
^{85}Kr	2.49E+2	2.56E+2	1.03
^{90}Sr	4.35E+3	4.76E+3	1.09
^{99}Tc	1.27E+0	1.56E+0	1.23
^{137}Cs	5.75E+3	6.39E+3	1.11
^{154}Eu	4.65E+1	4.58E+1	0.98
^{237}Np	2.65E−2	3.26E−2	1.23
^{238}Pu	5.31E+1	6.77E+1	1.27
$^{239,240}Pu$	1.58E+2	2.24E+2	1.42
^{241}Am	1.81E+2	2.56E+2	1.42

[a] Table 3.6 of Packer (1999).
[b] Table 6 of Schwinkendorf (2002).

Higher plutonium and fission product release also would occur due to the higher fuel dissolution rates early in the dissolver processing when exposed metal surface areas and nitrate concentrations are highest. However, the uranium metal slug dissolution rate is primarily affected by the nitrate concentration and the exposed metal surface area. Thus, the plutonium and particularly fission product non-homogeneity are relatively small effects in their release rate during fuel dissolution.

Enhanced plutonium concentrations likewise are found at the peripheries of irradiated UO_2 pellets with plutonium concentrations being up to 3 times greater at the cladding wall than in the pellet centerline. The enhanced plutonium concentration at the periphery falls off rapidly, to plateau values, within about 1 mm in an ~8-mm diameter pellet (Vaudano 2008b, Figure 25). For UO_2 in irradiated fuel pins, pellet cracking increases with irradiation, releasing the inert krypton and xenon fission product gas. For a nominal 55 gigawatt-day/metric ton exposure (equivalent to 5.5% fission of initial metal), about 80% of the inert gas escapes to the free volume within the pin (Vaudano 2008b, 170, Figure 28).

2.12 MOX DISSOLUTION

The dissolution of $(U, Pu)O_2$ (i.e., MOX) in HNO_3 is addressed in a recent review of nuclear material dissolutions (Desigan et al. 2018) with much of the overview provided by a key prior study (Uriarte and Rainey 1965). In that work, the dissolution of MOX in HNO_3 was found to be highly dependent on the microscopic degree of mixing of the constituent UO_2 and PuO_2, as determined by fuel preparation and irradiation, on the plutonium fraction in the fuel, and the use of hydrofluoric acid (HF) to enhance PuO_2 dissolution.

In these tests (described in Section 4.4 of Uriarte and Rainey 1965), MOX pellets were prepared both by mechanical blending of the separate oxides and by coprecipitation as "$Pu(OH)_4$" and "$(NH_4)_2U_3O_7$"; i.e., apparently by making acidic Pu(IV) and U(VI) nitrate solutions alkaline with ammonium hydroxide (NH_4OH) to form the coprecipitate. The green pellets of both types were formed by pressing into shape at ~0.6 cm in diameter and ~0.6 cm long and sintering under a reducing atmosphere (nitrogen gas containing 6% H_2) at 1,300°C–1,600°C for 1–48 hours. Dissolution tests were run with large stoichiometric HNO_3 excess. Some characteristic results are described here. Both mechanically blended and coprecipitated oxide pellets containing <20% PuO_2 dissolved completely in tests with 10 M HNO_3. Pellets with 20% PuO_2, mechanically blended and sintered at 160°C for 16 hours and pellets from coprecipitated oxide and sintered at 1,300°C for 2 hours, did not dissolve completely in boiling 10 M HNO_3 but did dissolve in 10 M HNO_3 containing 0.05 M HF. A comparative test of 5% PuO_2 pellets prepared by mechanical blending and by coprecipitation showed the coprecipitated pellet dissolved more slowly.[7] An increase

[7] The opposite trend was noted in another MOX dissolution study (Glatz, Bokelund, and Zierfuss 1990) conducted using 90°C 6 M HNO_3. The plutonium fraction in the MOX was not described. Pellets prepared by mechanical blending left 10% residue after 22 hours immersion, while the coprecipitated MOX dissolved completely after 4 hours even though the coprecipitated material began dissolution later (i.e., had a longer induction time).

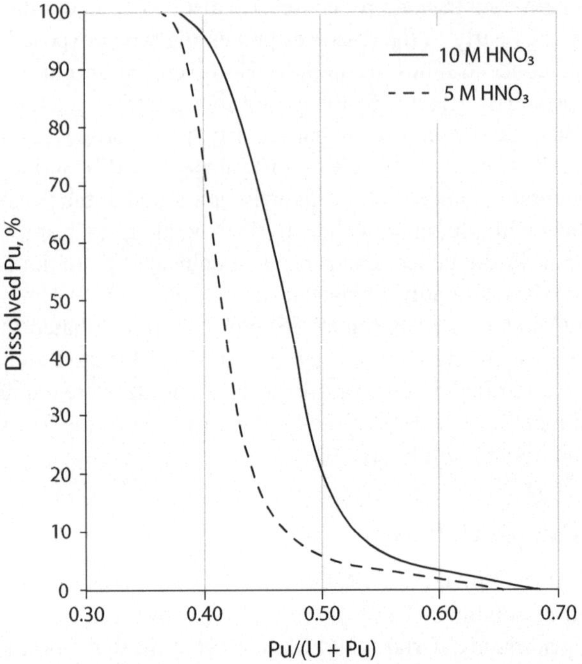

FIGURE 2.9 (U, Pu)O_2 MOX dissolution by six hours of boiling HNO_3 as a function of plutonium fraction. (After Figure 31 of Vaudano, A., The PUREX Process: Head-End Operations, in *Treatment and Recycling of Spent Nuclear Fuel: Actinide Partitioning–Application to Waste Management*, edited by M. Lecomte, 45–54, Commissariat à l'Énergie Atomique, Paris, France, 2008a.)

in the plutonium fraction increased the dissolution time and decreased the extent dissolved. Thus, coprecipitated pellets containing 0.5%, 5.0%, and 17.8% PuO_2 required 60, 240, and 360 minutes to dissolve in boiling 14 M HNO_3. The dissolution rates of irradiated pellets are about 5 times greater than similarly constituted non-irradiated pellets (Uriarte and Rainey 1965).

As shown in Figure 2.9, for irradiated fuel, residual plutonium solids begin to be found in the dissolver heel when the plutonium fraction in the fuel exceeds about 0.35; the dissolved plutonium goes practically to zero as the plutonium fraction reaches 0.7 (Vaudano 2008a). These findings mean that for pressurized water reactor MOX fuels, where the plutonium fraction is less than 0.10, or for fast reactor fuels, where the plutonium fraction is less than 0.30, dissolution should be straightforward. However, for lightly irradiated or non-irradiated fuels, which do not undergo annealing by fission and neutron exposure, harsh conditions should be maintained to assure complete dissolution.

In French practice, dissolution of low-enriched irradiated UO_2 fuels is conducted at >90°C using 3–4 M HNO_3 to attain 200–250 g of uranium/L (~0.84–1.05 M uranium), dissolving >99.9% of both plutonium and uranium. For light-water reactor MOX fuels, >90°C temperatures and 4–5 M HNO_3 are used to attain 200 g uranium/L, leaving plutonium dissolver fines at <0.1%. For fast reactor fuels irradiated up to 100 gigawatt-days per ton of heavy metal, dissolution in boiling >5 M HNO_3 for more than 6 hours is used to obtain 150–200 grams heavy metal per liter (~0.63–0.84 M U, Pu) but with 1% or more of the plutonium remaining in the heel. The extent of loss depends greatly on plutonium content in the fuel, fuel fabrication conditions, and the extent of irradiation (Vaudano 2008a, 47). Means to dissolve the plutonium in the heel include the use of low HF concentrations at the cost of incurring concomitant equipment corrosion or oxidative dissolution of the PuO_2 solid to PuO_2^{2+}. Oxidants used include trivalent cobalt, Co(III), tetravalent cerium, Ce(IV), or divalent silver, Ag(II). The use of such strong oxidants has the disadvantage of converting fission product ruthenium to volatile RuO_4 (Vaudano 2008a, 47).

REFERENCES

Alford, M. D. 1957. *I. And E. Fuel Element Dissolution Rate Studies* (Richland, WA: Hanford Atomic Products Operation). HW-52808. https://reading-room.labworks.org/Files/GetDocument.aspx?id=D8411254.

Allen, G. K., L. L. Jacobs, and D. W. Reberger. 1982. *PUREX Flowsheet–Reprocessing N Reactor Fuels* (Richland, WA: Rockwell Hanford Operations). PFD-P-020-00001.

Bement, A. L., and J. L. Swanson. 1957. *The Effects of Carbon Content on the Rate of Dissolution of Dingot Uranium in Nitric Acid* (Richland, WA: Hanford Atomic Products Operation). HW-52430. https://reading-room.labworks.org/Files/GetDocument.aspx?id=D8410862.

Benedict, M., T. H. Pigford, and H. W. Levi. 1981. *Nuclear Chemical Engineering*. 2nd ed. New York: McGraw-Hill Education.

Blaine, H. T. 1960. *Dissolution of Power Reactor Fuel Cores* (Richland, WA: Hanford Atomic Products Operation). HW-66320. https://digital.library.unt.edu/ark:/67531/metadc100595/m2/1/high_res_d/metadc100595.pdf.

Blanco, R. E., and C. D. Watson. 1961. "Head-End Processes for Solid Fuels." In *Reactor Handbook: Fuel Reprocessing*, edited by S. M. Stoller and R. B. Richard. New York: Interscience Publishers, Incorporated.

Chauve, H. J., J. Decours, R. Demay, M. Pelras, J. Simonnet, and G. Turluer. 1987. "Zirconium Use for Large Process Components." In *Proceedings of a Technical Committee Meeting on Materials Reliability in the Back End of the Nuclear Fuel Cycle*, pp. 165–192. Vienna, Austria: International Atomic Energy Agency. IAEA-TECDOC-421. https://www-pub.iaea.org/MTCD/Publications/PDF/te_421_web.pdf.

Colvin, C. A. 1959. *Partially Dissolved Fuel Slugs from REDOX Dissolvers* (Richland, WA: Hanford Atomic Products Operation). HW-58561. https://reading-room.labworks.org/Files/GetDocument.aspx?id=D8527530.

Cordara, T., S. Szenknect, L. Claparede, R. Podor, A. Mesbah, C. Lavalette, and N. Dacheux. 2017. "Kinetics of Dissolution of UO_2 in Nitric Acid Solutions: A Multiparametric Study of the Non-Catalysed Reaction." *Journal of Nuclear Materials* 496: 251–264. doi:10.1016/j.jnucmat.2017.09.038.

Cronin, J., and N. Collier. 2012. "Corrosion and Expansion of Grouted Magnox." *Mineralogical Magazine* 76 (8): 2901–2909. doi:10.1180/minmag.2012.076.8.05.

Culler Jr., F. L., R. E. Blanco, H. E. Goeller, and C. D. Watson. 1957. *An Analysis of Power Reactor Fuel Reprocessing* (Oak Ridge, TN: Oak Ridge National Laboratory). ORNL-2265(Del.). https://www.osti.gov/servlets/purl/4333047.

Decours, J., J. C. Decugis, R. Demay, M. Pelras, and G. Turluer. 1987. "Austenitic Stainless Steels: Assessment of Progress in Materials Performance for Reprocessing Applications." In *Proceedings of a Technical Committee Meeting on Materials*

Reliability in the Back End of the Nuclear Fuel Cycle, 117–127. Vienna, Austria: International Atomic Energy Agency. IAEA-TECDOC-421. https://www-pub.iaea.org/MTCD/Publications/PDF/te_421_web.pdf.

Desigan, N., N. P. Bhatt, N. K. Pandey, U. K. Mudali, R. Natarajan, and J. B. Joshi. 2017. "Mechanism of Dissolution of Nuclear Fuel in Nitric Acid Relevant to Nuclear Fuel Reprocessing." *Journal of Radioanalytical and Nuclear Chemistry* 312 (1): 141–149. doi:10.1007/s10967-017-5208-z.

Desigan, N., N. P. Bhatt, M. A. Shetty, G. K. P. Sreekumar, N. K. Pandey, U. K. Mudali, R. Natarajan, and J. B. Joshi. 2018. "Dissolution of Nuclear Materials in Aqueous Acid Solutions." *Reviews in Chemical Engineering*. doi:10.1515/revce-2017-0063.

DOE. 2002. *History of the Plutonium Production Facilities at the Hanford Site Historic District, 1943–1990* (Richland, WA: U.S. Department of Energy). DOE/RL-97-1047. https://www.osti.gov/servlets/purl/807939.

DuPont. 1944. *Hanford Engineer Works Technical Manual* (Richland, WA: General Electric, Hanford Atomic Products Operation). HW-10475 ABC. https://pdw.hanford.gov/arpir/pdf.cfm?accession=0076415H.

DuPont. 1957. *Savannah River Plant Engineering and Design History Volume III, 200°F and H Areas* (Wilmington, Delaware: E.I. Du Pont de Nemours and Company). DPE-972. https://www.osti.gov/servlets/purl/754230.

Evans, T. F. 1956. *Pilot Plant Dissolution of Unjacketed Fuel Elements* (Richland, WA: Hanford Atomic Products Operation). HW-46093. https://reading-room.labworks.org/Files/GetDocument.aspx?id=D199119856.

Evans, T. F. 1958. *Continuous Pilot Plant Dissolution of Aluminum-Jacketed Uranium Fuel Elements in Nitric Acid* (Richland, WA: Hanford Atomic Products Operation). HW-56403. https://reading-room.labworks.org/Files/GetDocument.aspx?id=D5766210.

Fairhall, G. A., and J. D. Palmer. 1992. "The Encapsulation of Magnox Swarf in Cement in the United-Kingdom." *Cement and Concrete Research* 22(2–3): 293–298. doi:10.1016/0008-8846(92)90068-7.

Fauvet, P., F. Balbaud, R. Robin, Q. T. Tran, A. Mugnier, and D. Espinoux. 2008. "Corrosion Mechanisms of Austenitic Stainless Steels in Nitric Media Used in Reprocessing Plants." *Journal of Nuclear Materials* 375 (1): 52–64. doi:10.1016/j.jnucmat.2007.10.017.

Gerber, M. S. 1982. *Past Practices Technical Characterization Study-300 Area-Hanford Site* (Richland, WA: Westinghouse Hanford Company). WHC-MR-0388. https://www.osti.gov/servlets/purl/10127106-FOxneb/.

Glatz, J. P., H. Bokelund, and S. Zierfuss. 1990. "Analysis of the Off-Gas from Dissolution of Nuclear Oxide Fuels and Carbide Fuels in Nitric-Acid." *Radiochimica Acta* 51(1): 17–22, https://www.degruyter.com/view/j/ract.1990.51.issue-1/ract.1990.51.1.17/ract.1990.51.1.17.xml?format=INT

Gnanamoorthy, J. B., and G. R. Balasubramanian. 1987. "Selection of Materials for Various Systems in Reprocessing Fast Reactor Fuels–Present Status and Future Programme." In *Proceedings of a Technical Committee Meeting on Materials Reliability in the Back End of the Nuclear Fuel Cycle*, 207–214. Vienna, Austria: International Atomic Energy Agency. https://www-pub.iaea.org/MTCD/Publications/PDF/te_421_web.pdf.

Gresky, A. T. 1952. *Recovery of Nitrogen Oxides and Rare Gas Fission Products from the Dissolution of Irradiated Uranium* (Oak Ridge, TN: Oak Ridge National Laboratory). ORNL-1208. https://www.osti.gov/servlets/purl/4116841/.

Groenier, W. S. 1971. *Equipment for the Dissolution of Core Materials from Sheared Power Reactor Fuels* (Oak Ridge, TN: Oak Ridge National Laboratory). ORNL-TM-3194. https://www.osti.gov/servlets/purl/4008629.

Guillet, J.-L., and Y. Guérin. *Nuclear Fuels*. Paris, France: Commissariat à l'Énergie Atomique, 2009. http://www.materials.cea.fr/en/PDF/MonographiesDEN/Nuclear-fuels-CEA-en.pdf.

Gupta, C. K. *Materials in Nuclear Energy Applications*. Vol. II. Boca Raton, FL: CRC Press, 2017.

Herrmann, B. 1984. *Auflösung Unbestrahlter UO_2-Pellets in Salpetersäure* (Karlsruhe, Germany: Kernforschungszentrum Karlsruhe). https://publikationen.bibliothek.kit.edu/200019712/3812698.

Herrmann, B., W. Bumiller, E. Henrich, and R. von Ammon. 1984. "Zur Chemie Der Auflösung Von UO_2 in Salpetersäure." *KfK Nachrichten* 16 (2): 87–93.

Haddeland, G. E., and T. W. Boyer. 1947. *The Dissolution of X Metal in Nitric Acid* (CRX-318).

Hyde, A. C., C. H. Holt, G. Bird, S. G. Thompson, and J. L. Dreher. 1944. *Process Development and Chicago Semi-Works Operation* (Chicago, IL: University of Chicago). CN-1751.

Hyder, M. L., W. C. Perkins, M. C. Thompson, G. A. Burney, E. R. Russell, H. P. Holcomb, and L. F. Landon. 1979. *Processing of Irradiated, Enriched Uranium Fuels at the Savannah River Plant* (Aiken, South Carolina: E. I. du Pont de Nemours & Company). DP-1500. https://www.osti.gov/servlets/purl/5819865.

IAEA. 1987. *Proceedings of a Technical Committee Meeting on Materials Reliability in the Back End of the Nuclear Fuel Cycle, IAEA-TECDOC-421*. Vienna, Austria: International Atomic Energy Agency. https://www-pub.iaea.org/MTCD/Publications/PDF/te_421_web.pdf.

Krieg, I. T., and R. E. Van der Cook. 1956. *Continuous Dissolution of Aluminum-Jacketed Fuel Elements in a Converted Batch Dissolver* (Richland, WA: Hanford Atomic Products Operation). HW-42666. https://reading-room.labworks.org/Files/GetDocument.aspx?id=D3161381.

Lacher, J. R., J. D. Salzman, and J. D. Park. 1961. "Dissolving Uranium in Nitric Acid." *Industrial and Engineering Chemistry* 53(4): 282–284. doi:10.1021/ie50616a023.

Larsen, R. P. 1959. "Dissolution of Uranium Metal and Its Alloys." *Analytical Chemistry* 31(4): 545–549. doi:10.1021/ac50164a026.

Laue, C. A., D. Gates-Anderson, and T. E. Fitch. 2004a. "Dissolution of Metallic Uranium and Its Alloys—Part 1. Review of Analytical and Process-Scale Metallic Uranium Dissolution." *Journal of Radioanalytical and Nuclear Chemistry* 261(3): 709–717. doi:10.1023/B:JRNC.0000037117.01721.f1.

Laue, C. A., D. Gates-Anderson, and T. E. Fitch. 2004b. "Dissolution of Metallic Uranium and Its Alloys—Part II: Screening Study Results: Identification of an Effective Non-Thermal Uranium Dissolution Method." *Journal of Radioanalytical and Nuclear Chemistry* 262(2): 517–524. doi:10.1023/B:JRNC.0000046790.80269.8f.

Long, J. T. 1967. *Engineering for Nuclear Fuel Reprocessing*. New York: Gordon and Breach Science Publishers.

Marc, P., A. Magnaldo, A. Vaudano, T. Delahaye, and É. Shaer. 2017. "Dissolution of Uranium Dioxide in Nitric Acid Media: What Do We Know?" *European Journal of Physics—Nuclear Science and Technology* 3(13). doi:10.1051/epjn/2017005.

Masson, M., V. Rovet, and A. Béziat. 2008. "Treatment of Advanced Fuels: Head-End Processes." In *Treatment and Recycling of Spent Nuclear Fuel: Actinide Partitioning—Application to Waste Management*, edited by M. Lecomte, pp. 97–99. Paris, France: Commissariat à l'Énergie Atomique. https://hal-cea.archives-ouvertes.fr/cea-01153306/file/cea6-en.pdf.

McNamara, B. K., E. C. Buck, C. Z. Soderquist, F. N. Smith, E. J. Mausolf, and R. D. Scheele. 2014. "Separation of Metallic Residues from the Dissolution of a High-Burnup BWR Fuel Using Nitrogen Trifluoride." *Journal of Fluorine Chemistry* 162: 1–8. doi:10.1016/j.jfluchem.2014.02.010.

Moore, J. D., R. L. Walser, and J. J. Fritch. 1980. *PUREX Technical Manual* (Richland, WA: Rockwell Hanford Operations). RHO-MA-116.

Moore, R. L., W. W. Schulz, and S. J. Walter. 1953. *Phosphate Catalysis in Nitric Acid Dissolutions of Uranium Metal* (Richland, WA: Hanford Atomic Products Operation). HW-28995. https://reading-room.labworks.org/Files/GetDocument.aspx?id=D198137244.

Moore, R. L., and R. A. Watts Jr. 1954. "Uranium Dissolution Studies." In *Quarterly Progress Report—Chemistry Unit—January, February, March, 1954, HW-31630*, edited by A. H. Bushey, pp. 14–14a. Richland, WA: Hanford Atomic Products Operation. https://reading-room.labworks.org/Files/GetDocument.aspx?id=D198167606.

Nakahara, M., Y. Sano, and H. Abe. 2018. "Electrochemical Properties of Zirconium in Highly Concentrated Plutonium Nitrate Solution." *Progress in Nuclear Science and Technology* 5: 52–55. doi:10.15669/pnst.5.52.

Noirot, J. 2009. "In-Reactor Behaviour of UO_2 and MOX Fuels." In *Nuclear Fuels*, edited by J.-L. Guillet and Y. Guérin, pp. 17–20. Paris, France: Commissariat à l'Énergie Atomique. http://www.materials.cea.fr/en/PDF/MonographiesDEN/Nuclear-fuels-CEA-en.pdf.

OECD. 2005. *The Safety of the Nuclear Fuel Cycle*. Paris, France: Nuclear Energy Agency, Organisation for Economic Co-Operation and Development. http://www.oecd-nea.org/nsd/pubs/2005/3588-safety-nuclear-fuel-cycle.pdf.

Packer, M. J. 1999. *105-K Basin Material Design Basis Feed Description for Spent Nuclear Fuel Project Facilities* (Richland, WA: Numatec Hanford, Inc). HNF-SD-SNF-TI-009, Rev 3. https://www.osti.gov/servlets/purl/798703.

Peper, S. M., L. F. Brodnax, S. E. Field, R. A. Zehnder, S. N. Valdez, and W. H. Runde. 2004. "Kinetic Study of the Oxidative Dissolution of UO_2 in Aqueous Carbonate Media." *Industrial & Engineering Chemistry Research* 43(26): 8188–8193. doi:10.1021/ie049457y.

Schneider, K. J. 1951. *Flow Sheets and Flow Diagrams of Precipitation Separation Processes* (Richland, WA: Hanford Atomic Products Operation). HW-23043. https://www.osti.gov/opennet/detail?osti-id=16430616.

Schulz, W. W. 1972. *Shear-Leach Processing of N-Reactor Fuel–Cladding. Fires* (Richland, WA: Atlantic Richfield Hanford Company). ARH-2351. https://www.osti.gov/biblio/4281991-shear-leach-processing-reactor-fuel-cladding-fires.

Schwinkendorf, K. N. 2002. *K-East Basin Sludge Characterization Using Exposure-Adjusted Radioisotope Source Terms for N Reactor Fuel* (Richland, WA: Fluor Hanford, Incorporated). HNF-8760, Rev 1.

Shabbir, M., and R. G. Robins. 1969. "Kinetics of Dissolution of Uranium Dioxide in Nitric Acid. II." *Journal of Applied Chemistry* 19 (2): 52–56. https://www.tandfonline.com/doi/abs/10.1179/000705990798269676.

Shaw, R. D. 1990. "Corrosion Prevention and Control at Sellafield Nuclear Fuel Reprocessing Plant." *British Corrosion Journal* 25 (2): 97–107. doi:10.1179/000705990798269676.

Slansky, C. M. 1961. "Preparation of Fuels for Processing." In *Chemical Processing of Reactor Fuels*, edited by J. F. Flagg. New York: Academic Press.

Soderquist, C. Z., A. M. Johnsen, B. K. McNamara, B. D. Hanson, J. W. Chenault, K. J. Carson, and S. M. Peper. 2011. "Dissolution of Irradiated Commercial UO_2 Fuels in Ammonium Carbonate and Hydrogen Peroxide." *Industrial & Engineering Chemistry Research* 50 (4): 1813–1818. doi:10.1021/ie101386n.

Soderquist, C. Z., B. K. McNamara, and B. Oliver. 2008. "Dissolution of Uranium Metal without Hydride Formation or Hydrogen Gas Generation." *Journal of Nuclear Materials* 378 (3): 299–304. doi:10.1016/j.jnucmat.2008.05.014.

Sokhina, L. P. 2000. *Pages of History and Radiochemical Plant Production Association "Mayak"*. Ozersk, Russia: Printing House Mayak. http://elib.biblioatom.ru/text/sohina_stranitsy-istorii-zavoda-mayak_2000/?bookhl=

Swanson, J. L. 1961. "The Zirflex Process." In *Progress in Nuclear Energy, Series III*, edited by F. R. Bruce, J. M. Fletcher and H. H. Hyman, In Process Chemistry. New York: Pergamon Press.

Swanson, J. L. 1988. *Recent Studies Related to Head-End Fuel Processing at the Hanford PUREX Plant* (Richland, WA: Pacific Northwest Laboratory). PNL-6609. https://www.osti.gov/servlets/purl/6868352.

Swanson, J. L., L. A. Bray, H. E. Kjarmo, J. L. Ryan, C. L. Matsuzaki, S. G. Pitman, and J. H. Haberman. 1985. *Laboratory Studies of Shear/Leach Processing of Zircaloy Clad Metallic Uranium Reactor Fuel* (Richland, WA: Pacific Northwest Laboratory). PNL-5708. https://www.osti.gov/servlets/purl/6315563.

Taylor, L. L. 2000. *N Reactor (U-Metal) Fuel Characteristics for Disposal Criticality Analysis* (Idaho Falls, Idaho: Idaho National Engineering and Environmental Laboratory). DOE/SNF/REP-056. https://www.osti.gov/biblio/911535-reactor-metal-fuel-characteristics-disposal-criticality-analysis.

Uriarte, A. L., and R. H. Rainey. 1965. *Dissolution of High-Density UO_2, PuO_2, and UO_2-PuO_2 Pellets in Inorganic Acids* (Oak Ridge, TN: Oak Ridge National Laboratory). ORNL-3695. https://www.osti.gov/servlets/purl/4652087.

Vaudano, A. 2008a. "The PUREX Process: Head-End Operations." In *Treatment and Recycling of Spent Nuclear Fuel: Actinide Partitioning–Application to Waste Management*, edited by M. Lecomte, 45–54. Paris, France: Commissariat à l'Énergie Atomique. https://hal-cea.archives-ouvertes.fr/cea-01153306/file/cea6-en.pdf.

Vaudano, A. 2008b. "Some Characteristics of Nuclear Fuels, Relevant for Their Treatment." In *Treatment and Recycling of Spent Nuclear Fuel: Actinide Partitioning–Application to Waste Management*, edited by M. Lecomte, pp. 37–41. Paris, France: Commissariat à l'Énergie Atomique. https://hal-cea.archives-ouvertes.fr/cea-01153306/file/cea6-en.pdf.

Wagman, D. D., W. H. Evans, V. B. Parker, R. H. Schumm, I. Halow, S. M. Bailey, K. L. Churney, and R. L. Nuttall. 1982. "The NBS Tables of Chemical Thermodynamic Properties: Selected Values for Inorganic and C_1 and C_2 Organic Substances in SI Units." *Journal of Physical and Chemical Reference Data* 11, no. 2, https://srd.nist.gov/JPCRD/jpcrdS2Vol11.pdf.

Watson, C. D., J. B. Adams, G. K. Ellis, G. A. West, F. L. Hannon, W. F. Schaffer, and B. B. Klima. 1959. "Mechanical Processing of Spent Power Reactor Fuel at Oak Ridge National Laboratory." *AEC Symposium for Chemical Processing of Irradiated Fuels from Power, Test, and Research Reactors*, Richland, WA. https://www.osti.gov/servlets/purl/4210048-pyJtx9/.

Watson, C. D., B. C. Finney, G. A. West, and B. A. Hannaford. 1961. "The Shear-Leach Process for Spent Nuclear Fuels." *Transactions of the American Nuclear Society* (6): 408–409.

Weakley, E. A. 1979. *Fuels Engineering Technical Handbook* (Richland, WA: United Nuclear). UNI-M-61.

Wymer, R. G., and R. E. Blanco. 1957. "Uranium-Aluminum Alloy Dissolution." *Industrial & Engineering Chemistry* 49(1): 59–61. doi:10.1021/ie50565a026.

Wymer, R. G., and D. L. Foster. 1956. "Nuclear Reactor Fuel Dissolution." In *Progress in Nuclear Energy, Series III, Process Chemistry*, edited by F. R. Bruce, J. M. Fletcher, H. H. Hyman and J. J. Katz. New York: McGraw-Hill Book Company.

Yamanouchi, T., A. Aoshima, N. Sasao, S. Takeda, and N. Ishiguro. 1987. "Experience of Corrosion Problems and Material Developments in the Tokai Reprocessing Plant." In *Proceedings of a Technical Committee Meeting on Materials Reliability in the Back End of the Nuclear Fuel Cycle*, pp. 129–146. Vienna, Austria: International Atomic Energy Agency. IAEA-TECDOC-421. https://www-pub.iaea.org/MTCD/Publications/PDF/te_421_web.pdf.

3 Precipitation and Crystallization Processes in Reprocessing, Plutonium Separation, Purification, and Finishing, Chemical Recovery, and Waste Treatment

Calvin H. Delegard and Reid A. Peterson

CONTENTS

3.1 INTRODUCTION

Definitions related to precipitation and crystallization processes and examination of precipitation and crystallization used in the discovery and early separations of the first two transuranium elements—neptunium (Np) and plutonium (Pu)—provide background to this chapter.

3.1.1 Definitions

Separations processes dependent on precipitation and crystallization rely on the preferential partitioning of a material of interest away from other materials present in the chemical system based on their respective distributions to solid and solution phases. According to the IUPAC "Gold Book" (IUPAC 2014), precipitation is "the sedimentation of a solid material (a precipitate) from a liquid solution in which the material is present in amounts greater than its solubility in the liquid." The same reference defines crystallization as "the formation of a crystalline solid from a solution, melt vapor or a different solid phase, generally by the lowering of the temperature or by evaporation of a solvent."

By these definitions, precipitation and crystallization share a degree of overlap. For example, a crystallization process, as achieved "by lowering of the temperature or evaporation of a solvent" would cause precipitation; i.e., "the sedimentation of a solid material from a liquid solution." In general, however, separations chemists understand precipitation and crystallization by the manners in which they are instigated. Thus, precipitation generally is understood more narrowly as the process in which the mixing of two clear solutions results in a solid forming (precipitating) from the mixture. For example, mixing a clear solution of silver nitrate with a separate clear solution of sodium chloride produces a solid precipitate of silver chloride with soluble salts remaining in solution. Crystallization is understood by separations chemists in the manner as defined by the IUPAC. For example, the salt pan evaporation of seawater results in sodium chloride crystallization. As might be inferred by these descriptions, precipitates generally have lower solubilities than solids formed by crystallization.

Precipitates may be allowed to age in their mother solutions to allow the solid particles to grow, improving their purity and making them easier to separate by mechanical means such as filtration or centrifugation. This process is called recrystallization, which the IUPAC (2014) defines as "the repeated dissolution and precipitation of a solid from a liquid solvent." Cyclic crystallization, dissolution, and recrystallization may also be used to effect separation of a solid from its mother solution, which might also contain other constituents. This purification process is called fractional crystallization.

The concentration of the material of interest in solution, such as plutonium present in dissolved irradiated uranium fuel, may be too low to be precipitated of itself. In this case, coprecipitation might be (and has been) used. The IUPAC (2014) defines coprecipitation as "the simultaneous precipitation of a normally soluble component with a macro-component from the same solution by the formation of mixed crystals, by adsorption, occlusion or mechanical entrapment." The macro-component is often called a carrier, and the coprecipitation called carrier precipitation. The IUPAC (2014) defines a carrier as "a substance in appreciable amount which, when associated with a trace of a specified substance, will carry the trace with it through a chemical or physical process."

The terms precipitation, crystallization, recrystallization, fractional crystallization, coprecipitation, and carrier precipitation will be used throughout this chapter, and their definitions refined through specific examples.

3.1.2 Discoveries of Plutonium and Neptunium and Introduction to Early Separations Processes

Plutonium was discovered in the products of cyclotron particle irradiation of uranium. The initial insights into the existence of plutonium, as ^{239}Pu, were inferred by McMillan and Abelson (1940; in May). These insights were based on observation of its ingrowth from neptunium, as ^{239}Np, produced by neutron capture on uranium, as ^{238}U, and subsequent beta decay:

$$^{238}U + n \rightarrow {}^{239}U \rightarrow \beta^{\nearrow} + {}^{239}Np \rightarrow \beta^{\nearrow} + {}^{239}Pu. \quad (3.1)$$

Plutonium, as ^{238}Pu, was discovered by Seaborg et al. (1946a) and reported secretly in January and March of 1941. These brief accounts provided initial information on plutonium's chemical properties. Further insights into the chemical and physical properties for ^{239}Pu, including its favorable slow-neutron fission properties, were published by Kennedy et al. (1946) and reported secretly in May 1941.

The initial separation of plutonium from cyclotron particle-irradiated uranium used carrier precipitation processes in which a compound present in macro concentration "carried," through incorporation within a chemically analogous compound, trace amounts of the plutonium precursor, neptunium. Direct carrier precipitation of plutonium also occurred. The early chemical methods used to isolate plutonium are outlined as follows:

1. Sequential sodium uranyl (tri)acetate and cerium fluoride coprecipitations of neptunium, followed by in-growth of ^{239}Pu (McMillan and Abelson 1940).
2. Coprecipitation of ^{238}Pu, suspected to be Pu(IV), from deuteron-irradiated uranium using the fluoride or iodate of a rare earth or thorium as carriers. Thorium peroxyhydrate also carried Pu(IV). The rare earth fluorides did not carry Pu, which had first been oxidized by silver-catalyzed persulfate. This Pu was later proven to be Pu(VI) (Seaborg et al. 1946b).
3. Preliminary diethyl ether extraction for uranium removal, followed by Cerium/lanthanum(Ce/La) fluoride coprecipitation for neptunium from the aqueous phase and passage of time to allow ^{239}Pu in-growth (Kennedy et al. 1946). In these experiments, the Np and subsequent in-grown Pu were present at about one part per billion of the source uranium.

The initial industrial processes used in the United States of America (US) and in the Union of Soviet Socialist Republics (Soviet Union or USSR) for plutonium separation from reactor neutron-irradiated uranium likewise used carrier precipitation processes for plutonium itself after decay of the 2.3-day half-life ^{239}Np precursor to ^{239}Pu:

1. In the United States, by coprecipitation of plutonium with bismuth phosphate carrier followed by coprecipitation of plutonium with lanthanum fluoride carrier (Du Pont 1944);
2. In the USSR, by sequential sodium uranyl acetate and lanthanum fluoride carrier coprecipitations of plutonium (Il'enko and Abramova 1999).

Further details on plutonium separation and purification by precipitation and crystallization methods are provided in later sections of this chapter. First, however, the historical and technological bases of the precipitation and crystallization methods in radiochemistry are examined.

3.2 HISTORY AND TECHNOLOGICAL BASES OF RADIOCHEMICAL PRECIPITATION AND CRYSTALLIZATION PROCESSES

Processes for plutonium separation from irradiated uranium by precipitation techniques are historically rooted in the deep scientific groundwork laid by pioneering chemical research into the separation of the similarly scarce natural radioelements polonium and radium present in uranium ores or in residues from uranium extraction.

This early work began with identification of polonium and radium by physicists Marie Skłodowska Curie and Pierre Curie and chemist Gustave Bémont (Curie and Curie 1898; Curie et al. 1898). Initial investigations were motivated by Marie Curie's observation that, although pure metallic uranium was radioactive, the radioactivity of the insoluble denuded residues from uranium extraction had 4.5 times higher specific radioactivity (Strutt 1904). Research by the Curie laboratory into the sources of the higher specific radioactivity was performed through heroic chemical fractionation techniques and led to the discoveries of polonium and radium.[1]

Polonium (primarily of isotope 210) is present in aged uranium ore in concentrations proportional to the ratio of the polonium-210 and uranium-238 half-lives ($t_{½}$ = 138 days and 4.47 × 10^9 years, respectively) or about one ten-billionth (8.5 × 10^{-11}) of the concentration of uranium. The Curie and Curie (1898) separation of polonium (Po) from uranium ore extraction residues was effected by coprecipitation with native (to the ore) bismuth (Bi), first as a sulfide (with copper and lead), the dissolution of the bismuth sulfide to form a sulfate in sulfuric acid, coprecipitation with bismuth as a hydroxide, and finally by its fractional dissolution away from bismuth in dilute acid (Figure 4 in Adloff and MacCordick 1995). Experiments also showed that Po could be separated from ore residues, which also have Bi, by sublimation of the sulfides (Adloff and MacCordick 1995). The contact reduction of Po on less noble metals, including Bi, also effects some separation of a purer Po fraction (Bagnall 1957, 34). However, the exceedingly small concentration and correspondingly miniscule amounts of Po as well as polonium's high specific activity make it difficult to achieve further purification, concentration, and thus chemical characterization by these techniques. Noteworthy is the fact that the technique and terminology of carrier precipitation was introduced by the Curies in their work to discover Po by their observations that the trace Po was chemically "carried" by the vastly more concentrated Bi as the sulfide and hydroxide.

Radium (Ra-226 or Ra) is present in the natural decay chain of aged uranium-238 at about 3.6 × 10^{-7} times the uranium concentration (the $t_{½}$ of ^{226}Ra is 1,602 years). Later in 1898, the year they discovered Po, the Curies, with their colleague Gustave Bémont (1898), identified Ra in a laboriously purified fraction isolated from uranium ore. According to the Kirby and Salutsky review of Ra radiochemistry written 66 years after this work (1964, 29), the method used by the Curie laboratory for extracting Ra from ore had not been improved.

In the Curie/Bémont process, the ore first is pulverized and then is digested in fused molten sodium carbonate. The cooled solid fusion is washed, first with water to dissolve the soluble carbonates (including the flux), then with dilute sulfuric acid to remove the soluble uranium. To bypass this initial uranium extraction sequence, extractions targeting radium, therefore, often began with the denuded uranium

---

[1] The residues investigated were from the sulfuric acid digest of sodium carbonate-roasted pitchblende (uraninite; primarily UO_2) and were retrieved from heaps discarded in the pine forests from uranium recovery processing near the Joachimsthal mine in Bohemia. Marie Curie processed tons of these residues in 20-kg batches, first by removing pine needles and other debris before boiling and manually stirring the sludges with an iron rod nearly her size in a drafty glass-roofed shed, to isolate, after thousands of recrystallizations, about 0.1 g of almost pure radium chloride (Fröman 1996).

extraction residues. The residues then are subjected to a series of acid and alkaline digestions and solid/solution separations, each time producing fractions enriched in the radioactive radium and its chemically analogous homolog carrier, barium (Ba). The final separation of radium from the Ba carrier takes place by fractional crystallization of radium chloride ($RaCl_2$) away from the accompanying slightly more soluble barium chloride ($BaCl_2$).

The chemical processing scheme utilized by the Curies to isolate radium from pitchblende (characterized by high uraninite, UO_2, concentrations) is shown in Figure 3.1 (after Strutt 1904). This diagram indicates, in bold, the fractions richer in

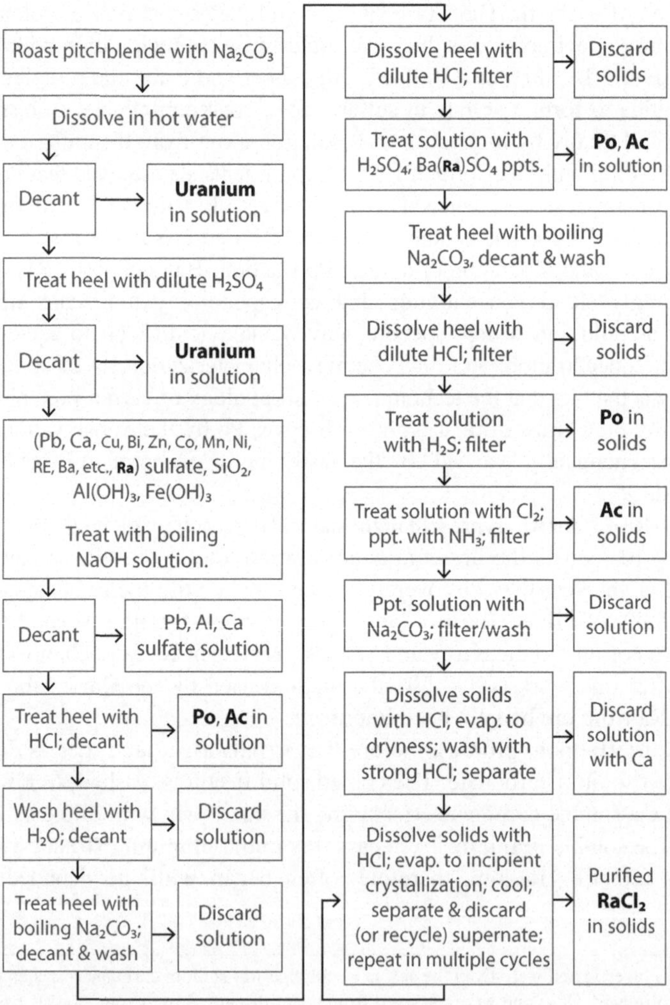

FIGURE 3.1 Curie/Bémont process for chemical separation of radium from uranium ore. (Strutt, R.J., The Treatment of Pitchblende Residues on a Large Scale, In *The Becquerel Rays and the Properties of Radium*, 37–41 and Appendix C, Arnold, Edward, London, UK, 1904.)

uranium (U), polonium (Po), and actinium (Ac), present as the uranium-235 decay-chain isotope, ^{221}Ac ($t_{1/2}$ 21.8 years). The U-, Po-, and Ac-rich fractions are collected as side streams. Radium is carried through nearly the entire process as a coprecipitate with Ba, occurring, in order, as the sulfate, chloride, carbonate, sulfate, carbonate, and, finally, chloride. The Ra separation concludes with the above-mentioned multiple fractional crystallizations of $RaCl_2$ away from $BaCl_2$.

Interposing treatments achieve further purifications by leaving the Ba/Ra combination, either as solution or solid, intact while precipitating or dissolving, respectively, other impurity elements. The treatments and separations are achieved by the following steps:

- Treatment with sodium hydroxide, which leaves Ra coprecipitated with Ba as sulfate, but dissolves lead and calcium sulfates and aluminum hydroxide;
- Treatment with hydrochloric acid, which leaves Ra coprecipitated with Ba as sulfate but partitions Po and Ac to the solution;
- Treatment with hydrogen sulfide, which captures some Po as a coprecipitate but leaves Ba and Ra in solution;
- Treatment with ammonia, which captures, by coprecipitation, Ac as the hydroxide but leaves Ba and Ra in solution.

The radium separation and isolation scheme developed by the Curies and Bémont features crystallization processes and technique enhancements for the trace element recoveries and chemical isolations subsequently used in large-scale industrial radiochemical processing. The types of chemical precipitation process steps used in the Curie/Bémont radium recovery scheme include:

- Use of a gross (macro) element (barium) to capture a trace element (radium) by carrier precipitation;
- Use of a variety of carrier compounds ($BaSO_4$, $BaCl_2$, and $BaCO_3$ for Ra) to effect separation from a corresponding variety of more soluble impurity elements;
- Use of solutions (sodium hydroxide and hydrochloric acid) to dissolve and rinse impurity elements away from solid carrier compounds of the desired element (Ra in Ba);
- Use of precipitating agents (hydrogen sulfide and ammonia) to carry impurities away from solutions of the desired element (Ba/Ra);
- Use of fractional crystallization to separate the target material (Ra, as purified radium chloride, $RaCl_2$) from its carrier element (Ba).

The fractional crystallization efficacy in the last step later was found to be improved by use of bromide instead of chloride, forming $BaBr_2$ and the favored $RaBr_2$ (Salutsky et al. 1953). Fractional *precipitation* of radium also has been practiced at the last step of the Ba/Ra separation process (Salutsky et al. 1953 and references therein, particularly Henderson and Kracek 1927). The precipitation agent, chromate (CrO_4^{2-}) is added to the purified dilute acid solution of Ba/Ra present at the penultimate step in the Curie/Bémont radium recovery scheme. The solution then is made alkaline. Radium chromate ($RaCrO_4$) precipitation is favored compared with that of barium as $BaCrO_4$. The radium-enriched precipitate is collected, dissolved in acid, and the process repeated by making the acidic solution alkaline. Homogeneous generation of alkaline conditions, as done by formation of ammonia through hydrolysis of urea or cyanate, and avoidance of recrystallization favors ideal separation factors of about 15 (Salutsky et al. 1953). The fractional precipitation method is increasingly attractive for concentrated radium solutions whose volumes become diminishingly small and difficult to handle by fractional crystallization methods.

3.3 PRINCIPLES OF CARRIER PRECIPITATION

Carrier precipitation of a trace element by a compound of a bulk element plausibly occurs by several mechanisms (Blaedel and Meloche 1963). These mechanisms are not necessarily mutually exclusive; one or more likely are operant in practically any "real" system. The mechanisms include:

- Isomorphic substitution, wherein the trace element and bulk element compounds have the same crystal structure and the trace element can substitute directly for the bulk element.
- Surface adsorption, wherein the trace element sorbs indiscriminately, with no unique structural similarity, onto the surface of the host solid phase.
- Internal adsorption and occlusion, wherein the trace element compound is not strictly isomorphous with the bulk element compound but is accommodated within the host crystal structure because of shared anions and nearness of structural fit.
- Anomalous mixed crystal formation, wherein the trace element compound is not isomorphous with the bulk element compound in three dimensions but is isomorphic or near-isomorphic along a two-dimensional plane and deposits preferentially upon that plane.
- Solution occlusion, wherein the dissolved trace element is only physically entrapped within the host bulk element compound as pockets of solution.

These various carrier precipitation mechanisms are sketched in Figure 3.2.

Carrier precipitation occurs most efficiently by the incorporation of the trace metal in a carrier compound that has very low solubility and is isomorphic with (i.e., has the same crystal structure and nearly the same unit cell dimensions) the corresponding trace element compound.[2] Such is the case for radium sulfate and its carrier, barium sulfate. For Pu(VI) as PuO_2^{2+} (the plutonyl cation), isomorphous carrier precipitation can be accomplished using sodium uranyl triacetate ($NaUO_2(O_2CCH_3)_3$) wherein PuO_2^{2+} can substitute at all proportions for U(VI) as UO_2^{2+}, the uranyl cation. Both

---

[2] Note that isomorphic coprecipitation does not occur if the isomorphs' solubilities differ greatly. For example, NaCl (0.563 nm lattice spacing) and AgCl (0.555 nm lattice spacing) are cubic isomorphs. However, sparingly soluble AgCl contains negligible freely soluble NaCl upon AgCl precipitation from NaCl solution (page 727 of Blaedel and Meloche 1963).

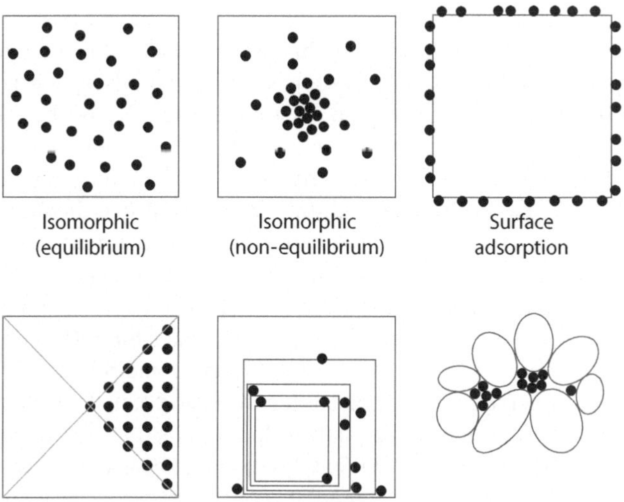

FIGURE 3.2 Distribution of trace elements in coprecipitation processes. The trace elements within the bulk host element compounds are depicted as dots.

equilibrium and non-equilibrium types of isomorphic coprecipitation are depicted in the upper and middle left of Figure 3.2. The differences between these two types will be described in subsequent paragraphs.

The bulk precipitate itself may also be very poorly crystalline (e.g., ferric hydroxide, $Fe(OH)_3$) and of high surface area. Such indiscriminate structure makes these ready hosts for a variety of similarly low-solubility trace metal hydroxides or oxides, including those of Pu(III), (IV), and even (VI), by attraction to their abundant surfaces. As such, metal hydroxides can serve to decontaminate solutions of dissolved plutonium by carrier precipitation processes. At the same time, the metal hydroxides generally provide little separation or purification from other low-solubility metal hydroxides (e.g., rare earths, most transition metals, other actinides). Even here, however, some regularities can be found. Depending on the pH, the trace component may exist in its low pH ionic state, wherein the trace ion is exchanged on the surface of the bulk precipitate electrostatically or may exist in a hydrolyzed state at higher pH, wherein chemisorption onto the bulk precipitate occurs (Grebenshchikova 1972). Surface adsorption is depicted in the upper right corner of Figure 3.2.

Carrier precipitation may also occur even if the respective carrier and trace element are not isomorphous but each form low solubility compounds sharing the same anion (DuPont 1944). In this case, the trace cation can bond onto the surface of the bulk precipitating carrier compound. The trace cation then becomes covered with the bulk carrier compound in a mechanism called internal adsorption or occlusion. The bulk crystal is distorted in the region of the engulfed trace cation to a greater or lesser degree, depending on the mismatch in the respective crystal structure or unit cell dimension occasioned by accommodating the trace cation. Internal adsorption and occlusion are depicted in the middle illustration of the bottom row in Figure 3.2.

Between pure isomorphic crystallization and internal adsorption is anomalous mixed crystallization. In this mechanism, the trace element is adsorbed as thin layers onto a particular plane or planes of the bulk crystal. The layer containing the trace element then becomes covered by an overlying layer of the bulk crystal (Aquilano and Pastero 2013). The resulting crystal, thus, consists of two distinct phases with the trace element compound forming lamellae on the structurally accommodating two-dimensional faces of the bulk crystal. At high relative concentrations, Pu(IV) has been inferred to form anomalous mixed crystals with lanthanum fluoride (LaF_3) based on the behaviors of Ce(IV) and U(IV) as surrogates for Pu(IV) because of the similar Pu(IV), Ce(IV), and U(IV) ionic radii (Schlyter and Sillén 1950). Likewise, the uptake of Pu(IV) onto particular bismuth phosphate precipitate crystal faces has been postulated based on the similar lattice dimensions on these faces for Pu(IV) phosphate and Bi(III) phosphate (DuPont 1944). This mechanism has been postulated even though the relatively low plutonium concentrations present under Bismuth Phosphate Process conditions would not have permitted anomalous mixed crystals to form extensively if at all. Anomalous mixed crystallization is sketched in the lower left corner of Figure 3.2.

Finally, simple occlusion of trace element-bearing solution within the bulk element compound precipitate can make carrier precipitation appear to be happening when, in fact, little or no chemical uptake is occurring. Hence, washing alone can remove the trace element. Simple solution occlusion is diagramed in the lower right corner of Figure 3.2.

The distribution of a trace element to the bulk element compound by isomorphous coprecipitation can be rationalized in at least two ways. In the first way, the trace element distributes itself homogeneously throughout the bulk precipitate. This condition is illustrated in the upper left diagram in Figure 3.2. Such a distribution would occur if the solid phase were digested (dissolved and reprecipitated) to assure that thermodynamic equilibrium of the trace element to the solid and solution phases would be established. Alternatively, a pseudo-homogeneous distribution of the trace element to the solid might occur if precipitation occurs instantaneously without allowing time for equilibrium to be established.

The mathematical expression for one type of isomorphous coprecipitation was developed by Henderson and Kracek (1927) based on their studies of the fractional precipitation of Ba/Ra chromate. This distribution indicates that the ratio of the amount of the trace element to the carrier element in the solid phase is proportional to the trace and carrier element concentrations in the solution phase as shown in Equation 3.2.

$$\left(\frac{\text{trace}}{\text{bulk}}\right)_{\text{solid phase}} = D\left(\frac{\text{trace}}{\text{bulk}}\right)_{\text{solution phase}} \tag{3.2}$$

The greater the value of the distribution coefficient, D, the greater the affinity of the trace element for the solid phase. If D > 1, the trace element concentrates preferentially in the solid phase. If D = 1, the trace element shows no distinct affinity for, or aversion to, the solid. If $D < 1$, the trace element prefers the solution phase. Equation 3.2, also known as Khlopin's law, clearly is analogous to Henry's law describing the solubility of gases in solution (Grebenshchikova 1972).

In the second type of isomorphic distribution, the trace element deposits on the bulk precipitate in proportion to the instantaneous concentration of the trace element in solution. Thus, at any given instant, the deposition follows Equation 3.2. However, as the depositions of the trace and bulk elements occur differentially over time, the concentration of the trace element in the continually formed outer precipitate layer changes. The trace element already adsorbed becomes incorporated within the bulk element solid phase, no longer able to exchange with the solution phase, and remains fixed. If the trace element favors deposition onto the solid phase surface, the solution becomes depleted in the trace element, leading, in turn, to decreased concentration of the trace element in the ensuing surface layers developed on the solid phase (as depicted in the upper middle row of Figure 3.2).

The mathematical form for this differentially changing second type of distribution was elucidated by Doerner and Hoskins (1925) and is described by Equation 3.3.

$$\log\left(\frac{\text{total trace}}{\text{trace in solution}}\right) = \lambda \log\left(\frac{\text{total bulk}}{\text{bulk in solution}}\right) \tag{3.3}$$

Achievement of the conditions implicit in Equation 3.3 means that dissolution and reprecipitation of the solid phase is necessarily slow relative to the initial precipitation; i.e., the solid phase is not digested or permitted to recrystallize. Such is the case for fractional precipitation of a sparingly soluble compound if the achievement of precipitation conditions in solution is done homogeneously, allowing time for the development of the differential solution concentrations (see Salutsky et al., 1953, for barium/radium chromate homogeneous precipitation). In contrast, the behavior implicit to adherence to Equation 3.2, with a homogeneous distribution of the trace element in the carrier compound solid, better fits the abrupt establishment of precipitation conditions or, alternatively, digestion of the solid (Salutsky et al. 1953).

Several generalized observations of the behavior of trace element uptake within (coprecipitation) or onto (adsorption) bulk solids are gathered under the Paneth-Fajans-Hahn rule(s) (see Gibson 1961, 19–25 and Vertés and Kiss 1987).

- The degree of uptake of the trace element in the bulk carrier by carrier precipitation is greater for trace element compounds having lower solubility.
- Adsorption is dependent on the surface area of the solid and its surface structure.
- Adsorption is more dependent on conditions such as concentration change than is coprecipitation.
- Adsorption is more dependent on the presence of competing cations than is coprecipitation.
- In both cases, behaviors may be strongly altered by the presence of cation complexation agents.

3.4 COPRECIPITATION FOR PLUTONIUM SEPARATION AND RECOVERY IN REPROCESSING

Following plutonium's discovery and early investigations into its chemical and physical properties at the University of California at Berkeley in 1941, highly ambitious and focused work into its chemistry took place at the Manhattan Project's Metallurgical Laboratory (Met Lab) at the University of Chicago beginning in February 1942. This work continued at the Met Lab until after the end of World War II when the Lab evolved and was relocated to the Argonne National Laboratory, south of Chicago, on July 1, 1946. Early studies at the Met Lab included research to identify means to separate Pu from pile-irradiated uranium. This work necessarily began by using Pu separated from cyclotron particle irradiated uranium; i.e., before Pu was produced at much greater rates in neutron-generating uranium piles/reactors. The demonstration of neutron multiplication (criticality) occurred on December 2, 1942, under Enrico Fermi's direction in mankind's first self-sustaining nuclear chain reaction at the Chicago Pile 1 (CP-1), also at the University of Chicago. The construction of the X-10 uranium pile (reactor) at the Clinton (Tennessee) Laboratory and its operation in late November 1943 then provided larger quantities of plutonium for physical and chemical research (Gosling 2010).

The Smyth Report (Smyth 1945), issued in August 1945, described to the world the $2,000,000,000 (1945 dollars) scope of the U.S. Manhattan Project to produce atomic bombs. This report provided a salient, but necessarily vague, overview of the investigations taken within the Manhattan Project to separate Pu from reactor-irradiated uranium and perspectives on the rationales behind the technical directions ultimately pursued. The Smyth (1945) Report outlined the nature of the chemical separations used in Pu processing within the Manhattan Project. In paragraph 8.21, it noted that volatility, absorption, solvent extraction, and precipitation were considered but that the Hanford Plant used precipitation. Paragraph 8.22 further elaborated details on repeated use of carrier precipitation and its impacts on plant design, while paragraph 8.23 described how the precipitation processes took advantage of differences in the behaviors of Pu(IV) and Pu(VI).

The urgency of the wartime work is evident by noting that the initial findings on the efficacy of bismuth phosphate in coprecipitating (cyclotron-produced) plutonium away from uranium in nitric acid (HNO_3) solution, at envisioned chemical processing plant concentrations, occurred on December 19, 1942 with key proof-of-principle experiments completed by March 29, 1943 (Kathren et al. 1993). Lanthanum fluoride already had been demonstrated in early 1941 to be effective in separating Pu(IV) from U(VI) (Seaborg et al. 1946). Although decontamination from fission products was still largely to-be-determined, the Hanford Site construction and operation authority contracted by the U.S. Army Corps of Engineers, DuPont, began coprecipitation plant engineering on June 7, 1943, using an ingeniously flexible design (discussed later) that allowed both processes, in either order, to be used and indeed allowed virtually any coprecipitation or chemical process to be implemented.

It is noteworthy that the initial bismuth phosphate tests occurred only days after the first self-sustaining nuclear chain reaction at the nearby Chicago Pile on December 2, 1942. Thus, the advanced Pu chemistry and separations research achieved by December

1942 exemplified not only the US's no-holds-barred multipronged attack in pursuit of nuclear weapons, but also the confidence that chain-reacting piles (nuclear reactors), the prospective source for military quantities of plutonium, would prove feasible.

Because of Pu(IV)'s marked propensity to hydrolysis, it was recognized that separations would necessarily take place in strong acid solution. Dissolution of the uranium metal or oxide target material occurs readily in oxidizing HNO_3 and less readily or conveniently in other common mineral acids that also are non-oxidizing (e.g., HCI, H_2SO_4, H_3PO_4). The resistance of stainless steel to corrosion in HNO_3 solidified its choice as the material of construction for chemical processing.

The Met Lab plutonium coprecipitation investigations cast a wide net. Coprecipitation agents found useful to capture the Pu(III), (IV), and (VI) oxidation states readily accessible in HNO_3 are shown in Table 3.1 (Hyde and Seaborg 1956; see also Hyde 1954). Plutonium(V) is not stable in HNO_3 solution.

TABLE 3.1
Coprecipitation Agents for Plutonium

Carrier Compound	Pu(III)	Pu(IV)	Pu(VI)
Insoluble hydroxides	C	C	C
Lanthanum fluoride	C	C	NC
Phosphates			
Zirconium phosphate	NC	C	NC
Bismuth phosphate	C	C	NC
Thorium pyrophosphate	NC	C	NC
Thorium hypophosphate	—	C	NC
Oxalates			
Thorium oxalate	C	C	NC
Uranium(IV) oxalate	C	C	NC
Bismuth oxalate	C	C	NC
Lanthanum oxalate	C	C	NC
Iodates			
Zirconium iodate	—	C	NC
Thorium iodate	—	C	NC
Ceric iodate	—	C	NC
Sodium uranyl acetate	NC	NC	C
Zirconium phenylarsonate	NC	C	NC
Thorium peroxide	—	C	—
Bismuth arsonate	—	C	—

Source: Hyde, E.K., and G.T. Seaborg, *The Transuranium Elements*, University of California, Berkeley, CA, UCRL-3312, https://pubarchive.lbl.gov/islandora/object/ir:146622, 1956; Hyde, E.K., Radiochemical Separation of the Actinide Elements, In *The Actinide Elements*, edited by G. T. Seaborg and J. J. Katz, In National Nuclear Energy, McGraw-Hill Book Company, New York, 1954.

Note: C means coprecipitation is complete—i.e., nearly quantitative—under proper conditions; NC means the coprecipitation is incomplete and can be made <1%–2% under proper conditions

The availability of multiple plutonium oxidation states also afforded a powerful chemical handle to affect its separation from uranium, fission products, process chemical impurities, and stainless steel corrosion products (iron [Fe], chromium [Cr], nickel [Ni]) in coprecipitation and other chemical processes. Such is not the case for radium, which has available only one oxidation state in solution.

An example will illustrate the influence of Pu oxidation state in achieving separation. As seen in Table 3.1, Pu(III) and Pu(IV) are carried by coprecipitation with lanthanum fluoride (LaF_3), but Pu(VI) is not. Thus, a LaF_3 coprecipitation cycle beginning with dissolved Pu in reduced states—Pu(III) and/or Pu(IV)—will carry the Pu. Fission product rare earths (e.g., radiocerium) also are carried by the LaF_3, but hexavalent uranium, U(VI), is not and remains in solution. Solid-liquid separation by filtration or centrifugation and washing of the precipitate partitions the plutonium, with the LaF_3 and rare earths, to the solid phase and the uranium to the solution phase. Metathesis of the LaF_3 to $La(OH)_3$ followed by HNO_3 treatment will dissolve the La and its captive Pu and rare earth hydroxides. Treatment of the dissolved Pu with a strong oxidant will form Pu(VI) but leave La and the rare earths in their (III) states. The bulk La can then be reprecipitated with fluoride to again form LaF_3. The LaF_3 will carry the fission product rare earths but leave the Pu(VI) in solution. Separation of the dissolved Pu from the trace fission product rare earths held in the bulk LaF_3 solids, via filtration or centrifugation and washing, is thus achieved. Parallel separations of other unwanted contaminants from the Pu are also accomplished because most elements do not share Pu's two different oxidation state precipitation behaviors with respect to the bulk LaF_3.

3.4.1 The Bismuth Phosphate Process at Hanford

The first industrial-scale reprocessing to extract plutonium from irradiated uranium occurred in the United States at the Hanford Site T Plant on December 6, 1944, using slugs irradiated at the Clinton (now known as Oak Ridge) X-10 reactor and dummy non-irradiated slugs. Subsequent runs, on December 14, 1944, using some slugs irradiated in the Hanford Site B Reactor (with dummy non-irradiated slugs), and on December 26, 1944, using a complete load of irradiated slugs, then occurred. The B Plant, twin to T Plant, began operation on April 13, 1945 (DuPont 1946a, 64–65). Plutonium recovery was 60%–70% in the first two months of Bismuth Phosphate Process operation, 90% in the third, 93% after six months, and then above 95% with a decontamination factor (DF) of 10^8 (Seaborg 1958). As will be seen in later experiences in the USSR, the early low recoveries may reflect plutonium uptake onto process equipment surfaces.

The initial steps in Hanford reprocessing are dissolution of the cladding followed by dissolution of the irradiated uranium metal fuel itself. Chemical dissolution of the aluminum cladding was accomplished by treatment with sodium hydroxide (NaOH) solution, which vigorously corrodes Al metal but has little effect on the underlying uranium metal core. Sodium nitrate ($NaNO_3$) is added to the NaOH as a hydrogen radical scavenger to suppress flammable hydrogen gas formation (Schneider 1951). Dissolution of the uranium metal fuel cores is accomplished in hot concentrated HNO_3. Dissolved uranyl nitrate is the principle product along with Pu and myriad fission product nitrates (see Chapter 2, Section 2.2).

Oxidation state adjustments are necessary in all Pu separations processes. In the Bismuth Phosphate Process, chemical reduction to the Pu(IV) state used $NaNO_2$ for the dissolver solution; ferrous ammonium sulfate [$Fe(NH_4)_2(SO_4)_2$] and, later, ferrous sulfate ($FeSO_4$) were used in subsequent steps. Ferrous ions rapidly reduce Pu to the (III) state, which then oxidizes to (IV) in HNO_3. Adjustment to the Pu(VI) state used sodium bismuthate ($NaBiO_3$) and sodium dichromate ($Na_2Cr_2O_7$) (DuPont 1944; Schneider 1951).

The Bismuth Phosphate Process used at the Hanford Site employed first $BiPO_4$ and then LaF_3 in a carrier precipitation sequence and alternated the Pu oxidation state between (IV) and (VI) in the process with the (IV) state carried and the (VI) state remaining in solution for both $BiPO_4$ and LaF_3 carrier precipitates (Table 3.1). The sequence is shown in Figure 3.3, with reagents introduced from the left, the Pu-rich stream proceeding down the center column, and the waste streams indicated in the right column. Although not shown in Figure 3.3, these waste streams, if not already alkaline, were made alkaline with sodium hydroxide or sodium carbonate solution so that they could be held in the mild steel-lined reinforced concrete underground waste storage tanks on the Hanford Site.

As shown in Figure 3.3, the Bismuth Phosphate Process involves four major coprecipitation cycles:

1. *Extraction*, using $BiPO_4$ to separate the trace Pu, present at about 250 g per metric ton of irradiated uranium (2.5×10^{-4} g Pu/g U) during the Manhattan Project era, from nearly all of the irradiated uranium and 97% of the fission product activity.
2. *First Decontamination Cycle*, using $BiPO_4$ to effect purification of Pu from fission products (aided by added zirconium, Zr, and cerium, Ce, scavengers targeting these fission product elements, and the other fission product rare earths), process chemicals, and process impurities (such as stainless steel corrosion products).
3. *Second Decontamination Cycles*, using $BiPO_4$ (but without added Zr and Ce) to further purify Pu from fission products, process chemicals, and process impurities.
4. *Concentration* or *Crossover and Metathesis*, using LaF_3 to achieve further purification, but primarily to concentrate the Pu with minimal accompanying carrier.

In the Extraction step, $BiPO_4$ precipitation carries Pu(IV). The uranium is prevented from precipitating as uranyl phosphate by the addition of sulfuric acid (H_2SO_4). The sulfate complexes the uranium(VI) to keep it in solution. The Pu-bearing $BiPO_4$ solids are removed from the accompanying solution by centrifugation and washing. The dissolved uranium(VI) and the fission products in the solution are made alkaline and collected in the tank farms as metal waste for future processing to recover uranium values.

In the subsequent First Decontamination Cycle, the $BiPO_4$ cake with contained Pu is dissolved in strong acid, the Pu oxidized to the (VI) oxidation state, and $BiPO_4$ reprecipitated in the presence of added Zr and Ce, leaving the non-carried Pu(VI) in

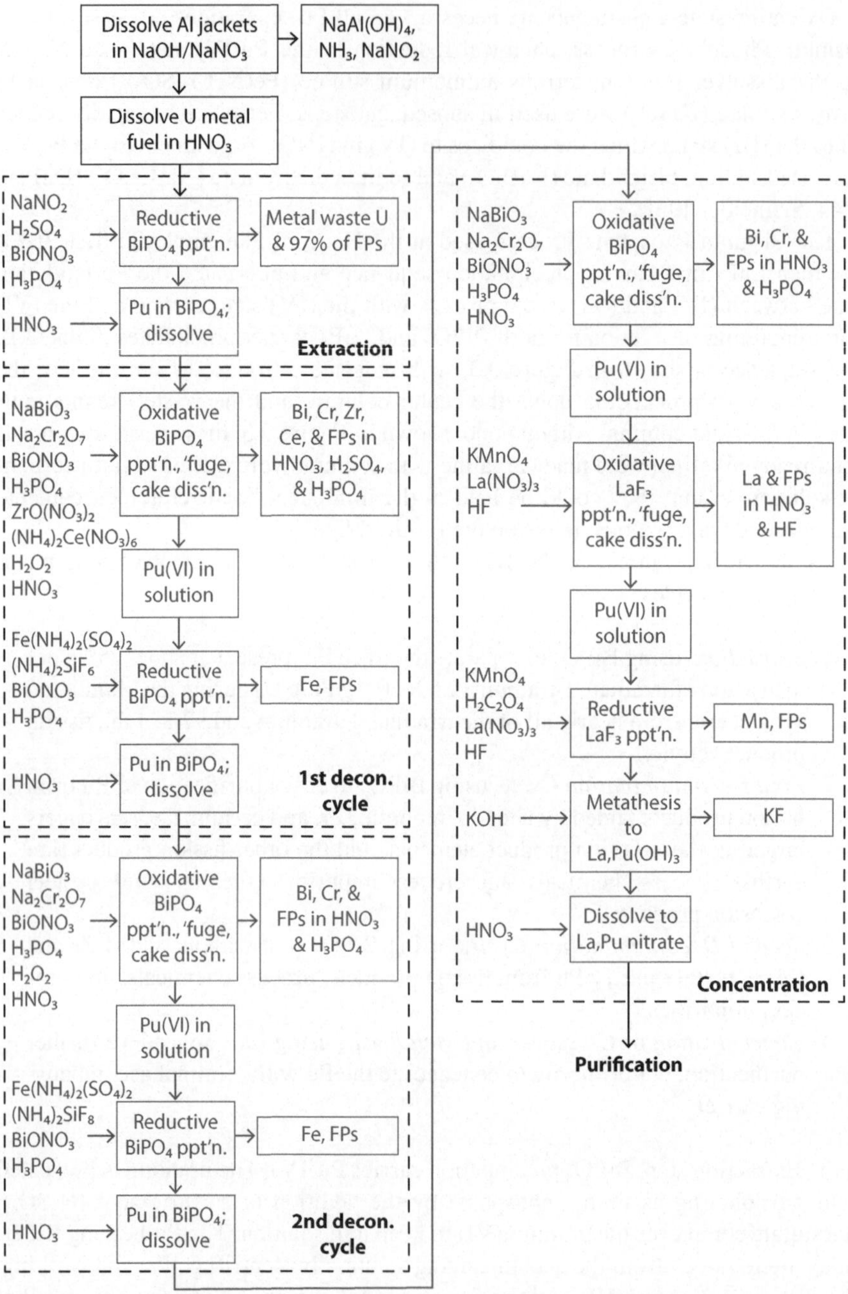

FIGURE 3.3 Hanford Bismuth Phosphate Process flowchart. (Derived from DuPont, *Hanford Engineer Works Technical Manual*, General Electric, Hanford Atomic Products Operation, Richland, WA, HW-10475 C, https://pdw.hanford.gov/arpir/pdf.cfm?accession=0076415H, 1944.)

solution, while $BiPO_4$ carries other impurities and fission products, particularly Zr, Ce, and the rare earths chemically analogous to Ce. As will be seen, the added Zr and Ce dilute the chemically trace fission product Zr and rare earths to sweep them from solution. The $BiPO_4$ cake with fission products and contaminants is dissolved, made alkaline, and discarded as first cycle (1C) waste to the tank farms.

The Pu(VI) in solution is then chemically reduced to Pu(IV) and the $BiPO_4$ precipitation process sequence repeated, first with Pu(IV) and then with Pu(VI) in the Second Decontamination Cycle. Successively smaller quantities of $BiPO_4$ are used in the First and Second Decontamination Cycles to further concentrate the Pu. As in the first cycle, the second cycle (2C) wastes are made alkaline and discarded to the tank farms.

Precipitation using LaF_3 follows, first with non-carried Pu(VI)—primarily to remove fission product rare earths with the LaF_3 cake—and then, after reduction to Pu(IV), another LaF_3 precipitation to carry the plutonium. The Pu-bearing LaF_3 is then metathesized with potassium hydroxide to form $La(OH)_3$. The Pu remains with the La while the metathesis eliminates corrosive fluoride—as dissolved potassium fluoride—by it being sent to the waste tank farms.

Because these latter LaF_3 precipitations not only achieve further decontamination but also chemically concentrate the Pu, this sequence is called the Concentration Cycle. The La:Pu mole ratio in the $La(OH)_3$ and the resulting solution is about 8.7:1, an ~460-fold Pu concentration increase in the solid phase compared with the ~4,000:1 U:Pu mole ratio in the irradiated fuel.

The Pu-bearing $La(OH)_3$ cake is then dissolved in HNO_3. This dissolution product proceeds to a sequence of carrier-free Pu purification steps to form increasingly concentrated and decontaminated Pu compounds. Discussion of these plutonium purification processes, most involving precipitation, will be taken up in subsequent sections of this chapter.

The First Decontamination Cycle differs from the Second Decontamination Cycle by the addition of Zr and Ce as zirconyl nitrate, $ZrO(NO_3)_2$, and ceric ammonium nitrate, $(NH_4)_2Ce(NO_3)_6$, respectively, to the oxidative $BiPO_4$ precipitation strike. The amounts of Zr and Ce added are appreciable, about 10% of that for bismuth. These reagents chemically dilute the radiologically prominent but chemically trace Zr and Ce (and other rare earth) fission products. The added Zr and Ce nitrates accompany the analogous fission products to make the total sufficiently concentrated to precipitate as their respective phosphates rather than be carried by the $BiPO_4$. The resulting phosphates also have high specific surface area, which can lead to some Pu product loss by adsorption. Although the zirconium phosphate is bulky and difficult to centrifuge, its handling in centrifugation is improved by the coincident ceric phosphate. Removal of the $BiPO_4/ZrO(H_2PO_4)_2/Ce_3(PO_4)_4$ cake from the centrifuge bowl for subsequent dissolution and discharge to waste is aided by the addition of hydrogen peroxide (H_2O_2), which reduces the ceric (Ce(IV)) phosphate to the more readily acid-soluble cerous (Ce(III)) phosphate; oxygen gas bubbling from peroxide decomposition also helps loosen the compacted cake (DuPont 1944, 604 and 608).

Figure 3.3 also shows that ammonium silicofluoride, $(NH_4)_2SiF_6$, is added in reductive $BiPO_4$ precipitation steps in both the First and Second Decontamination Cycles. The fluoride contained in this reagent complexes the highly hydrolyzable carrier and fission product Zr and fission product niobium (Nb) to keep them in solution and not carried with the plutonium in the $BiPO_4$ cake (DuPont 1944, 151).

The efficacy of the Bismuth Phosphate Process, including both the $BiPO_4$ and LaF_3 carriers, in retaining the Pu delivered from the reactors, separating the Pu from bulk uranium and various process chemicals and carriers, and decontaminating the Pu from fission products is illustrated in Figure 3.4. The process quantity basis is the dissolver charge of 1.5 metric tons of irradiated uranium metal. Under the flowsheet outlined in the Hanford Technical Manual (DuPont 1944), this amount of uranium is accompanied by 375 g of Pu and contains 3×10^5 curies of fission products. The nominal amount of Pu remaining in solution after the concentration step as plutonium nitrate (1.48 moles) in the presence of lanthanum nitrate (10.27 moles) and HNO_3 is ~354 g, a Pu yield of ~94.4%. The Pu-bearing stream, thus, is decreased from 1.5 metric tons in uranium metal to just over 4 kg as La/Pu nitrate salts, a concentration factor of (1,500/4 =) 375. The accompanying fission products are decreased to 0.03 curies, a decontamination factor (DF) of ($3 \times 10^5/0.03 =$) 10^7; process experience in the first three months of B Plant operation (to 1 July 1945) demonstrated average DFs of ~2×10^7 (DuPont 1946a, 95).

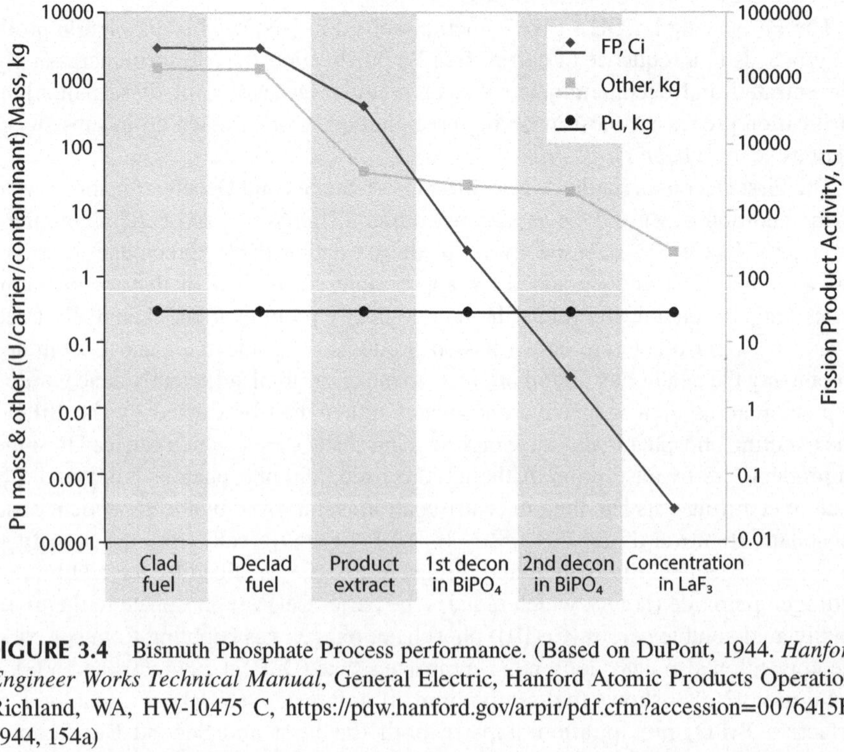

FIGURE 3.4 Bismuth Phosphate Process performance. (Based on DuPont, 1944. *Hanford Engineer Works Technical Manual*, General Electric, Hanford Atomic Products Operation, Richland, WA, HW-10475 C, https://pdw.hanford.gov/arpir/pdf.cfm?accession=0076415H, 1944, 154a)

The Bismuth Phosphate Process developed at the Met Lab from 1942 to 1944 and used at the Hanford Site from 1944 until 1956 implemented a broad arsenal of carrier precipitation techniques to achieve high selectivity, product recovery, and purification:

- Two different Pu carriers, $BiPO_4$ and LaF_3, in decreasing quantities in successive Pu coprecipitation steps to increase Pu concentration in the processed solids;
- Two different Pu oxidation states, (IV) and (VI), for both $BiPO_4$ and LaF_3 carriers with Pu(IV) carried and Pu(VI) not carried in the respective reduced and oxidized precipitation steps;
- Uranium(VI) complexation by sulfate to prevent it following Pu in $BiPO_4$ precipitation;
- Fission product Zr and Nb complexation by fluoride to prevent them following Pu in $BiPO_4$ precipitation;
- Addition of macro quantities of Zr and Ce to carry trace fission product Zr and rare earths as their respective phosphate precipitates, while Pu(VI) remains in solution;
- Metathesis of Pu-bearing LaF_3 in strong hydroxide solution to eliminate corrosive fluoride from ensuing process steps.

Subsequent process improvements quickly occurred, mostly in decreasing the reagent quantities and resulting waste volumes. Waste generation rates in 1945 were 64 L/kg of irradiated uranium metal but decreased to an average ~30 L/kg uranium over the process lifetime (Barney and Delegard 1999).

The $BiPO_4$ processing took place in the identical 221-T and -B Buildings, each 247-m long, thickly constructed reinforced concrete "canyons" consisting of a line of 40 separate identical rectangular cells, each cell being about 4.0 × 5.4 × 6.7 (high) meters. Each cell contained 42 connector flanges exactly arrayed on the front (10) and two side walls (16 each) about 5.3 m above the cell floor. The flanges, in the standard arrangement, provided 16 water, steam, and chemical feed lines; 13 electrical and instrument lines; and 13 process fluid lines. The process equipment—e.g., tanks, dissolvers, precipitators, centrifuges—was held in jigs that could be located precisely within each cell using guides built into the cell floor. A paired set of cells containing a "standard section" for a single precipitation and decantation step is shown in Figure 3.5. The first cell contained a precipitator vessel and a solution tank (for the solids dissolved from the centrifuge), while the second cell contained a centrifuge and a supernatant solution catch tank.

The precise relative geometric locations of the guides and flanges within the cells required unprecedented concrete construction tolerances but allowed process equipment and associated jigs, shop-fabricated to similarly close tolerances, to be reliably installed in the process cells remotely, using an overhead bridge crane run by a skilled and practiced operator within a heavily shielded cab. The process equipment could then be connected to the fluid and electrical services and the process lines using appropriately dimensioned "jumpers," likewise shop-fabricated and having

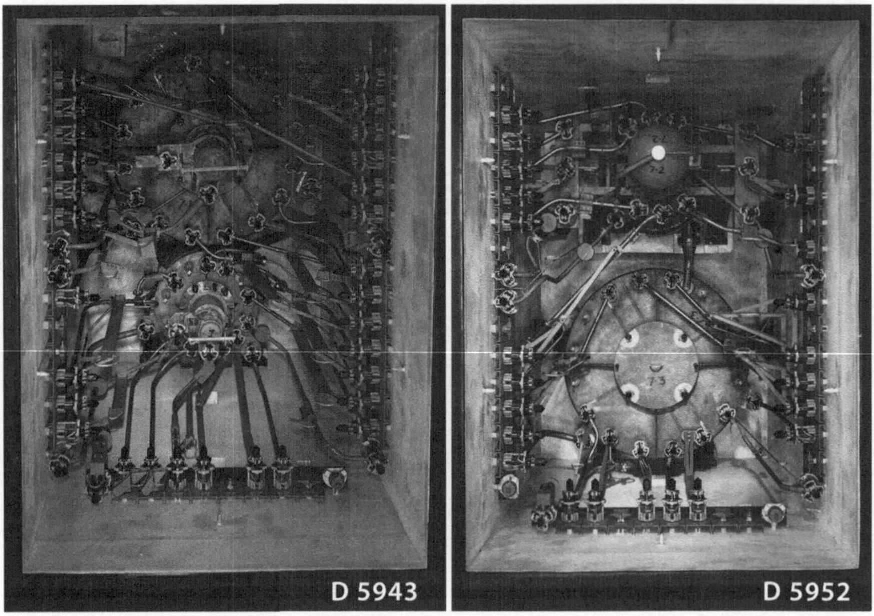

FIGURE 3.5 Overhead views of $BiPO_4$ process cells. Left – Precipitator (top) and solution tank. Right – Centrifuge (top) and supernatant catch tank. (From Figures 3 and 4, respectively, Courtesy of DuPont 1944.)

strategically located bales to allow their proper orientation when hung from a cable hook for installation by the overhead crane. The jumpers were secured to the appropriate cell flanges and process equipment with mechanical screw-and-cam connectors that could be opened and closed remotely using a cable-slung impact wrench, likewise, operated from the crane. A jumper with three connectors, one at the top and two near the bottom, and a sketch of a connector are shown in Figure 3.6. For radiological safety, the operators visually monitored their actions from the shielded crane cabs using periscopes and early application of television. This highly flexible process equipment design allowed radiologically safe remote maintenance and operation. In fact, to assure that failed or upgraded equipment could be replaced after "hot" operations commenced (i.e., operations with highly radioactive materials), the crane operators first became practiced and proficient by installing the original equipment under nonradioactive conditions using the remote-handling protocols. Once the T and B Plants became operational and "went hot," this training occurred in the reserve identical, but not operational, U Plant.

3.4.2 The Sodium Uranyl Acetate Process in the USSR

The "secrets" of the U.S. Manhattan Project did not remain secret for long, largely because of the efforts of sympathetic workers within the Manhattan Project, willing

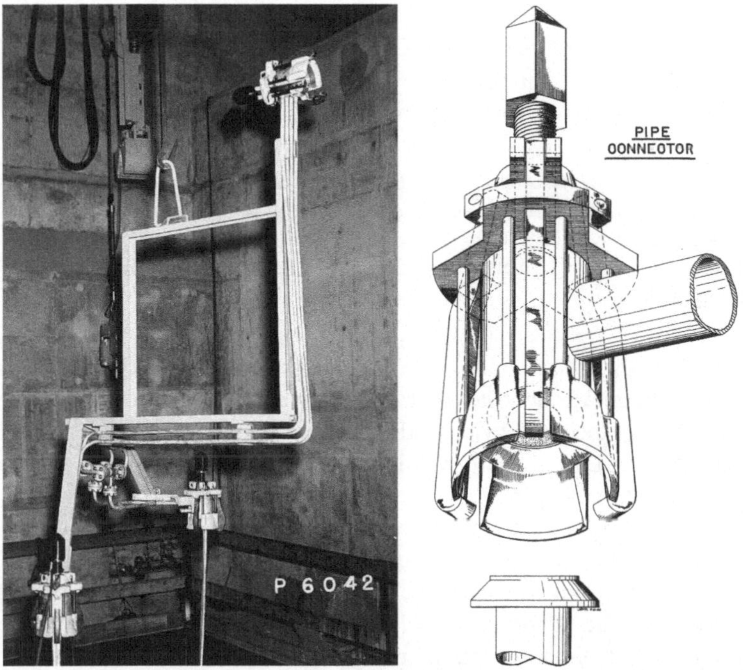

FIGURE 3.6 Hanford jumper and pipe connector (installation of a piping jumper in T Plant canyon cell, 23 August 1944, and Figure 36 of DuPont 1944).

to share the sensitive technical information with USSR espionage agents (Hawkins 2013). As a result of this intelligence and the massive USSR resources devoted to matching the demonstrated US nuclear capability shown in the events at Hiroshima and Nagasaki in August 1945, the USSR's major milestones in atomic fission weapons development occurred following cessation of World War II hostilities, almost exactly four years after their occurrences in the United States. Thus, the first nuclear criticality in the Chicago Pile on December 2, 1942, was mirrored by the first criticality in the Soviet reactor Ф-1 near Moscow on December 26, 1946. The first production reactor criticality, at Hanford's B Reactor, occurred on September 26, 1944; Mayak's A Reactor first reached criticality on June 10, 1948. Similarly, the US Trinity Test of the plutonium-based "gadget," occurring on July 16, 1945, was followed with the USSR's first atomic explosive test РДС-1 or "Joe-1" using an identical plutonium-fueled design at the Semipalatinsk test site on August 29, 1949 (Gosling 2010, 63–65; Kruglov 2002, 38–39, 55, 92).

Thus, it is noteworthy that the USSR did not follow the United States in using the Bismuth Phosphate Process of sequential $BiPO_4$ and then LaF_3 carrier precipitations for plutonium separation from irradiated uranium. Instead, the USSR used sequential $NaUO_2(O_2CCH_3)_3$ and LaF_3 carrier precipitations abetted by Pu(VI) nitrate extraction in diethyl ether (Il'enko and Abramova 1999). However, even here, the USSR was sufficiently aware of U.S. efforts in plutonium separation that on May 10, 1946, M. G. Pervukhin, the USSR Minister of Chemical Industry, could inform L. P. Beria, head of the USSR People's Commissariat for Internal Affairs and head of the USSR atomic effort (Ryabev 2000):

> Of all the known [emphasis added] methods for the separation of plutonium: a) bismuth phosphate, b) lanthanofluoride, c) ether, d) acetate-lanthanofluoride, the acetate-lanthanofluoride method has the greatest practical importance at present as the most studied theoretically and to a greater extent than all the rest that have been checked on the laboratory scale.

This was when the USSR only possessed plutonium from cyclotron production (Il'enko and Abramova 1999) to verify the very same coprecipitation separations methods, including the non-obvious $BiPO_4$ carrier, undertaken four years earlier by the secret U.S. Met Lab research. While the USSR did not use $BiPO_4$, the carriers selected were strikingly akin to those used by McMillan and Abelson (1940) in the isolation and identification of neptunium.[3]

Information on the development and workings of the initial USSR processes to separate plutonium from irradiated uranium is neatly summarized by Il'enko and Abramova (1999). The authors aver that $BiPO_4$ coprecipitation was tested (by Alexandr A. Grinberg of the Radium Institute in Leningrad) and even, unlike Hanford's Bismuth Phosphate Process, included uranium recovery measures. However, the key bismuth reagent was stated to be unavailable in the USSR. Therefore, preference was given to processes, such as those selected, that could be undertaken using domestically available chemicals. Interestingly, doubts about the availability of bismuth were also voiced in planning for US operations. In fact, bismuth reagent delivery delays nearly forced suspension of Hanford T and B Plant processing in early 1945 (DuPont 1946a, 52).

The "acetate-lanthanofluoride" process developed under the direction of Vitaly G. Khlopin by key chemists Boris A. Nikitin and Alexandr P. Ratner of the Radium Institute in Leningrad, is outlined in Figure 3.7 and was implemented at Plant Б at the Mayak site in the Ural Mountains (Il'enko and Abramova 1999; Kruglov 2002, 78–79).[4] In this diagram, the process reagents are introduced from the left, the plutonium-rich streams proceed down the center columns, and the waste and side streams are indicated in the right columns within each cycle. The sodium uranyl triacetate side stream, indicated as "$NaUO_2Ac_3$ ppt. for U recovery," is directed to uranium recovery and not to the waste. As at Hanford, the waste streams, if not already

[3] Both used sodium uranyl triacetate in the initial separation steps. Lanthanum(III) fluoride then was used in the USSR to carry Pu(IV) rather than the chemically analogous Ce(III) fluoride, which was used by McMillan and Abelson (1940) to carry Np(IV).

[4] The Mayak site was within a closed region of the USSR and was originally named Chelyabinsk-40 and then Chelyabinsk-65. The letters A, Б, and B (equivalent to A, B, and V in English) are the first three in the Cyrillic alphabet and were used in the Mayak plutonium production project to represent the reactor (Plant A), the reprocessing plant (Plant Б), and the plant for plutonium finishing and shaping to metal parts (Plant B). Shop No. 9, an open laboratory with few fume hoods and no gloveboxes except for plutonium metal fabrication, was used for initial plutonium finishing processes and metal shaping in the period of February–August 1949 before Plant B started operations (Sokhina et al. 2003).

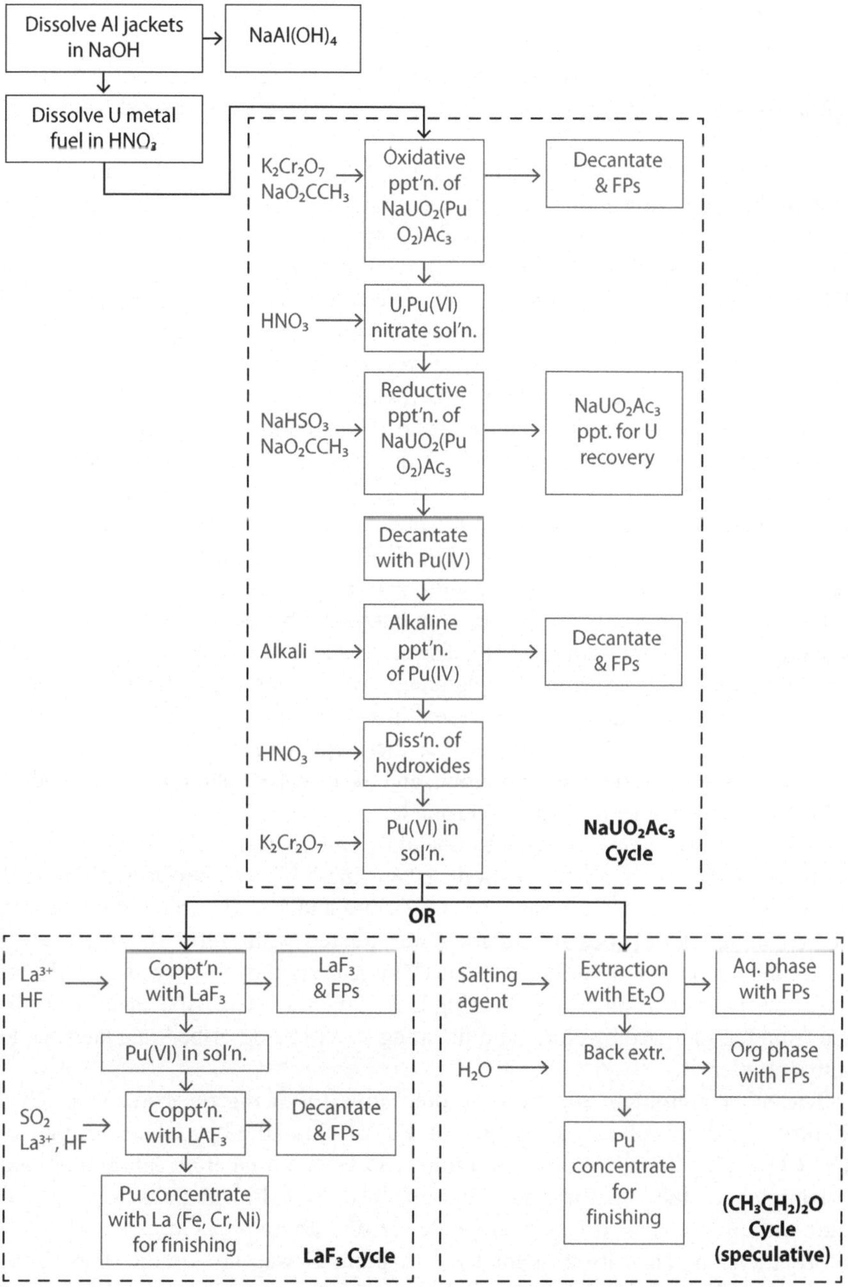

FIGURE 3.7 Acetate-lanthanofluoride Mayak reprocessing flowchart. (From Ilenko, E.I., and Abramova, N.A., *Radiochemistry*, 41, 101–107, 1999; Kruglov, A., *The History of the Soviet Atomic Industry*, Taylor & Francis Group, London, UK, 2002; Sokhina, L.P. et al., *Plutonium in Girl's Hands*, South Ural Institute of Biophysics, Ekaterinburg, Russia, 2003.)

alkaline, were made alkaline so they could be discarded to concrete underground waste storage tanks at Mayak.

As shown in Figure 3.7, three different cycles were used in plutonium separation at Mayak—the $NaUO_2Ac_3$ Cycle and two alternative subsequent cycles: the LaF_3 Cycle and the $(CH_3CH_2)_2O$ Cycle.

As at Hanford, aluminum-clad uranium metal fuel was used at Mayak. The first steps in the fuel reprocessing were also similar—the chemical removal of the aluminum jackets (cladding) in strong NaOH solution containing $NaNO_3$ as a hydrogen radical scavenger and the dissolution of the uranium metal fuel cores in strong HNO_3. However, nitric acid dissolution of the cladding, abetted by mercury(I) or (II) catalysis, was initially used for decladding (Sokhina 2000, 23; see Chapter 2).

The $NaUO_2Ac_3$ Cycle's first step coprecipitated Pu(VI) isomorphically with $NaUO_2(O_2CCH_3)_3$. The plutonium was oxidized using potassium dichromate ($K_2Cr_2O_7$) before the sodium acetate strike. The solution, containing most of the fission products, was discarded, and the solids containing Pu(VI) and nearly all the uranium were retained for further processing. This was a necessarily large precipitate volume because very little of the dissolved fuel bulk was removed. The sodium uranyl acetate cake was then dissolved in HNO_3, the plutonium reduced to Pu(IV) using sodium bisulfite ($NaHSO_3$) and sodium uranyl acetate re-precipitated. The precipitate, carrying virtually all the uranium but negligible plutonium, was processed in another plant for uranium recovery. The separated solution containing the Pu(IV) was made alkaline, carrying the plutonium with other metal hydroxides, chief among them residual uranium(VI) present in solution at or near the solubility limit for $NaUO_2(O_2CCH_3)_3$. The precipitate was then separated from the fission product-bearing solution, acidified for dissolution, and the dissolved plutonium was oxidized to Pu(VI), again using potassium dichromate.

The Pu(VI) was next processed by one of two alternative cycles. In the first alternative, the LaF_3 Cycle, a precipitation strike using LaF_3 was performed. The LaF_3 precipitate carried fission products and impurities and was discarded to waste while the remaining Pu(VI)-bearing solution was treated with sulfur dioxide (SO_2) to reduce the plutonium to Pu(IV). The Pu(IV) was carried by a succeeding LaF_3 precipitation strike. This plutonium-bearing LaF_3 cake was then dispatched for separation from La, followed by plutonium finishing as will be described in a later section of this chapter.

The other plutonium purification alternative following the $NaUO_2Ac_3$ Cycle (Figure 3.7) was solvent extraction of Pu(VI) into diethyl ether through the $(CH_3CH_2)_2O$ Cycle (Il'enko and Abramova 1999; Sokhina et al. 2003). Although distribution to the organic phase is enhanced by salting agents (e.g., ammonium nitrate was used at Los Alamos Scientific Laboratory; Gergen et al. 1946), information on the salting agent used in Plant Б was not found. The solvent extraction achieved significant purification for plutonium compared with the LaF_3 alternative as few metals present in fuel reprocessing streams, except uranium, are extractable by ether. The plutonium could then be stripped from the ether using water.

Although the $(CH_3CH_2)_2O$ Cycle product was much purer than the LaF_3 Cycle product, the ether cycle was soon abandoned after several small explosions and a

TABLE 3.2
Quality of Reprocessing Plant Б and ББ/ДБ Product Solutions

	Product Solution Compositions		
	Plant Б		Plant ББ/ДБ
Component	**Acetate-Fluoride**	**Acetate-Ether**	**Double Acetate**
Pu, g/L	10	17	17
U, g/L	10	1.9	7.7
Fe, g/L	3.5	0.8	0.8
Cr, g/L	3.5	0.6	0.8
La, g/L	8–10	–	–
mg-equivalent Ra/L[a]	350	100	250
Rare earths, % of initial activity	26	0.95	44

Source: Table 2.5 of Sokhina (2000).

[a] To a first approximation, mg-equivalent Ra/L ≈ mCi/L. The irradiated fuel dissolver solution activity was ~100 Ci/L and contained ~100 mg Pu/L (Sokhina et al. 2003, 23).

large one occurred (Sokhina 2000, 39). A third alternative, consisting of a second acetate precipitation step, was later instituted to replace the corrosive LaF_3 precipitation step. This evolved process started in 1959 (Cochran and Norris 1995) and took place in a new ББ (or ДБ, double Б) Plant, while Plant Б ceased operation in 1960. The product solution compositions emerging from these three alternative processes, each of which began with sodium uranyl acetate precipitation, are shown in Table 3.2.

The acetate step in Plant Б processing was confounded by radiolytic reactions. For example, hydrogen peroxide, formed radiolytically in nitric acid, chemically reduced the Pu(VI), probably to Pu(IV), causing process losses because $NaUO_2(O_2CCH_3)_3$ does not carry Pu(IV). Higher dichromate concentrations were used subsequently to overcome this effect. The chemical products of radiolysis also could increase corrosion. Organic compounds such as acetate decomposed and process solutions heated by radiolysis (Sokhina 2000, 21). These radiolytic manifestations were unanticipated or underestimated because they did not occur in the early laboratory studies that used lightly irradiated uranium and smaller volumes. In addition, plutonium deposited on surfaces of the process apparatus as colloid and polymer, especially in the alkaline metathesis steps of the process, led to unexpected losses (Sokhina 2000, 36–37). As a result of process upsets, equipment corrosion exacerbated by fluoride attack, seal and valve failure, loss of process materials to exhaust ventilation (Kruglov 2002, 79), and the vexing unexplained plutonium losses, manned entries into the Plant Б process cells became routine. These manned entries—for equipment repair and replacement, tracing and troubleshooting material losses, and understanding process upsets—resulted in uncommonly large personnel radiological exposure.

TABLE 3.3
Early Personnel Radiological Dose in U.S. and USSR Separations Plants

US—Hanford		USSR—Mayak		
	Dose Rate		Average Dose Rate	
Period	Weekly, mrem	Period	Weekly, mrem	Annual, rem
December	0.10% 100–200	1949	900	48.0; 6.9% >100
1944–July 1945	0.003% 200–600	1950	1800	94.0; 36.5% >100
	0% >600	1951	2200	113.3; 44.6% >100
DuPont (1946b).		From Table 6 of Kruglov (2002).		

In addition, although the Radium Institute developers had predicted a DF of 10^6 for the combined sodium uranyl acetate and lanthanum fluoride steps in the pioneering 1949 operations, the initial DF was only about 860 (Reed and Kramish 1996, 32). Even with process improvements, the routine Plant Б acetate-fluoride process DF only reached ~29,000 based on the values presented in Table 3.2. These DFs were much less than the 10^7 DF, soon attained through $BiPO_4$ and LaF_3 processing at Hanford (Figure 3.4). Thus, it fell to plutonium finishing steps at Mayak's Shop No. 9 (February–August 1949) and succeeding Plant B, at great radiological hazard given the high radioactivity of the Plant Б product (Table 3.2), to improve the plutonium purity through supplemental processing. The radiological safety impacts can be appreciated in Table 3.3 by comparing the personnel dose records during early operations at Mayak with the vastly smaller personnel doses seen in early Hanford operations.

3.5 PROCESSING OF URANIUM AND PLUTONIUM BY PRECIPITATION AS PURE COMPOUNDS

Plutonium chemical forms of technological importance for application as nuclear fuel and explosives are generally used as the dioxide, PuO_2, for fuel and as a metal precursor and as the metal for weapons and occasionally as fuel. The first step in PuO_2 and metal preparation is conversion from the concentrated nitrate solutions obtained by reprocessing or plutonium scrap recovery processes to solid chemical forms, usually by chemical precipitation. Ideally, these precipitation processes achieve supplemental purification from accompanying fission and corrosion products (e.g., Fe, Cr, and Ni from stainless steel) and from reagent impurities in the nitric acid streams. Plutonium conversion processes done without precipitation include direct thermal denitration of the plutonium nitrate solution and calcination of plutonium-loaded ion exchange resin (IAEA 2012). Uranium product from reprocessing is usually crystallized as uranyl nitrate hexahydrate ($UO_2(NO_3)_2 \cdot 6H_2O$) (UNH) as a prelude to its precipitation by ammonium diuranate and calcination (IAEA 2012) or direct calcination to UO_3 for shipment or storage (IAEA 1999). Therefore, most of the discussion in this section focuses on plutonium precipitation processes.

3.5.1 General Considerations

As noted in the Hanford Technical Manual (803 of DuPont 1944, 803), compounds selected for plutonium precipitation and purification should satisfy certain conditions:

- The compound should have low solubility in the precipitation medium (to minimize process recycle).
- Mechanical separation of the compound from the supernatant solution should be simple (e.g., large or readily separable crystals should form that can be readily washed free of the mother solution).
- Decontamination from key impurity cations (e.g., La, Fe, Zr) should occur (i.e., the precipitate should be highly selective for plutonium).
- The precipitated compound should re-dissolve readily (for recycling or further processing).

Candidate compound types investigated for coprecipitation of plutonium are shown in Table 3.1. Similar compounds would likewise be of interest in the processing of pure plutonium compounds. Thus, it is seen that hydroxides of all plutonium oxidation states available in nitric acid ((III), (IV), and (VI)), fluorides, phosphates, and oxalates ((III) and (IV)), peroxides (IV), and sodium acetate double salts (VI) are promising precipitates for plutonium processing based solely on solubility.

However, as noted previously, the hydroxides precipitate non-selectively because many metal hydroxides, including those of prominent impurities (e.g., Fe, Cr, La), have low solubility. Plutonium hydroxides (or, in the case of Pu(IV), the hydrated oxides $PuO_2 \cdot xH_2O$), in general, are also difficult to handle, forming finely particulate precipitates that filter and settle slowly and readily retain contaminated interstitial solution. In the absence of a holding reductant, Pu(III) hydroxide ($Pu(OH)_3$) rapidly oxidizes to $PuO_2 \cdot xH_2O$ in alkaline solution, while Pu(VI) hydroxide itself is much more soluble and slowly reverts to $PuO_2 \cdot xH_2O$, likely through radiolytic influences (Delegard and Jones 2015). While the hydroxide precipitate purity can be poor, dependent as it is on the purity of the starting solution, the low solubility of the hydroxide commends its use in decontaminating solutions of plutonium activity and recovering plutonium values, as will be seen. Carriers can be used to enhance the extent of decontamination (Christensen et al. 1988).

Phosphates also have relatively low selectivity, though not as indiscriminate as hydroxide, and likewise can be difficult to separate because of small particle size.

Of the remaining plutonium precipitate compound types, the trifluoride ($PuF_3 \cdot 0.75H_2O$), which can be dehydrated to PuF_3; the Pu(III) and (IV) oxalates $Pu_3(C_2O_4)_3 \cdot 10H_2O$ and $Pu(C_2O_4)_2 \cdot 6H_2O$; the Pu(IV) peroxide, approximately $Pu_2O_7 \cdot 6H_2O$; and the double salt $NaPuO_2(O_2CCH_3)_3$ attracts the most interest as compounds for plutonium processing as they provide readily handled solid forms that achieve appreciable decontamination.

Slurries of various plutonium precipitates, shown in Figure 3.8, illustrate the vivid colors of many of these precipitates. The colors of the $Pu_3(C_2O_4)_3 \cdot 10H_2O$, $Pu(C_2O_4)_2 \cdot 6H_2O$, and PuF_3 filter cakes are shown in Figure 3.9. In plant operations, attentive observation of color is often key to process control as it indicates proper plutonium oxidation state and functioning of the precipitation process.

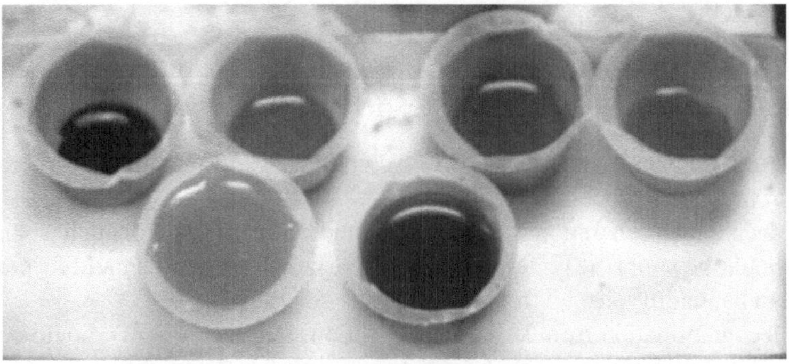

FIGURE 3.8 Precipitated plutonium compound slurries. Top, left to right: Blue Pu(III) oxalate, beige/tan Pu(IV) oxalate, green Pu(IV) peroxide, pink/red Pu(IV) fluoride. Bottom, left to right: Pink Pu(IV) sulfate, lavender Pu(III) fluoride. (Courtesy of Susan A. Jones, PNNL.)

FIGURE 3.9 Plutonium compound filter cakes. Left to right: Blue Pu(III) oxalate, tan Pu(IV) oxalate, and lavender Pu(III) fluoride. (Left – Courtesy of Matthew K. Edwards, PNNL; middle – Courtesy of Susan A. Jones, process support labs, Hanford Site; right – SRS production.)

The oxalates and peroxide attract process interest because they can be thermally decomposed to generate PuO_2 itself. The PuO_2, while being a product for reactor fuel or for long-term Pu storage (DOE 2012), can also serve as a precursor to Pu metal by, first, hydrofluorination to form PuF_4 by the reaction:

$$PuO_2 + 4\,HF \rightarrow PuF_4 + 2\,H_2O \tag{3.4}$$

as was done in the United States (Baker 1946). The PuF_4, dehydrated PuF_3 from precipitation (Mainland et al. 1961), or precipitated PuF_3, oxidized by roasting in the presence of oxygen to form ¼ PuO_2 + ¾ PuF_4 (Orth 1963), can be metallothermically reduced, generally with calcium, to produce Pu metal according to the following respective reactions:

$$PuF_4 + 2\,Ca \rightarrow Pu + 2\,CaF_2 \tag{3.5}$$

$$2\ PuF_3 + 3\ Ca \rightarrow 2\ Pu + 3\ CaF_2 \tag{3.6}$$

$$\tfrac{1}{4}\ PuO_2 + \tfrac{3}{4}\ PuF_4 + 2\ Ca \rightarrow Pu + 3/2\ CaF_2 + \tfrac{1}{2}\ CaO. \tag{3.7}$$

The precipitated double salts, $CaPuF_6$ (Harmon and Reas 1957) and Cs_2PuCl_6 (Muscatello and Killion 1990), have also been investigated as immediate precursors to Pu metal reduction but have not come into routine production.

Los Alamos used $NaPuO_2(O_2CCH_3)_3$ precipitation in early operations to purify Hanford-origin Pu of accompanying La. The $NaPuO_2(O_2CCH_3)_3$ precipitation followed an initial Pu(III) oxalate precipitation/purification step at the Los Alamos Scientific Laboratory (LASL), which, of course, did not remove La but achieved decontamination from other constituents such as Fe, Cr, and Ni. The rinsed Pu(III) oxalate solids were dissolved in heated HNO_3 containing sodium bromate ($NaBrO_3$) as an oxidant. The bromate oxidized the oxalate to CO_2 and the Pu to Pu(VI). The $NaPuO_2(O_2CCH_3)_3$ was precipitated from an initial solution containing 0.14 M PuO_2^{2+} (and 0.004 M La^{3+}), 2.75 M Na^+, 0.24 M acetic acid, 0.91 M acetate ($CH_3CO_2^-$), 2.00 M NO_3^-, and 0.12 M BrO_3^-. The mixed acetate/acetic acid buffered the solution to about pH 5. The $NaPuO_2(O_2CCH_3)_3$ solubility in 20°C water is 19.7 g Pu/L but decreases to 0.07 g Pu/L in 4 M sodium. In the 65°C 2.75 M sodium process solution, the solubility is 0.33 g Pu/L (Carritt et al. 1946, 23–24 and 30–31).

In analogy with the precipitation of ammonium uranyl carbonate ($(NH_4)_4UO_2(CO_3)_3$), ammonium plutonyl carbonate ($(NH_4)_4PuO_2(CO_3)_3$) can also be precipitated, forming U_3O_8 and PuO_2, respectively, upon calcination (Roepenack et al. 1984).

Co-crystallization of Pu(VI) nitrate with U(VI) nitrate has been demonstrated in the development of the NEXT (New EXtraction System for TRU Recovery) Process, a potential reprocessing technology developed in Japan (Hommaa et al. 2008). Steps in this process include: (1) fuel dissolution in nitric acid, (2) oxidation of plutonium to the (VI) state by heating, (3) isomorphous co-crystallization of Pu(VI) with U(VI) as $(U,Pu)O_2(NO_3)_2 \cdot 6H_2O$ to effect separation from fission products and minor actinides that do not crystallize in this system and thus remain in solution, (4) dissolution of the $(U,Pu)O_2(NO_3)_2 \cdot 6H_2O$ salt, (5) reduction of Pu(VI) to Pu(IV), and (6) crystallization of $UO_2(NO_3)_2 \cdot 6H_2O$ (UNH) to effect separation from the Pu(IV), which remains in solution with the mother liquor containing still-dissolved U(VI) nitrate. Recrystallization of the UNH achieves further purification with the purified product suitable for calcination to UO_3 or U_3O_8, re-enrichment, or fabrication into fuel. The purified mixed Pu(IV)/U(VI) nitrate solution from step 6 can be blended and processed as feed for mixed uranium/plutonium oxide (MOX) fuel. The actinyl nitrate hexahydrate salts' crystallization and recrystallization occur by cooling.

The NEXT Process is, thus, a potential competitor to the PUREX Process, the current standard method for irradiated fuels reprocessing. The NEXT Process advantages over the PUREX Process include: (1) organic solvents and redox reagents are not required because Pu(IV) can be oxidized to Pu(VI) by heating and reduction to Pu(IV) can occur by radiolysis, and (2) because residual uranium remains in the mother liquor from UNH crystallization, plutonium cannot be cleanly separated from uranium, offering nuclear weapons proliferation resistance. Eliminating organic solvents/diluents and redox reagents enhances operational safety, eliminates the solvent recovery cycle, and diminishes waste handling cost and complexity. However, the NEXT Process does not achieve, in a single step, the separation factors achievable by the PUREX Process and leaves relatively high U and Pu quantities, with fission products and minor actinides, in the mother liquor for recycle to the feed stream. Of the fission products and minor actinides built-up in these liquors, cesium (Cs) and Ba are the most problematic as they can report to the $(U,Pu)O_2(NO_3)_2 \cdot 6H_2O$ product crystals. This build-up of fission products and minor actinides in the mother liquor is overcome by splitting the mother liquor and incorporating a secondary $(U,Pu)O_2(NO_3)_2 \cdot 6H_2O$ crystallization, rejecting the secondary mother liquor to the high activity waste and recycling the secondary $(U,Pu)O_2(NO_3)_2 \cdot 6H_2O$ crystals to the feed stream (Hommaa et al. 2008). Decontamination factors of up to ~100 for 0.85/0.15 U/Pu nitrate solution mixtures from La, Ce, and neodymium (Nd) in well-washed $(U,Pu)O_2(NO_3)_2 \cdot 6H_2O$ crystals from a single crystallization have been demonstrated (Volk et al. 2018). The NEXT Process steps are illustrated in Figure 3.10.

Various methods to precipitate plutonium initially present in HNO_3 solution were surveyed at Rocky Flats as alternative means to convert anion exchange-purified Pu(IV) eluate to a solid form (Martella et al. 1984). Precipitation products investigated were Pu(IV) peroxide, Pu(III) fluoride, Pu(III) and Pu(IV) oxalate, Pu(III) and Pu(IV) carbonates, mixed Pu(IV)/(VI) carbonate, Pu(VI) carbonate, and Pu(IV) and Pu(VI) hydroxide. The precipitation conditions tested on a laboratory scale were

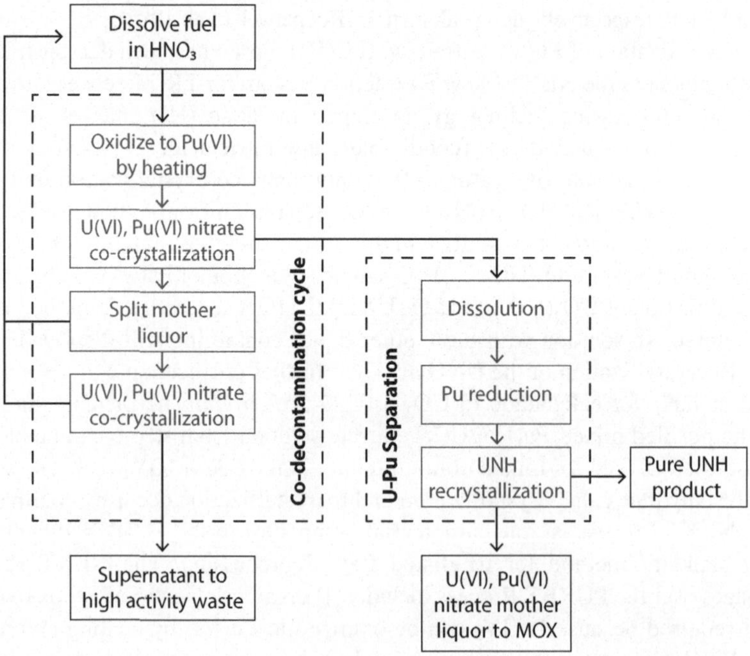

FIGURE 3.10 NEXT process flowsheet. (After Hommaa, S. et al., *J. Nucl. Sci. Technol.*, 45, 510–517, 2008.)

derived from a survey of the technical literature. Simplicity in envisioned operation, large particle size formation, rapid filtration, and low filtrate losses were sought-after process features. On these bases, Pu(IV) peroxide (used on a process scale at Rocky Flats), Pu(III) fluoride, Pu(III) and (IV) oxalate, and Pu(IV) hydroxide were judged to give satisfactory results, while Pu(III), (IV), and (VI) carbonates required long filtration times and Pu(VI) hydroxide resulted in high filtrate losses and post-precipitation.

An earlier survey at Rocky Flats examined seven alternative methods to prepare PuO_2 from Pu(IV) nitrate solution—Pu(III) oxalate, Pu(IV) oxalate, and Pu(IV) peroxide precipitations, sol-gel, and three direct denitration methods (Lehmkuhl 1977). The sol-gel process involved dispersion of Pu(IV) nitrate solution droplets into heated immiscible alcohol suspensions to affect the required dehydration and denitration. Process flow sheets and equipment design considerations were provided for all seven alternatives. Preparations of PuO_2 by various methods, including beginning with Pu(III) and Pu(IV) oxalate and Pu(IV) peroxide precipitation, were surveyed by Lerch and Norman (1984).

Review of the scientific aspects of actinide (including Pu(III) and (IV)) oxalate precipitation is provided by Abraham et al. (2014) and an overview of the process chemistry of Pu(III) and (IV) oxalate processes is given by Silver (2010). Preparation of mixed actinide oxides using the (III) and (IV) actinide oxalates and an overview of their structures and precipitation theory are provided by Sarsfield (2015).

An excellent synopsis of industrial precipitation processes that provide precursors to dry Pu(IV) fluoride for plutonium metal production has been prepared by Cleveland (1967). The precursor precipitates examined include Pu(IV) peroxide, Pu(III) fluoride, Pu(III) oxalate, Pu(IV) oxalate, and calcium Pu(IV) hexafluoride ($CaPuF_6$).

Formation of readily separable Pu compound crystals that achieve good decontamination from solution impurities is optimized by the control of precipitation conditions. Therefore, extensive laboratory and process tests of the effects of process conditions must be undertaken to maximize product quality and throughput. Such processing parameters may include:

- Reagent (including acid) concentrations
- Order and type of reagent addition
- Direct strike (addition of the precipitating agent solution to the plutonium solution)
- Reverse strike (addition of the plutonium solution to the precipitating agent solution)
- Co-addition of the plutonium and precipitating agent solutions
- addition of the precipitation agent as a solid
- Homogeneous generation of the precipitating agent (e.g., hydrolysis of oxalate esters to produce oxalate)
- Speed of reagent addition
- Temperature
- Stirring/agitation rate
- Digestion time

Plutonium precipitations as pure compounds usually occur by batch processes in which a set quantity of Pu is precipitated in a critically safe vessel (limited in at least one dimension to 12–15 cm). The batch precipitate is then settled, centrifuged, or filtered to effect solid-liquid separation, and washed to rinse out contaminant-bearing interstitial solutions. Alternatively, continuous precipitation processes have also been employed in which the Pu and precipitating agent solutions are added concurrently, with agitation, to critically safe dimensioned vessels for aging and crystal growth. The suspension overflow arising from the continuous addition is continuously withdrawn for phase separation and rinsing.

The major industrial Pu conversions by batch precipitation as Pu(IV) peroxide, Pu(III) fluoride, Pu(III) oxalate, and Pu(IV) oxalate and continuous precipitation as Pu(IV) oxalate are examined in further detail.

3.5.2 Plutonium Peroxide Precipitation Processes

Plutonium(IV) peroxide precipitates have uncertain stoichiometry and often incorporate solution anions in variable quantities, particularly sulfate that assists in Pu(IV) crystallization and that cannot be removed by washing. The peroxide component of the precipitate is likewise variable but approximated by the incorporation of three peroxides per two Pu(IV) and is what is thought to be a bridging oxide (Leary et al. 1959). The nominal precipitation of Pu(IV) peroxide from nitrate solution can be written:

$$\begin{gathered} 2\ \mathrm{Pu(NO_3)_4} + 3\ \mathrm{H_2O_2} + 0.5\ \mathrm{H_2SO_4} + 7\ \mathrm{H_2O} \rightarrow \\ \mathrm{Pu_2O_7 \cdot 0.5(H_2SO_4) \cdot 6H_2O} + 8\ \mathrm{HNO_3}. \end{gathered} \tag{3.8}$$

This reaction illustrates the incorporation of sulfate in the Pu(IV) precipitate and generation of four moles of HNO_3 per mole of Pu(IV) precipitated. An overview of peroxide interactions with plutonium, in redox reactions, complexation, and precipitation, is provided by Fugate and Navratil (2006).

The first industrial-scale Pu conversion process in the United States (and the world) occurred at Hanford with the precipitation of Pu(IV) peroxide, expressed simply as Pu_2O_7, from the mixed La(III)/Pu(IV) nitrate solutions arising as product from the LaF_3 *Concentration* or *Crossover and Metathesis* step (Figure 3.3). Plutonium peroxide precipitation has since been used in the United States in another Hanford application as well as at Los Alamos, the Savannah River Site, and at Rocky Flats. Implementation of Pu(IV) peroxide precipitation in these various US processes is examined in further detail in this section of this chapter.

The flowchart for Hanford's Pu peroxide precipitation following the Bismuth Phosphate Process (called the *Isolation Process*) is shown in Figure 3.11.

As seen in Figure 3.11, two precipitation cycles occurred (as derived from Section C of DuPont 1944, 805–815). The ~30-L feed solution to the initial plutonium peroxide precipitation strike contained 0.4 M La, 0.004–0.01 M Fe, 0.1–0.2 M K, 1.8–2.2 M HNO_3, 0.03 M Pu, variable low concentrations of Zr, Ca, and Mn, and some suspended MnO_2. Thus, the feed solution contained about 13 times more

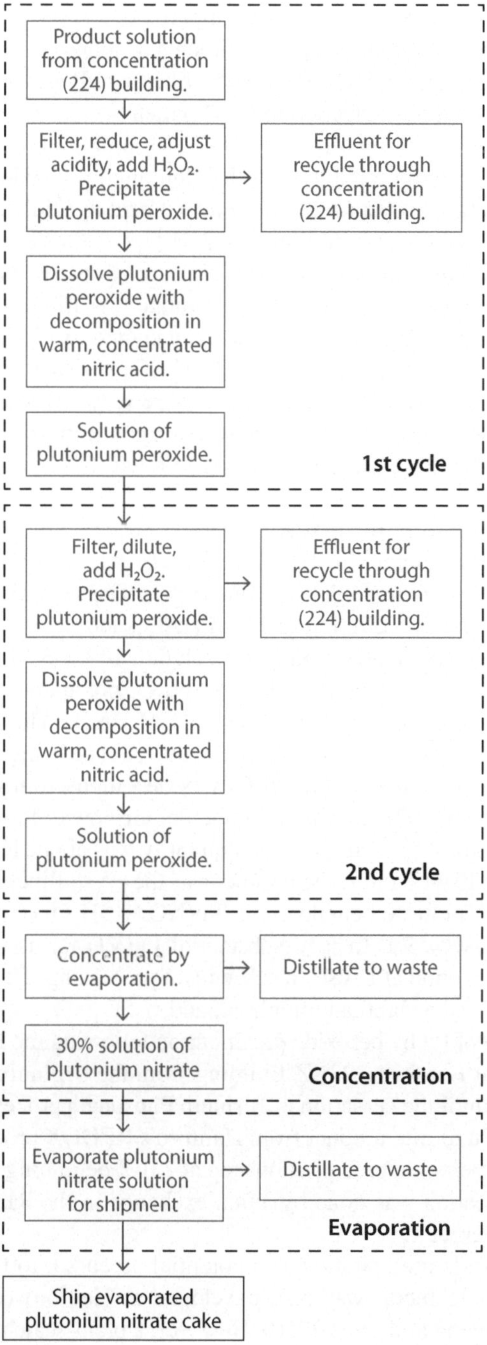

FIGURE 3.11 Hanford plutonium peroxide precipitation (*Isolation*) flowchart (Section C of DuPont 1944, 154).

La than Pu on a mole basis. The selectivity of peroxide to precipitate Pu allowed the Pu_2O_7 to form while leaving the other constituents in solution.

The process began with MnO_2 removal by filtration. Because some Pu(VI) was present in the feed solution, reduction to Pu(IV) then occurred using 0.05 M ammonium sulfite, $(NH_4)_2SO_3$, in the presence of 0.2 M ammonium sulfate, $(NH_4)_2SO_4$. The sulfate aided in later precipitation of the plutonium peroxide by forming larger and more readily settling crystals. A minor adjustment to achieve 2.0 M HNO_3 was then done followed by the slow addition of 30% H_2O_2 (~9.8 M) to make the final solution 10% H_2O_2 while maintaining 20°C to minimize iron-catalyzed decomposition of peroxide to oxygen gas and water. The precipitate was allowed to digest, growing the Pu_2O_7 particle size. The suspension was then cooled to ~5°C, settled, decanted, and washed with three 5-L portions of 0.4 M H_2SO_4 with settling and decantation between the ensuing washes. The decantates were recycled to the *Crossover* step (Figure 3.3) to conserve Pu values. The Pu solution concentration under these conditions, and those of the subsequent second precipitation strike, was about 0.02 g/L or $\sim 8 \times 10^{-5}$ M.

The product Pu_2O_7 from the first strike (about 80% Pu and 20% La on a weight basis or 70 mole% Pu compared with 7 mole% in the starting solution) was then dissolved by heating to 60°C in 1.7 L of ~13 M HNO_3. The second peroxide precipitation strike occurred by first diluting to 1.5 M HNO_3, adding concentrated H_2SO_4 to 0.5 M concentration, and again making the solution 10% (~3 M) in H_2O_2, with the total solution volume being about 20 L. The Pu_2O_7 precipitate was again washed three times, this time by using 5-L portions of 1 M HNO_3. This product contained about 93.4 mole% (~97 wt%) Pu as total metal, the balance being predominantly La. As in the first precipitation step, the Pu_2O_7 was decomposed in hot HNO_3, and the solution concentrated by 125°C evaporation under air flow to form a gummy plastic mass for shipment to LASL. The shipment quantity was about 160 g of Pu in 400 g total mass or about 40 wt% Pu. If the product was the crystalline product from gentle evaporation—Pu(IV) nitrate pentahydrate ($Pu(NO_3)_4 \cdot 5H_2O$) (Ryan et al. 1985)—it would contain ~42 wt% Pu. In fact, substantial Pu(VI) forms under these heated HNO_3 conditions as found at LASL in subsequent processing. The product also contained the La impurity, as lanthanum nitrate, and H_2SO_4.

The purification of Pu by peroxide precipitation following the Bismuth Phosphate Process occurred at Hanford's 231-Z Building from the beginning of processing in the spring of 1945 until the cessation of Bismuth Phosphate processing at Hanford in March 1956. The Pu nitrate product from Hanford's REDOX process was also purified by a single Pu peroxide precipitation step at 231-Z beginning January 1952 until adequate Pu purification was done by anion exchange in the REDOX Plant beginning April 1955 (Gerber 1997).

Plutonium peroxide precipitation as a potential precursor to form PuO_2 for subsequent processing to metal was also developed at Hanford (Harmon and Reas 1957; see also Kennedy and Lyon 1950). However, a plant-scale process test proved that the Pu peroxide had substantial drawbacks—low density (thus high cake volume), difficulty in drying the fine product filter cake, which spattered upon heating even with drying using methanol rinse (a safety hazard in itself and particularly in the presence of peroxide), and erratic Pu metal yields in subsequent processing

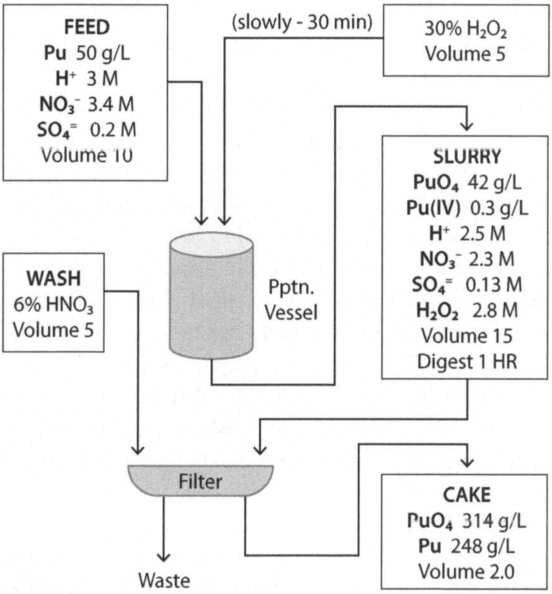

FIGURE 3.12 Plutonium peroxide flowchart. (Relative volumes shown; from Figure 3 of Harmon, K.M. and Reas, W.H., *Conversion Chemistry of Plutonium Nitrate*, General Electric, Hanford Atomic Products Operation, Richland, WA, HW-49597-A, http://www.osti.gov/scitech/biblio/4364697, 1957.)

(Kerr 1952, 4). A flowchart developed for producing Pu peroxide at Hanford, shown in Figure 3.12, is very similar to that outlined in the previous paragraphs for the *Isolation* step following the Bismuth Phosphate Process and as practiced elsewhere. Plutonium peroxide precipitation was also used in early (1949) processing in the USSR to purify Pu (Sokhina et al. 2003), but specific detail for the USSR process has not been located.

Leary and colleagues (1959) at LASL systematically investigated Pu peroxide precipitation and found two different Pu peroxide phases that arose depending on the acid concentration—hexagonal at ≥2 M HNO_3 and cubic at ≤1 M HNO_3. The structures were not altered by varying H_2O_2 or sulfate concentrations. The cubic phase had a Pu:peroxy oxygen mole ratio of 3.03, indicating inclusion of another anion in the structure, thought to be a bridging oxide, to satisfy the Pu(IV) valence, while the hexagonal phase's ratio in the wet precipitate was 3.37:1 and 3.03 in the dry, again indicating incorporation of a supplementary anion. Sulfate was present in both structures at relatively steady amounts irrespective of preparation—0.25 moles sulfate per mole of Pu in the cubic structure and 0.38 moles of sulfate per mole Pu in the hexagonal. The cubic formed a much denser filter cake than did the hexagonal. The cubic phase showed the apparent formula $Pu_2O_7 \cdot 6H_2O$ plus 0.5 moles of sulfate to yield a 0.640 gravimetric factor in Pu; the hexagonal phase gravimetric factor was 0.467. It was also found that although neither Fe(III) nor peroxide individually oxidized Pu(IV) to Pu(VI), together they did. The product Pu(VI) could then be reduced by peroxide to form Pu(IV) to close a catalytic peroxide decomposition cycle. Therefore, solutions containing too much dissolved iron cannot be processed by Pu peroxide precipitation.

A Pu peroxide precipitation sequence comparable to that shown in Figure 3.12 was used at LASL (Figure 47 in Christensen and Maraman 1968) based on the research of Leary et al. (1959). The LASL procedure started with 400 g Pu in nitrate solution that was first adjusted to 3–4 M HNO_3, using water or more concentrated HNO_3 solution with H_2SO_4 added to make the solution 0.15 M in sulfate. Based on the work of Leary et al. (1959), these conditions would, thus, favor the hexagonal form, as would the process outlined in Figure 3.12, while the original Hanford *Isolation Process* likely would have produced a mixture of cubic and hexagonal. The solution was cooled to 14°C and 100 mL of 30% H_2O_2 added to reduce any Pu(VI) present to Pu(IV), allowing one hour for this to occur. The precipitation then occurred by adding more 30% H_2O_2, slowly and to a stoichiometric excess, while ensuring the solution remained below 20°C. The precipitate was digested for one hour, filtered, and washed five times with 150–300 mL portions of 5% H_2O_2 in 0.5 M HNO_3. The inclusion of HNO_3 in the rinse, not mentioned in Leary et al. (1959), prevented unwanted peptization of the filter cake solids, while instead, Leary et al. (1959) used an ethanol rinse for drying.

The initial Pu precipitation process employed at the Savannah River Site (SRS), near Aiken, South Carolina, was adapted from the peroxide precipitation process used for Pu purification at LASL (Moore et al. 2000). The SRS use of Pu peroxide precipitation in routine processing to prepare Pu metal began in 1952 and continued until 1959, being supplanted in 1960 by Pu(III) fluoride precipitation (Mainland et al. 1961; Orth 1963).

The SRS peroxide process began with purified Pu(III) cation exchange eluate containing 65 g Pu/L in about 0.3 M sulfamic acid (NH_2SO_3H) and 4.7 M HNO_3 (Mainland et al. 1961). The valence adjustment and precipitation strike used 50% H_2O_2 (~17.5 M H_2O_2). The first portion of H_2O_2 was added slowly to overcome the excess sulfamic acid, slowly oxidize the Pu(III) to Pu(IV), and initiate and complete precipitation. Enough H_2O_2 was added initially to make the solution 2 M in peroxide and dilute to 4 M acid, with the solution chilled to 15°C to decrease peroxide decomposition. Additional H_2O_2 was then added more rapidly, all with stirring and continued chilling, to provide 6 M total H_2O_2, about 8 times the stoichiometric amount. The tan-colored slurry (and olive-green precipitate; Section 7.5.3[b][iii] of Clark et al. 2006) was chilled further to 6°C and digested for 30 minutes before filtration. The precipitation was done in glass vessels. Excess peroxide in the filtrates was destroyed by heating in stainless steel vessels. Peroxide precipitation and filtration/drying equipment material and construction details are also described by Mainland et al. (1961).

Interestingly, sulfate or sulfuric acid was not used in the SRS process (Mainland et al. 1961). Filterable crystals formed without this addition through judicious selection of precipitation conditions. The fact that the Pu(IV) was not initially present but was instead slowly generated as the sulfamic acid was consumed, apparently allowing larger hexagonal-phase crystals to form. The sulfate-free hexagonal-phase cakes produced were about six times denser and filtered six times faster than those formed under less satisfactory conditions but in the presence of sulfate. Agitation rates were also important; if stir speed were too high, the crystals would break into

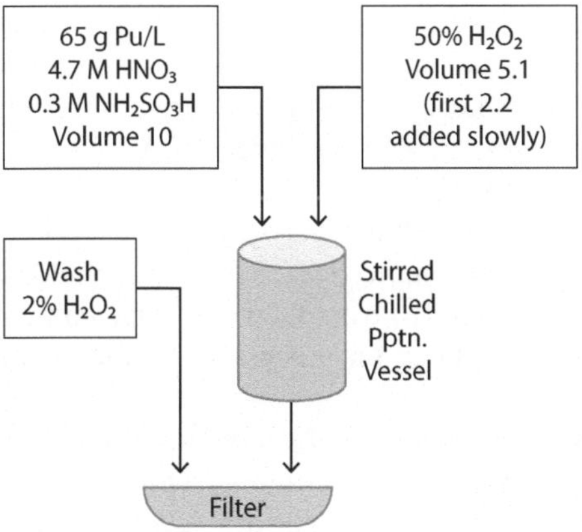

FIGURE 3.13 Savannah River Site plutonium peroxide precipitation process flowchart. (After Mainland, E.W., *Ind. Eng. Chem.*, 53, 685–694, 1961.)

fines, while low speeds did not promote proper mixing, leading to zones, where the H_2O_2 was added, being too dilute in acid, thus forming the less dense cubic phase. The process used at SRS is outlined in Figure 3.13.

SRS improvements to process conditions included a redesigned precipitator, reagent addition rate and agitator speed adjustments to eliminate localized low acid regions, better drying procedures, increased batch sizes, higher H_2O_2 concentration, and upgraded materials of construction (Moore et al. 2000).

The Rocky Flats Plant (RFP) near Golden, Colorado, used Pu(IV) peroxide precipitation throughout its operations in the period of 1965–1989 (extrapolated from Baldwin and Navratil 1983) to convert Pu nitrate solutions to a pure solid Pu compound. A parametric study of the effects of Pu(IV) peroxide precipitation conditions on the filtration rate and Pu concentration in the filtrate was undertaken at the RFP (Hagen and Miner 1980). Six variables were investigated—HNO_3 concentration (1.8, 2.5, and 3.1 M in the precipitating medium), H_2O_2 concentration (2.9, 3.9, and 4.9 M in the precipitating medium), the presence or absence of impurities in the 43 g Pu/L solution (0.31 g/L Fe, 0.04 g/L Cr, 0.06 g/L Ni, 0.03 g/L Cu, and 0.02 g/L Pb), digestion time (30, 60, and 90 minutes), rate of H_2O_2 addition (~5%, 3.2%, and 1.6%, relative to the Pu solution volume, per minute), and temperature (22°C, 14°C, and 6°C).

Hagen and Miner's (1980) parametric study generated the following process observations:

- Higher acid concentration and lower H_2O_2 addition rate favored faster filtration.
- Lower acid concentrations and higher final H_2O_2 concentrations decreased Pu filtrate loss.
- The other variables had lower impacts that were statistically insignificant.
- Because the effects of acid concentration on process efficiency went in opposing directions regarding filtration time and loss to filtrate, 2.3–2.7 M HNO_3 was recommended, midway in the range studied.

- The recommended H_2O_2 concentration in the precipitating solution provided 22 moles H_2O_2 per mole Pu—about seven times the stoichiometry of the product Pu_2O_7, which, according to Leary et al. (1959), contains three moles of peroxide per mole of Pu. Evidently, no higher and more soluble anionic Pu-peroxide complexes exist.
- The recommended H_2O_2 addition rate—2 moles H_2O_2 per minute per mole of Pu—would attain the recommended 22:1 H_2O_2:Pu mole ratio in 11 minutes.
- Higher temperature slightly favored lower Pu concentrations in the filtrate, but 14°C was recommended for processing to diminish parasitic/catalytic loss of H_2O_2.
- Because digestion time had little effect on filtration time and Pu loss to the filtrate, the lowest tested digestion time—30 minutes—was recommended.
- The effect of the presence of impurities (generally unavoidable in most plant processing) was small but beneficial to the filtration rate and plutonium filtrate loss.

3.5.3 Plutonium Trifluoride Precipitation Processes

The precipitation of Pu(III) fluoride is straightforward, producing three moles of HNO_3 from the three moles of HF required to form PuF_3, with variable amounts of weakly bound non-crystalline water:

$$Pu(NO_3)_3 + 3\ HF + x\ H_2O \rightarrow PuF_3 \cdot xH_2O + 3\ HNO_3. \tag{3.9}$$

Hanford demonstrated a small-scale continuous PuF_3 precipitation process (Harmon et al. 1961) but did not implement it for production. The 35 g Pu(III)/L feed stream, in 3.5–5 M HNO_3, was prepared using either ascorbic acid ($C_6H_8O_6$) plus sulfamic acid or sulfur dioxide (SO_2) gas plus sulfamic acid as reductant and nitrite scavenger, respectively. The precipitating agent was 3 M HF. The Pu and HF streams entered simultaneously into a vigorously stirred reactor with 1 M HF excess maintained at rates to give a 15- to 20-minute residence time, presumably at ambient temperature, although reaction heating occurred. The filtrate contained <0.05 g Pu/L and DFs were >10 for Fe, 20 for Ni, 10 for Cr, and 4 for Mg.

Early Hanford studies found that PuF_3, in a form assayed to be $PuF_3 \cdot 0.40H_2O$, could be prepared as a Pu laboratory standard (Jones 1953). The $PuF_3 \cdot 0.40H_2O$ preparation began by dissolving pure Pu metal in HCl, adjusting any Pu(IV) to the (III) state using SO_2 gas, and performing a reverse strike addition of the 5–85 g Pu(III)/L solution to HF solution such that the final free HF concentration was 2–3 M. The lavender-colored precipitate was digested 15 to 60 minutes with stirring (apparently at room temperature; temperature not stated), centrifuged, and washed sequentially with water, methanol, and ether.

Savannah River investigated two alternative production methods to precipitate PuF_3 (Harmon and Reas 1957; Harmon et al. 1961). In the first method, a direct strike of 2.7 M hydrofluoric acid (HF) was added over 30 minutes and at ≥40°C to an equal volume of 50 g/L Pu(III) nitrate solution in 5.7 M HNO_3. In the second method, the same 2.7 M HF and Pu solution volumes were added simultaneously over 30 minutes to half the volume of 0.1 M HNO_3 at 25°C. Nitrite scavengers, though not specified, must have been present to maintain Pu(III) in 5.7 M HNO_3 and the resulting precipitating mixtures. Both methods required vigorous agitation and a digestion of 30 minutes to produce easy-to-filter crystals, but the latter yielded larger crystals. In both cases, the solids were filtered and washed with 0.8 M HF to prevent oxidation of the solids by residual HNO_3 and were dried, with or without alcohol wash, by suction. The precipitates could then be dehydrated in a flow of 200°C helium.

The PuF_3 precipitation process developed and implemented on a production scale at SRS in the late 1950s supplanted the Pu(IV) peroxide process in the conversion steps to Pu metal production. Compared with Pu(IV) peroxide, the PuF_3 product was more readily dried with less loss to the filtrate and eliminated problems with hydrogen peroxide reagent instability. However, it provided less feed solution impurity decontamination, particularly for Al, and shifted the fluoride corrosion problems from the hydrofluorinator to the precipitation vessel (Moore et al. 2000).

The PuF_3 precipitation process at SRS employed two sequential batchwise precipitation stages (Orth 1963). The first stage created the initial precipitate from an ~50 g/L Pu(III) feed stream in 4 HNO_3 containing 0.2 M NH_2SO_3H as reductant and 0.3 M hydroxylammonium nitrate ($NH_2OH \cdot HNO_3$) as nitrite scavenger. Ascorbic acid was added to the Pu feed to ensure complete reduction to Pu(III) and thus diminish loss to the filtrate occasioned by the precipitation of the more soluble Pu(IV) fluoride hydrate ($PuF_4 \cdot 2.5H_2O$).

Because HF is a weak acid, the free fluoride and the PuF_3 solubility depends on the HNO_3/HF mole ratio. Hence, the 4 M HF precipitant was added to a HNO_3/HF mole ratio of about 7 (Orth 1963). Because these conditions allowed relatively high Pu concentration, rapid PuF_3 crystal particle growth occurred. It was found that if the HNO_3/HF ratio was too high, precipitates plated on the vessel walls. The crystals were allowed to grow to the point that still allowed particles to be readily slurried out of the precipitation vessel. The second precipitation stage occurred by adding more 4 M HF until a HNO_3/HF mole ratio of about 3 was attained, further growing the PuF_3 crystals but minimizing dissolved Pu concentration to leave a 0.05%–0.1% loss to the filtrates and the ensuing 0.8 M HF wash. The Pu filtrate concentration would correspond to an expected PuF_3 solubility of about 0.1 g Pu/L (Mandleberg et al. 1961). Lab studies at SRS showed that much of the Pu in the filtrate could be recovered as PuF_3, PuF_4, and double salt precipitates upon solution neutralization using NaOH or KOH. Alternatively, the Pu could be coprecipitated using bismuth trifluoride (BiF_3) as the carrier (Orth 1963).

The two-stage precipitation process as described by Orth (1963) had decreased Pu filtrate loss by moving the initial PuF_3 slurry to a second precipitation vessel for the supplemental HF addition. Burney and Tober (1965) further developed and refined the SRS process. They found that PuF_3 contains no crystalline water but had interstitial water in the filter cake. Like Orth (1963), Burney and Tober (1965) used ascorbic acid addition just prior to the precipitation step to remove the Pu(IV) traces in the feed solution that would produce higher solubility, muddy, lower density, and thus less filterable Pu(IV) fluoride precipitate. High stirring speeds (2,500 RPM for 2-cm diameter blades) during precipitation produced ~2–3 μm particles, whereas lower (600 RPM) stir speeds allowed much more readily settled and filterable 10–15 μm particle agglomerates. Once formed, stirring of the larger agglomerates at rapid rates did not reduce their size. Burney and Tober's (1965) research showed that, at HNO_3/HF ratios higher than about 5, the initial Pu solution concentration was greater and precipitation to a steady Pu concentration substantially delayed compared with those found at lower HNO_3/HF ratios. Thus, filtrate losses at HNO_3/HF mole ratios less than 3 were 0.01–0.015 g Pu/L. This work also showed that the filtrate solution volume could be sharply decreased by using concentrated (28 M) HF, provided it was adjusted to 0.5 M HNO_3 to suppress free fluoride concentration; i.e., move the $HF \rightleftharpoons H^+ + F^-$ equilibrium to the left. This adjustment avoided localized high free fluoride regions during precipitant addition.

Burney and Tober (1965) also showed that, while decontamination from Fe was good, decontamination from Al, U(VI), and Zr was poor. The general SRS PuF_3 process flow, including the primary and secondary precipitation vessels, is shown in Figure 3.14.

A parametric study of the effects of PuF_3 precipitation conditions as practiced at SRS was performed to optimize the process by minimizing solids plating on the precipitation vessel walls, maximize particle size, and minimize losses to the filtrate (Luerkens 1983). Twelve experiments were conducted, all at 22°C. Six had high (27.5 g/L) Pu and low (2.4 M) HNO_3 concentrations, and the other six had low (7.5 g/L) Pu and high (6.5 M) HNO_3 in the feed solutions. The as-mixed total HF concentrations

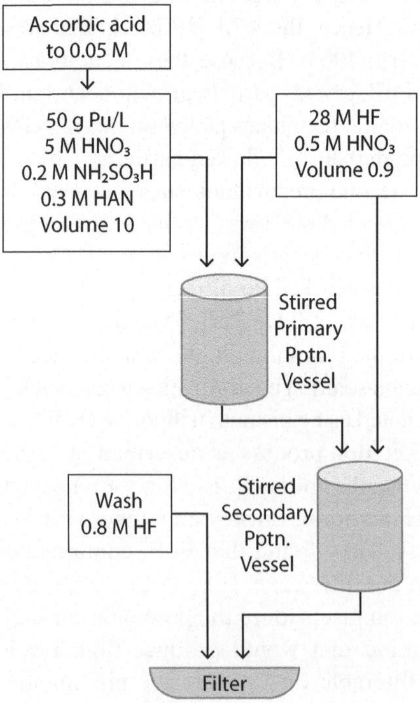

FIGURE 3.14 Savannah River Site plutonium trifluoride precipitation process flowchart. (After Burney, G.A. and Tober, F.W., *Ind. Eng. Chem. Process Des. Dev.*, 4, 28–32, 1965.)

for the six 27.5 g Pu/L tests ranged from 0.1–2 M; the HF concentrations for the six 7.5 g Pu/L tests ranged from 0.4 to 2.5 M. Based on the test observations, the conditions to be used in the primary (precipitation) vessel should be 25–35 g Pu/L in 2–3 M HNO_3 containing 0.03 M ascorbic acid and 0.22 M sulfamic acid, and the precipitation should occur by direct strike addition of HF solution to the Pu solution. The testing showed that plating was minimized and crystal growth maximized when the total as-mixed fluoride in the primary precipitation vessel was 0.10–0.15 M. In the secondary (crystal growth) precipitator, the as-mixed total fluoride should range from 0.7–1.0 M and the residence time should be at least 10–15 minutes to minimize the PuF_3 product solubility and thus limit loss to the filtrate.

Several plutonium fluoride precipitation processes were also tried at LASL (Section 8.4.4 of Christensen and Maraman 1968) but were abandoned because of inadequate decontamination. The best, a double salt method to produce dipotassium Pu(IV) fluoride (K_2PuF_6) was used primarily to separate Pu from uranium. The starting solution was made ~1 M HNO_3 and potassium nitrate (KNO_3) added to a 5:1 K:Pu mole ratio. Then, 28 M HF solution was added to attain a 1–2 M HF excess over that needed to form the K_2PuF_6.

3.5.4 Plutonium(III) Oxalate Precipitation Processes

The precipitation of Pu(III) oxalate from Pu(III) solutions in HNO_3 proceeds with the following reaction stoichiometry to produce three moles of HNO_3 per mole of Pu(III) precipitated:

$$2Pu(NO_3)_3 + 3H_2C_2O_4 + 10H_2O \rightarrow Pu_2(C_2O_4)_3 \cdot 10H_2O + 6HNO_3. \quad (3.10)$$

Complete precipitation of 100 g Pu/L (~0.42 M Pu), therefore, will increase HNO_3 concentration by 1.26 M for the mixture, assuming no dilution from the added oxalic acid solution.

Plutonium(III) oxalate ($Pu_2(C_2O_4)_3 \cdot 10H_2O$) is brilliant green (Pittman et al. 1944), though this color might indicate the presence of some beige/tan Pu(IV) oxalate with the teal/blue Pu(III) oxalate ordinarily observed (Figure 3.9). It may be precipitated from HNO_3 solution by either direct, reverse, or concurrent strike and even strike by dissolution of solid oxalic acid ($H_2C_2O_4 \cdot 2H_2O$) (Christensen et al. 1988). Valence adjustment to Pu(III) is necessary and, like the case for PuF_3, a holding reductant is generally required to suppress or scavenge nitrite generated by nitrate reduction or by radiolysis. Nitrite oxidizes Pu(III) to Pu(IV), thus leading to Pu(IV) oxalate precipitation.

As is the case for Pu(IV) peroxide and Pu(IV) oxalate, the objectives of Pu(III) oxalate precipitation are twofold: (1) purify Pu nitrate streams in HNO_3 solution from common metal impurities and (2) use as an intermediate in converting Pu nitrate solution to PuO_2. The PuO_2, in turn, can be a product in its own right for nuclear reactor MOX fuel, be fabricated into $^{238}PuO_2$ heat sources (e.g., Stoll et al. 2012), be a form for storage (DOE 2012), or itself be an intermediate on the way to preparing Pu metal through subsequent hydrofluorination and calciothermic reduction steps as was done, by way of Pu(III) oxalate, in early LASL and Hanford Site processing for weapons production.

A survey of the technical literature describing very early Pu(III) oxalate precipitation within weapons metal production processes at LASL (Christensen and Maraman 1968; Pittman et al. 1944; Carritt et al. 1946; Gergen et al. 1946) showed Pu(III) concentrations ranging from ~43 to 81 g/L in the precipitation liquor itself. The reductant used in the cited cases was hydriodic acid, HI. Other chemical reductants (e.g., sulfur dioxide, SO_2; formaldehyde, HCHO; hydrogen sulfide, H_2S; hydrogen, H_2, on platinum black) and electrolytic reduction were tested and found successful in early research but were rejected due to lack of confidence in their consistent performance (Gergen 1946, 48–49). The Pu(III) feed concentrations during oxalate precipitation, according to a 1944 flowsheet, was ~48 g/L (Figure 44 of Christensen and Maraman 1968 based on Pittman et al. 1944).

Within the first year of Pu production, i.e., from early 1945 until early 1946, LASL used, sequentially, three different processes—"A," "B," and "C"—all featuring Pu(III) oxalate precipitation steps (Carritt et al. 1946). In all cases, 5.5 M HI was added, at the rate of 8 moles HI per mole of Pu, to chilled (<17°C) feed solution to chemically reduce, to Pu(III), diluted Hanford product nitrate paste, containing a 70:30 to 30:70 ratio mixture range of Pu(IV):Pu(VI). The low temperature and excess HI were used to favor formation of non-volatile triiodide (I_3^-) thus minimizing elemental iodine and its escape to the vapor phase. The HNO_3 concentration in the valence-adjusted feed solutions ranged from ~0.6 to 2.5 M. Supporting studies at LASL showed that excess oxalic acid had to be limited to less than 0.4 M to minimize formation of soluble anionic Pu(III) oxalate complexes, while HNO_3 concentrations should be >0.5 M to enhance crystal growth (Gergen et al. 1946).

The "A" process used two Pu(III) oxalate precipitation steps with intermediate sodium plutonyl acetate precipitation followed by diethyl ether extraction: the first to purify the initial Hanford product and the second to prepare the Pu(III) oxalate as a PuO_2 precursor (Carritt et al. 1946, 28–29 and 34). The Pu(III) concentrations in the feeds for the first "A" process precipitation were 81 g/L, or ~31 g/L in the precipitating mixture for the first Pu(III) oxalate precipitation, and 43 g/L in the second feed, or ~23 g Pu/L in the precipitating mixture, in the second precipitation. In the "B" process, the initial Pu(III) oxalate precipitation was abandoned and the initial step became purification of the Hanford plutonium nitrate, first oxidized to Pu(VI) using sodium bromate, by diethyl ether extraction. Following this purification, the Pu(III) solution was prepared (using HI) and its concentration, when fed to precipitation, was 50 g/L, or ~25 g Pu/L in the precipitating mixture, with the Pu(III) oxalate again used as the PuO_2 precursor (Carritt et al. 1946, 58). In the "C" process, the Pu(III) feed concentration was 81 g/L, or ~31 g Pu/L in the precipitating mixture (Carritt et al. 1946, 76).

All the early LASL Pu(III) oxalate precipitations occurred by direct strike addition of 0.67 M $H_2C_2O_4$ solution, with stirring, to the valence-adjusted Pu solutions. Twenty-minute digestion time produced large crystals that were separated using a 15-minute settling time. The decontamination factor from uranium via the Pu(III) oxalate process was 20, but, of course, no separation occurred from the abundant La (30,000 parts per million parts Pu; i.e., 3%) present in the original Hanford feed material; by the summer of 1945, the La impurity in the Hanford product decreased to ~1% (Gergen et al. 1946). Decontamination (or purification) factors for other elements are provided in Table VI of Gergen et al. (1946).

Hanford adopted the LASL Pu(III) oxalate precipitation process when it assumed routine Pu metal and weapons part processing from LASL in July 1949 (Gerber 1997). Improvements in Pu(III) oxalate processing were implemented at Hanford, but HI continued to be used as reductant until Pu(IV) oxalate processing began in June 1952. A Hanford lab study investigated avoiding HI to produce Pu(III) oxalate, adding 0.3 M sulfamic acid as a nitrite scavenger to feed solution containing 25–70 g Pu/L and 2–4.5 M HNO_3. This solution was then treated with sufficient H_2O_2 to precipitate Pu(IV) peroxide. Continued addition of peroxide reduced the Pu to the (III) state, dissolving the Pu(IV) peroxide to make the soluble Pu(III) peroxide. At this point, oxalic acid was added to precipitate dense crystals of Pu(III) oxalate (Brown and Reas 1952). This process, however, was not implemented.

To avoid corrosion of iron-based materials by the HI, the precipitation vessel used in Hanford Pu(III) processing was constructed of glass and the process tubing of tantalum and saran (Work 1950). Removal of HI from filtrate solutions, by fuming, was also necessary before the solutions could be sent to Hanford's 224-B and -T Buildings for Pu recovery. The Pu(III) precipitation feed solution was about 70–80 g Pu/L in ≤2 M HNO_3 in a late Hanford process test (Kerr 1951). The reagent for reduction to Pu(III) was 47 wt% (~5.6 M) HI. Precipitation took place using direct strike addition of 0.67 M $H_2C_2O_4$, and the precipitate washed using 0.1 M HNO_3 containing 0.1 M $H_2C_2O_4$ (Kerr 1951). The flowchart for the Pu(III) oxalate process, *circa* 1951 at Hanford, is shown in Figure 3.15.

Further studies to optimize Pu(III) oxalate precipitation took place at the SRS (Porter and Symonds 1965). These systematic investigations showed that coarse, dense, and readily filtered Pu(III) oxalate crystals could be produced by slow

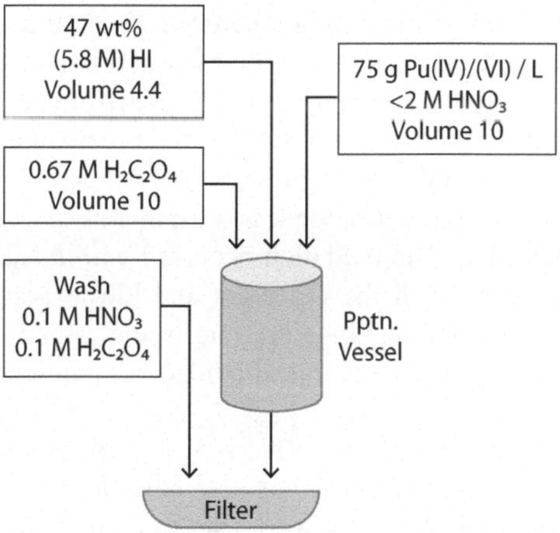

FIGURE 3.15 Hanford plutonium(III) oxalate precipitation process flowchart. (After Kerr, W.B., *Final Report – Production Test 231-10 – Supplement B – Process Evaluation-Production of Concentrated Plutonium Nitrate Solution in Building 231 – Transfer to and Processing in Building 234-5*, General Electric, Richland, WA, HW-22333. http://www.osti.gov/scitech/biblio/86307, 1951.)

addition of 1 M oxalic acid to 5–100 g Pu(III)/L solutions in 1 M HNO_3 at room temperature. The adjustment to Pu(III) was accomplished using ascorbic acid with hydrazine (N_2H_4) or sulfamic acid as nitrite scavengers. Under these conditions, Pu filtrate loss was <0.02 g/L. The Pu(III) oxalate product solubility, determined as functions of HNO_3 and excess oxalic acid concentrations (Porter and Symonds 1965; see also Burney and Porter 1967), was found to have an ~3.6-order direct dependence on HNO_3 concentration above 1 M HNO_3. For example, with 0.095 M excess oxalic acid, the Pu(III) oxalate solubility in 1 M HNO_3 is about 0.03 g Pu/L; at 3 M HNO_3, the solubility is ~1.15 g Pu/L (Porter and Symonds 1965). Similar trends but somewhat different values were found in the latter study (Burney and Porter 1967). Although solubility decreases with increasing excess oxalic acid in the tested range of 0.095–0.285 M, the filtrate volumes likewise increase, negating the diminished solubility by increased volume leading to similar Pu quantity reporting to the filtrates.

Porter and Symonds (1965) recommended that the Pu(III) oxalate precipitation process proceed by the following steps:

1. Adjust the 5–100 g Pu/L feed solution to <1 M HNO_3, add 0.05 M hydrazine or sulfamic acid to scavenge nitrite, and stir for 10 minutes.
2. Add 1 M ascorbic acid in an amount sufficient to reduce to Pu(III) and include a 0.03 M excess. Stir for 10 minutes to complete the reduction.
3. While maintaining thorough mixing, add a stoichiometric amount of 1 M oxalic acid, plus 0.2 M excess, over a 30- to 45-minute period.
4. Digest for 30 minutes and filter.
5. Wash with about three cake volumes of 0.5–1 M HNO_3 containing 0.2 M oxalic acid.

The product particles from this method range from 15 to 40 μm. Smaller particles, on the order of 1–10 μm, can be produced by faster addition of the oxalic acid solution, reverse strike (Pu solution added to the oxalic acid), or use of 0.2 M oxalic acid.

Porter and Symonds (1965) also showed that the precipitate left in contact with HNO_3 would oxidize to Pu(IV) causing filtrate losses due to the greater solubility of the resulting Pu(IV) oxalate. Filtration time also increased because of the smaller Pu(IV) oxalate particle size. The oxidation occurred within stagnant layers of precipitate that became depleted in the reductant and nitrite scavengers available to greater extent within the solution. Once dry and free of HNO_3, the Pu(III) oxalate precipitate becomes more stable to chemical oxidation though not immune to radiolytic oxidation.

Later SRS research examined the effects of precipitation parameters on the size and shape of the Pu(III) and Pu(IV) oxalate particles, though with more focus on Pu(III), exclusively so in the latter work (Smith et al. 1976; Burney and Smith 1984). These two complementary reports seemingly describe aspects of the same study. The work was undertaken to determine the effects of precipitation and calcination conditions on the physical qualities of the product plutonium dioxide (PuO_2) for use in $^{238}PuO_2$ heat sources. Such qualities included particle size and shape, specific surface area, and the sinterability of the compacted powders. The former

(Smith et al. 1976) described experiments into the effects of precipitation parameters on the size and xyz dimensions of the lath-like Pu(III) oxalate and the ball-like clumps of blades for the Pu(IV) oxalate. The latter study (Burney and Smith 1984) reduced these findings to parametric equations to describe how precipitation variables affect Pu(III) oxalate solubility and the size and shape of the product $^{238}PuO_2$ and how desired particle size, shape, and degree of agglomeration can be obtained by managing the precipitation conditions.

Because the focus was on preparation of $^{238}PuO_2$, the Pu feed solution concentrations were limited to 2–8 g/L in 0.2–5 M HNO_3 in the latter study. The lower concentrations limited the radiolytic effects which can alter the precipitation temperature and oxidation state distribution. Adjustment of the Pu(IV) anion exchange-purified feed to Pu(III) was done using 0.05 moles of excess ascorbic acid with 0.05 M hydrazine as a nitrite scavenger. Both reverse and direct strike precipitation were tested using 0.9 M oxalic acid with the final mixtures containing 0.20 M excess oxalic acid. The Pu(III) or oxalic acid was combined by constant rate addition over 12 minutes into a vessel paddle-stirred at 550 RPM. Temperatures were controlled to set values (22°C–55°C) in this parametric study. Because the particle size developed quickly, filtration could occur within five minutes of completion of mixing.

As is usually observed for both Pu(III) and Pu(IV) oxalate, the precipitated particles retained their morphology upon calcination to PuO_2 meaning the Pu(III) or Pu(IV) oxalate precipitation conditions determine the calcined product PuO_2 particle shapes and sizes (Smith et al. 1976). Among the studied Pu(III) precipitation parameters, direct/reverse strike mixing order, reactant concentrations, and temperature had the greatest effects on the particle morphologies whereas these properties were independent of agitation or solution addition rates, presence of impurities or suspended matter, and digestion time (Burney and Smith 1984). Increased precipitation temperature favored larger primary particles but less particle agglomeration. Higher HNO_3 concentration likewise favored larger particles, perhaps because of greater solubility of the plutonium oxalates. Direct strike mixing favored higher initial supersaturation in the Pu than reverse strike mixing and thus caused greater agglomeration of the laths. Prediction equations for the PuO_2 particle size (length, width, and thickness of the lath and the particle mode size) and the degree of agglomeration were developed for Pu(III) reverse strike precipitation parameters in 0.2 M excess oxalic acid in the slurry for the 20°C–60°C, 0.7–2 M HNO_3, and 2–8 g Pu/L ranges of precipitation conditions.

Smith et al. (1976) studied homogeneous precipitation of Pu(III) oxalate using dimethyl oxalate ($H_3CO_2CCO_2CH_3$) as oxalic acid precursor. Oxalic acid and two moles of methanol arise slowly and homogeneously in the acid solution by hydrolysis of the parent dimethyl ester. This incremental and evenly distributed oxalic acid generation allows uniform approach to precipitation and crystal growth conditions, thus contrasting strongly with direct or reverse strikes, which generate high differential solute concentrations within the mixing vessel. However, no further discussion of these experiments is provided here nor in the companion paper (Burney and Smith 1984).

Subsequent studies of the use of homogeneous generation of oxalic acid for Pu(III) oxalate precipitation were performed at the Los Alamos National Laboratory

(Yarbro et al. 1992). In this case, diethyl oxalate ($H_3CCH_2O_2CCO_2CH_2CH_3$) was the oxalic acid precursor. The Pu(III) arose by elution, using a chemical reductant, from an anion exchange purification process. To decrease solution volume, the original Pu(III) precipitation process operated using solid oxalic acid. However, introduction and metering of the solid was difficult to control. Therefore, the use of liquid diethyl oxalate to homogeneously generate oxalic acid was investigated. The diethyl oxalate process was perceived to have some advantages over the solid oxalic acid alternative. First, the liquid, available at 7.3 M concentration, would produce nearly the same filtrate volume as the solid oxalic acid reagent but be easier to meter. In comparison, the aqueous solubility of oxalic acid itself is only ~1 M at room temperature. Second, the precipitation rate would be more readily controlled by simply adjusting temperature (acid concentration also affects the diethyl oxalate hydrolysis rate). Finally, the product Pu(III) oxalate would be easier to filter, lowering Pu loss to the filtrate. One drawback was the presence of ethanol in the HNO_3-bearing filtrate. Ethanol fumes are flammable, while reaction of the alcohol with HNO_3 causes gassing and can engender runaway autocatalytic reactions. Parametric studies of temperature (50°C and 70°C) and HNO_3 concentration in the feed (0.5 M and 1 M) showed an increase in either showed an increase in either temperature or HNO_3 concentration increased the diethyl oxalate hydrolysis rate. The Pu(III) oxalate formed by this process was coarse, filtered almost as rapidly as water itself, and, when run at higher hydrolysis temperature, gave half the filtrate loss compared with use of solid oxalic acid.

3.5.5 Plutonium(IV) Oxalate Precipitation Processes

The reaction of Pu(IV) nitrate from HNO_3 with oxalic acid proceeds according to the following reaction to produce Pu(IV) oxalate hexahydrate and four moles of HNO_3 per mole of Pu(IV) precipitated:

$$Pu(NO_3)_4 + 2\,H_2C_2O_4 + 6\,H_2O \rightarrow Pu(C_2O_4)_2 \cdot 6H_2O + 4\,HNO_3. \qquad (3.11)$$

For example, the complete precipitation of 100 g Pu/L (~0.42 M Pu) will increase HNO_3 concentration by 1.68 M for the mixture, assuming no dilution from the added oxalic acid. An overview of Pu(IV) oxalate precipitation processing, with the ensuing steps to produce PuO_2 and an engineering flowsheet to accomplish these operations on a production basis, is provided by Greintz and Neal (1978).

Use of Pu(IV) oxalate precipitation to accomplish Pu purification from impure reprocessing plant product and provide, in a subsequent step, precursor to PuO_2 occurred in initial Pu industrial production at Mayak's Shop No. 9 in the USSR in 1949 (Sokhina et al. 2003). Details on these processes are sparse. However, it is known that the initial Pu refining processes were hampered by the great compositional variability of the starting reprocessing plant product solutions, some containing more impurity (e.g., La, Fe, U) than Pu, and some containing more Pu(VI) than Pu(IV). Although the initial precipitation step was to be Pu(IV) peroxide, the Fe and Mn impurities made this approach untenable. The high complement of impurity metal ions also complicated use of Pu(IV) oxalate as the starting step. When the high impurity content was recognized, higher oxalic acid quantities were used to overcome the oxalate consumed by complexation with Fe, U, La, and others. The Pu(VI) present in the reprocessing plant product was reduced to Pu(IV) using SO_2 gas; H_2O_2 was used in later processing. Thus, Pu(IV) oxalate, containing the La(III) oxalate impurity, precipitated, while substantial oxalate was consumed by Fe and uranyl complexation while leaving these impurities in solution. The Pu(IV) within the impure oxalate precipitate, then, was leached using ammonium carbonate to form soluble anionic Pu(IV) carbonate (or perhaps carbonate-oxalate) complexes (see pages 135–138 of Gelman et al. 1962 for underlying chemistry), leaving La(III) oxalate in the solid residue. Acidification of the Pu-bearing carbonate solution allowed further purification of the Pu, now free of Fe, U, and La, by Pu(IV) peroxide precipitation. A second Pu(IV) oxalate precipitation later in the Pu finishing process sequence produced feed for calcination to PuO_2. Although not a precipitation reaction, to complete the Pu conversion sequence, chlorination of the PuO_2 using hot carbon tetrachloride (CCl_4) vapor took place. Then, the product $PuCl_3$ was calciothermically reduced to form Pu metal (pages 40–60 and 78 of Sokhina et al. 2003).

Use of Pu(IV) oxalate for industrial-scale Pu processing in the United States did not occur until 1952 at Hanford. Laboratory investigations at Hanford of Pu(IV) oxalate precipitation, proof that hydrogen peroxide could adjust plutonium valence from the (III) and (VI) states to Pu(IV), and inventions associated with these findings were reported in October 1951 (Brown and Reas 1951). Thus, the Pu(III) oxalate precipitation step, with its attendant use of corrosive HI as a plutonium reductant, was supplanted in June 1952 by Pu(IV) oxalate precipitation in all subsequent Hanford processing to form PuO_2, either as a product in itself or as a route to metal. At Hanford, the Pu(IV) oxalate initially was produced batch-wise using a noble metal-lined filter that also functioned as a calcination boat to form PuO_2 with simultaneous or ensuing hydrofluorination to PuF_4 (Section 7.4 of Gerber 1997). Improvements in oxalate filter cake drying by heat lamp and by through-cake air flow were implemented in October 1952.

The industrial batch-wise preparation of Pu(IV) oxalate was described in a book chapter written ~1959 by five Hanford Site scientists and engineers (Harmon et al. 1961). The valence adjustment, precipitation, filtration, and wash operations, shown in Figure 3.16 and as described by Harmon and Reas (1957), are identical to those same steps outlined several years later (Harmon et al. 1961) and may reflect Hanford's highest state of development for the batch Pu(IV) oxalate process. Besides the valence adjustment, precipitation, and filter cake washing steps, the latter batch flowsheet showed that filtrates (designated "Waste" in Figure 3.16) and solid and dissolved oxalate were oxidized by potassium permanganate ($KMnO_4$) with H_2O_2 reduction of excess permanganate to divalent manganese (Mn(II)) to produce a Pu-bearing solution for recycle. The filtrate contained ~0.4 g Pu/L; the treated filtrates were 0.32 g Pu/L in 2 M HNO_3 containing 0.05 M each of potassium and manganese (Harmon et al. 1961).

Batch Pu(IV) oxalate precipitation process parameters at Hanford are provided by Facer Jr and Harmon (1954). The Pu feed solution could be 10–550 g Pu/L, with upper concentrations typically about 200–300 g Pu/L for the plant feeds of the mid-1950s. The acid concentration in the precipitating slurry, taking into account the

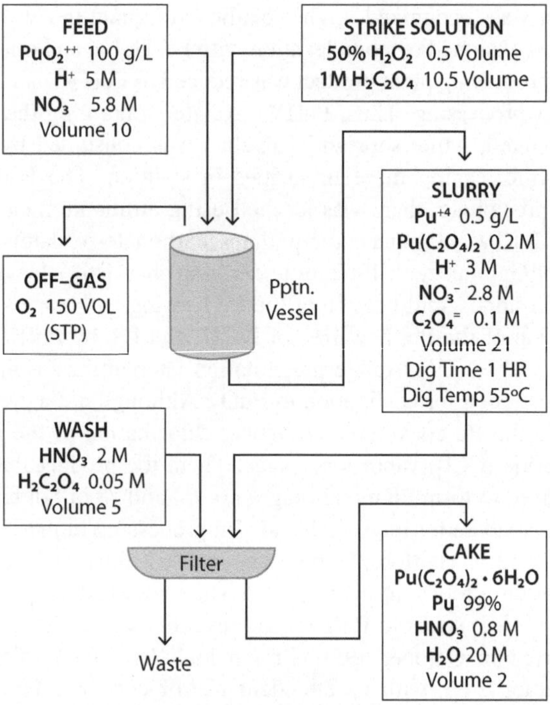

FIGURE 3.16 Plutonium(IV) oxalate flowchart (Figure 4 of Harmon and Reas 1957).

production of four moles of H^+ per mole of Pu(IV) precipitating and the dilution caused by added oxalic acid solution, could range from 1.5 to 4.5 M H^+. To diminish Pu filtrate loss, the H^+ concentration normally was ~2 M as shown in Figure 3.17.

With low residual contaminant in the purified Pu nitrate feed solutions, the H_2O_2 solution can be used to adjust Pu valence to (IV) for Pu(IV) oxalate precipitation (Facer Jr and Harmon 1954). The valence adjustment can be observed by the initial formation of a brown Pu(IV) peroxy complex followed, in turn, by a red Pu(IV) peroxy complex, blue Pu(III) solution, and then, finally, the dissolved Pu(IV) as oxidation of the Pu(III) by nitrous acid (HNO_2) occurs. The Pu(IV) solution is brown in more dilute ($\lesssim$2 M) HNO_3, becoming olive or emerald green in more concentrated HNO_3. Excess H_2O_2 precipitates Pu(IV) peroxide, but this solid decomposes and dissolves by reduction to Pu(III) as well as by heating in nitric acid.

If Pu(VI) remains in the feed solution, the Pu oxalate precipitation filtrate can contain as much as 2.3 g Pu/L, as limited by Pu(VI) oxalate solubility (Jenkins et al. 1965), to give the filtrate a characteristic orange color (Stubbs and Durnil 1991). The precipitated Pu(VI) oxalate also may pose hazards in subsequent calcining because it reportedly "decrepitates" when heated (Jenkins et al. 1965). Hanford operations likewise found that Pu(VI) oxalate decomposes too vigorously upon heating (page 8 of Facer Jr and Harmon 1954).

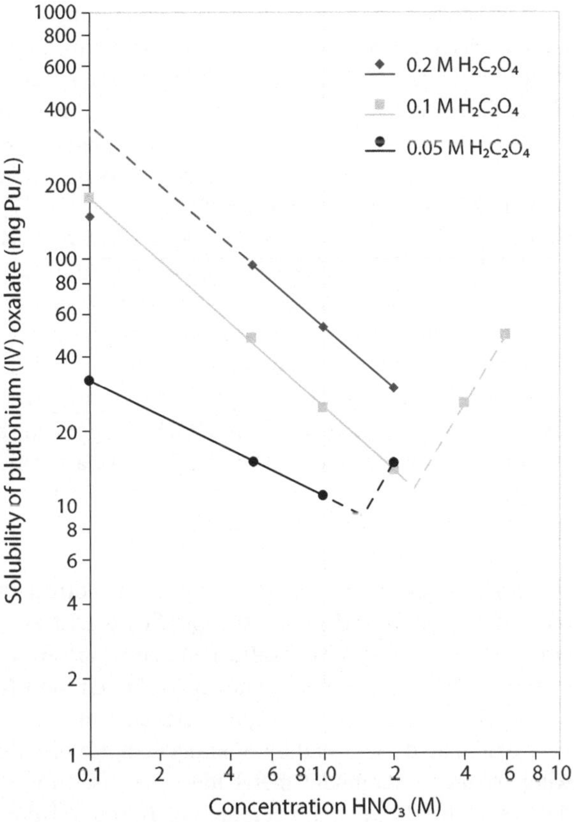

FIGURE 3.17 Pu(IV) oxalate solubility in HNO_3-$H_2C_2O_4$ solutions (Figure 2 of Facer Jr and Harmon 1954).

The conversion of Pu(VI) to Pu(IV) using hydrogen peroxide (H_2O_2) (Myers 1956; Marsh and Gallegos 1987) works reliably, leaves no residue, and acts on Pu(VI) nitrate according to the following reaction stoichiometry:

$$PuO_2(NO_3)_2 + 2\ HNO_3 + H_2O_2 \rightarrow Pu(NO_3)_4 + 2\ H_2O + O_{2\ \text{gas}} \tag{3.12}$$

consuming two moles of HNO_3 per mole of Pu(VI). Hydrogen peroxide disproportionation also occurs, catalyzed by dissolved Fe and Pu:

$$2\ H_2O_2 \rightarrow 2\ H_2O + O_2. \tag{3.13}$$

Because of the latter parasitic reaction, excess H_2O_2 must be used for valence adjustment. Fortunately, the decomposition residue is water, meaning dilution is the only consequence of superfluous H_2O_2 addition.

Valence adjustment at room temperature requires ~30 minutes but less than 10 minutes at 50°C if 0.5–1 wt% H_2O_2 (~0.15–0.3 M) is used. Because H_2O_2

decomposition rates increase with temperature to cause bubbling and foaming, solutions containing excess H_2O_2 are heated with caution. Failure to fully reduce Pu(VI) to Pu(IV) is shown by orange Pu(VI) oxalate in the ordinarily tan Pu(IV) oxalate precipitate (see Figure 3.9) to give a salmon color and accompanying high filtrate losses. Alternatively, if insufficient time is allowed for oxidation from Pu(III), the mixed Pu(III)/Pu(IV) oxalate is olive green in color, a combination of blue/teal Pu(III) oxalate and tan Pu(IV) oxalate (Figure 3.9).

As seen in Figure 3.17, HNO_3 concentrations above ~2 M increase Pu(IV) oxalate solubility steeply (by forming higher anionic oxalate complexes). Facer Jr and Harmon (1954) also found that precipitating at >4.5 M acid concentrations produced thixotropic Pu(IV) oxalate precipitates of low porosity. The low porosity cake inhibits ready reaction with air during calcination to form PuO_2 such that when reaction ultimately gets underway at more elevated temperatures, it proceeds so rapidly that the cake heats excessively, producing refractory PuO_2. In general, at <1.5 M H^+, coprecipitation of impurities is favored and the Pu(IV) oxalate solids become too finely divided for quick filtration. However, at H^+ >4.5 M, Pu(IV) oxalate solubility is excessive and the slurry becomes thixotropic; i.e., the slurry becomes shear-thinning and gels without agitation.

Oxalic acid solution was added to the Pu(IV) solution in a direct strike, sufficient to form stoichiometric $Pu(C_2O_4)_2 \cdot 6H_2O$ and still provide 0.1 M excess oxalic acid (Facer Jr and Harmon 1954). The excess oxalic acid could vary from 0.04 to 0.2 M for pure Pu solutions. With ordinary but unknown Al, Fe, Cr, and Ni contaminant concentrations, filtrate losses could be 100 times higher using 0.05 M rather than with 0.10 M excess oxalic acid. As found by Sokhina et al. (2003), this implied that oxalate complexing by the contaminant metal ions decreased the amount of free oxalic acid available for Pu(IV) oxalate precipitation. To minimize recycle solution volume, the Hanford plant used 1 M oxalic acid as the highest concentration conveniently available without requiring heat-jacketed apparatus (Facer Jr and Harmon 1954). Solid oxalic acid was not used because of difficulty in adding and mixing, slow dissolution, and its occlusion by Pu(IV) oxalate precipitating on the oxalic acid solid surface.

At Hanford, the oxalic acid strike was performed at ≥50°C with addition rate sufficiently slow to enhance crystal growth, typically 45–90 minutes per batch (Facer Jr and Harmon 1954). Precipitation at 50°C–60°C allowed greater crystal particle growth and filtration rates and decreased Pu filtrate losses. Higher precipitation temperatures, however, yielded "gummy" precipitates that were difficult to transfer, filter, or wash. After all the oxalic acid was added, the slurry was chilled to 10°C to decrease the Pu(IV) oxalate solubility and Pu filtrate loss in post-precipitation and filter cake washes. The precipitating slurry was agitated gently to avoid the slow-to-filter fine particles incurred by vigorous mixing.

The Pu(IV) oxalate particles produced in the above manner were 5–7 μm in diameter (Facer Jr and Harmon 1954). The slurry was filtered on sintered metal, plastic, or ceramic filters with an average pore diameter of <25 μm and maximum pore diameter of 40 μm. In plant processes, platinum filters with 15 μm pore diameter were used, showing improved filtration rates over the 8 μm diameter filters they replaced. Water or mixed $H_2C_2O_4/HNO_3$ solutions could be used for filter cake washing. Best results were obtained with 2 M HNO_3 containing 0.05 M $H_2C_2O_4$, but water or 0–4 M HNO_3 containing 0.05 M $H_2C_2O_4$ was nearly as good and showed little difference (0.03–0.06 g Pu/L) in filtrate loss. Twenty grams of wash were used per gram of Pu.

The freshly precipitated Pu(IV) oxalate was cream to tan in color (Figure 3.9) and lighter for finely divided precipitate (Facer Jr and Harmon 1954). The precipitate acquired a green color on standing, likely due to Pu(III) oxalate formation from self-radiolysis. The suction-compacted filter cake density was 0.4–0.7 g Pu/cm^3, while the solids that only gravity-settled in the mother solution were about half as concentrated. The settle/decant rinsing of Pu(IV) oxalate precipitates was practiced in early (1952) Hanford processing using triplicate contacts containing 1.5 M HNO_3/0.05 M $H_2C_2O_4$ (Collins and Isaacson 1952).

By the end of batch Pu(IV) oxalate processing at Hanford's 234-5 Building in 1957, 70–90 g Pu(IV)/L was precipitated in a glass vessel with sufficient 1 M $H_2C_2O_4$ to have 0.1 M excess (Beede 1956). The slurry was digested at 55°C and cooled to 20°C before the Pu(IV) oxalate solids were separated in the boat filters.

A continuous Pu(IV) oxalate precipitation process was developed at Hanford to supplant the batch process. It was envisioned to be superior because it would generate a more uniform Pu(IV) oxalate product having improved filtration characteristics while eliminating the batch process requirements for heating and cooling (Beede 1956). Heating was provided by the proximate precipitation vessel location to the calciner. The cooling step was obviated by the better mixing and improved particle size control. Once underway, the overflow from the continuous precipitation vessel was withdrawn at the same rate as the solution additions. This continuous precipitation arrangement is called a "mixed suspension, mixed product removal" (MSMPR) reactor. The slurry withdrawal occurring near the upper outer wall of the precipitator minimized the short-circuiting of the finer and more easily suspended Pu(IV) oxalate particles before they had time to grow and ripen. The continuous process provided greater process throughput rates and much lower hands-on manipulation with consequently lower personnel radiation exposure compared with the prior batch operations.

According to the development studies, feeds containing 10–100 g Pu(IV)/L in 1.0–6.5 M HNO_3 could be precipitated by the concurrent addition of near-saturated 1 M $H_2C_2O_4$. Initial testing showed that including Pu(VI) valence adjustment using H_2O_2 within the same process vessel as the continuous Pu(IV) oxalate precipitation gave high filtrate losses because of inadequate Pu(VI) reduction (Beede 1956). The valence adjustment, thus, was shifted to before the precipitation step, generally with heating that not only hastened the valence adjustment, but also warmed the Pu(IV) solution in anticipation of the oxalate precipitation.

The influence of precipitation temperature on the particle size of Pu(IV) oxalate produced in the continuous process was examined at ambient (~25°C) and 50°C–60°C temperatures using 180–200 g Pu/L solutions in 3–5 M HNO_3 and 0.1 M excess $H_2C_2O_4$ (Mudge 1959). Although good filter cakes were found at both temperatures, larger particle sizes, with 50 wt% of the particles >33 μm, were found at the higher precipitation temperature compared with 50 wt% >19 μm at ambient temperature.

Batch operations in Hanford's 234-5 Building ceased in July 1957 with adoption of the Remote Mechanical A (RMA) Line continuous process. The 1957 flowsheet, Figure 3.18, shows that 76 g/L Pu(IV) nitrate in 3.3 M HNO_3 was added to

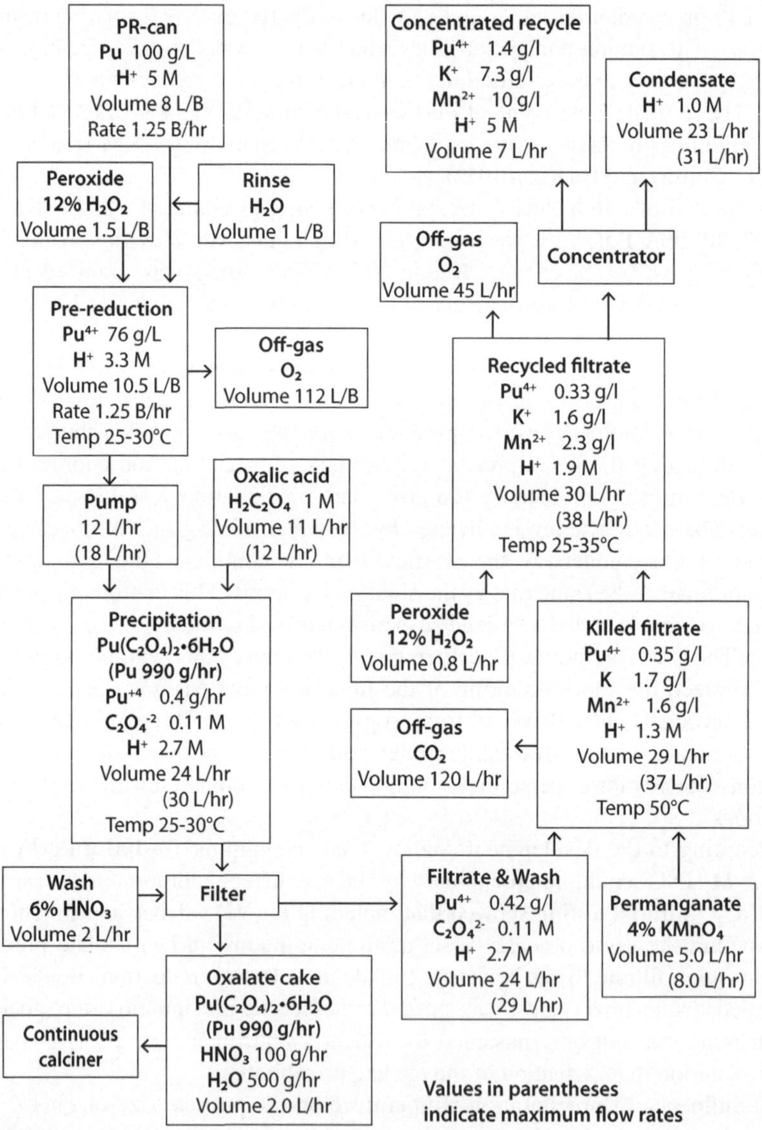

FIGURE 3.18 Continuous 1957 Pu(IV) oxalate flowsheet for the Hanford RMA line. (Maffei, H.P., *Continuous Task I–II*, General Electric, Hanford Atomic Products Operation, Richland, WA, HW-51006, https://www.osti.gov/opennet/detail?osti-id=16441914, 1957.)

the precipitation vessel simultaneously with 1 M $H_2C_2O_4$. The excess $H_2C_2O_4$ concentration was 0.11 M, and the digestion temperature was 25°C–30°C. Given the combined 24-L/hour output flow rate and 36-L reactor volume, the average digestion time was 1.5 hours.

The Pu(IV) oxalate slurry from the reactor vessel then overflowed to a 4-L capacity pan in which a cylindrical vacuum drum filter was partially immersed (Maffei 1957). Agitation within the pan kept the Pu(IV) oxalate solids suspended while the vacuum drum filter rotated on its horizontal axis partly immersed in the pan. The vacuum drew the slurry toward the drum, where the solids were collected on the outer filter cloth surface, and the filtrate was removed by suction through the drum filter. As the drum rotated like a Ferris Wheel, the filter cake rose out of the slurry on the drum. Near the top of the rotation, the cake was sprinkled with a rinse solution of 6% (1 M) HNO_3 with the rinsate drawn through the cake by suction and gravity. The rinsed Pu(IV) oxalate cake continued around the drum, was scraped from the drum by a "doctor blade," and the scrapings directed into a hopper leading to the calciner. The mixed filtrate/rinsate contained 0.42 g Pu/L in 2.7 M H^+ and 0.11 M $H_2C_2O_4$ (Figure 3.18).

Plutonium(IV) oxalate precipitation for the production of PuO_2 for fuel also took place at Hanford using the same continuous process (Lerch and Norman 1984). A relatively strict control of conditions was required to generate PuO_2 of suitable purity and particle size distribution for its intended use in MOX fuel. The process occurred by the simultaneous addition of nominally 250 g Pu(IV)/L and 1.0 M oxalic acid solutions to give a 0.1 M oxalic acid excess in 2.5–3.5 M HNO_3 in the precipitating slurry. The acid concentration was selected to decrease impurities and produce a finely divided and easily filterable product. The Pu(IV) valence was ensured by prior treatment with 12% (~3.5 M) H_2O_2. The Pu(IV) oxalate precipitation and digestion occurred at 45°C–55°C with digestion lasting 30–60 minutes. The solids were filtered and washed with 2 M HNO_3 containing 0.06–0.4 M oxalic acid. The 0.4 M oxalic acid in the rinse seems unusually high considering concentrations used in other Hanford flowsheets. For example, a somewhat earlier Hanford flowsheet is otherwise similar, but the rinse solution was 2 M HNO_3 containing 0.005 M oxalic acid (White 1977).

The feed solution in the continuous Pu(IV) oxalate precipitation process in the Hanford RMC Line in the 1985 campaign to make PuO_2 feed for Pu metal was 200 g Pu(IV)/L in 6.1 M HNO_3, and the precipitation agent was 0.95 M $H_2C_2O_4$ at 0.09 M concentration excess. The precipitation residence time and temperature were unstated, though warmed by the calcination furnace (Rodgers 1986) and likely 35°C–60°C as described in subsequent operations six years later (Stubbs and Durnil 1991). The precipitation reactor slurry overflow drained to a filter pan, was agitated by an oscillating paddle, and the solids separated by a vacuum drum filter as described for the RMA Line (see text describing Figure 3.18). The precipitation vessel and vacuum drum filter are diagramed in Figure 3.19. The filter cake wash solution was 0.05 M $H_2C_2O_4$ in 2 M HNO_3. The filtrate, including the wash solution, contained 0.6 g Pu/L, 3.2 M HNO_3, and 0.1 M $H_2C_2O_4$ (Rodgers 1986).

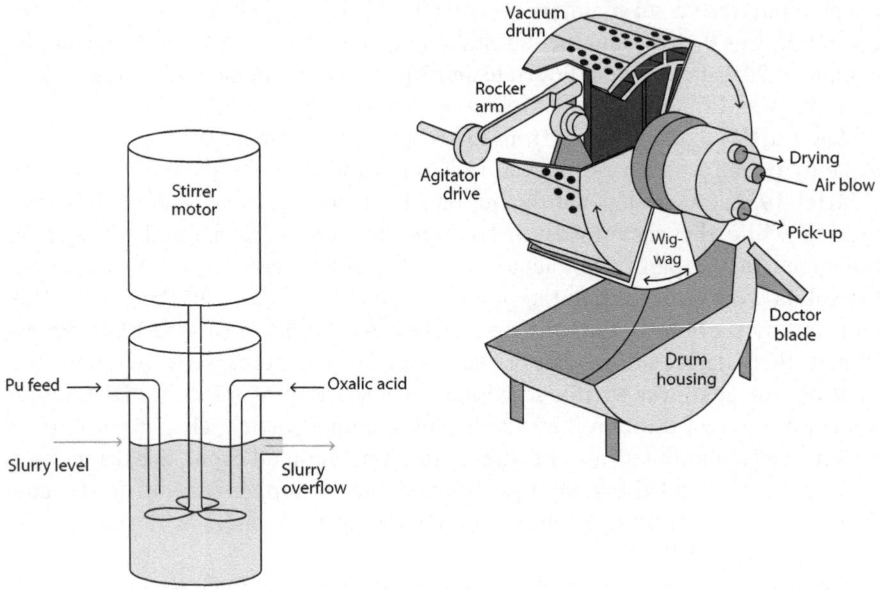

FIGURE 3.19 Hanford continuous Pu(IV) oxalate precipitation vessel (left) and vacuum drum filter. (Figures 6 and 7, respectively, of Rodgers, L.H., *1985 Remote Mechanical C (RMC) Line Campaign Report*, Rockwell Hanford Operations, Richland, WA, RHO-RE-EV-85, 1986.)

In the continuous Pu(IV) oxalate precipitation process operational in the RMC Line in 1991, the feed solution containing 150 g Pu(IV)/L in 5.5 M HNO_3 and the 0.92 M $H_2C_2O_4$ solution were added simultaneously at the optimum digestion temperature of 35°C–60°C (Stubbs and Durnil 1991). The slurry had an 8- to 10-minute precipitation reactor residence time, a much shorter digestion and higher temperature than the 90-minute residence time in the RMA Line reactor run at 25°C–30°C (Maffei 1957). The RMA Line in 1991 used 0.15 ± 0.05 M stoichiometric excess oxalic acid. As was the case for the 1950s RMA Line system (Maffei 1957), the overflow from the 1991 RMA Line precipitation reactor drained to a filter pan with an oscillating paddle immersed in the slurry to prevent solids from settling. The 11¾-inch (~30-cm) diameter and 5-inch (~13-cm) wide vacuum drum filter rotated into this suspension to pick up the Pu(IV) oxalate solids by suction from the slurry. As the drum rotated, the filter cake moved upwards and out of the pan, and the filter cake scraped from the outer wall of the drum by the doctor blade and into the calciner (Figures 3.19 and 3.20). As in the 1950s RMA Line system, the filter cake was washed by sprinkling with a rinse solution of dilute HNO_3 sometimes also containing dilute $H_2C_2O_4$. The filtrate contained 0.94 g Pu/L in 2.72 M HNO_3 and 0.15 M $H_2C_2O_4$ (Stubbs and Durnil 1991).

Conversion of a variety of pure and impure Pu-bearing solutions, such as mixed Pu/U nitrate solutions used in critical mass laboratory experiments and concentrated Pu(IV) oxalate filtrate solutions remaining from the 1980s operations, to PuO_2 suitable for long-term storage under the DOE-STD-3013 criteria (DOE 2012) occurred

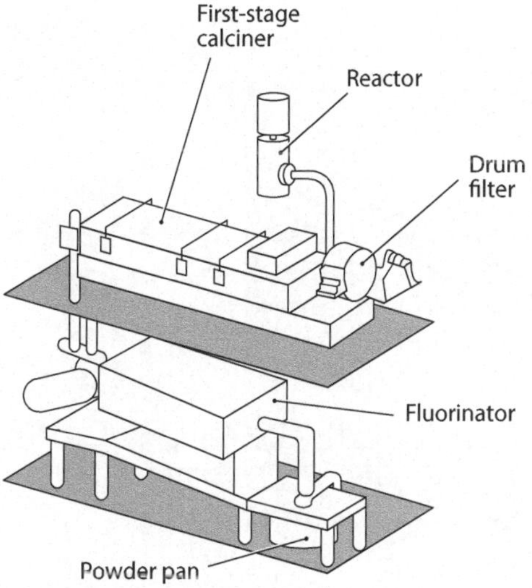

FIGURE 3.20 Hanford RMC line precipitation reactor, calciner, and hydrofluorination apparatus. (Figure 4.8 of Rodgers, L.H., *1985 Remote Mechanical C (RMC) Line Campaign Report*, Rockwell Hanford Operations, Richland, WA, RHO-RE-EV-85, 1986.)

from July 2001 to July 2002 in anticipation of the decommissioning of the Plutonium Finishing Plant at Hanford. A simple batch Pu(IV) oxalate precipitation process, followed by calcination to form PuO_2, was employed. Multi-year solution storage meant that the Pu had become exclusively tetravalent making valence adjustment unnecessary. Rather than the direct strike addition of oxalic acid solution, as was the practice in prior Hanford batch processing, the oxalic acid was added directly to the Pu solution as crushed crystalline $H_2C_2O_4 \cdot 2H_2O$ with amounts adjusted based on the Pu quantities. Stoichiometric control for uranium-rich solutions targeted a 2.1:1 to 2.3:1 $H_2C_2O_4$:Pu mole ratio, equivalent to 0.02–0.06 M excess $H_2C_2O_4$ in the ~50-g Pu/L feed solution, to avoid excessive uranium precipitation. For other solutions whose impurities did not form insoluble oxalates, the addition of excess oxalic acid was of less concern other than assuring that oxalic acid be enough to limit Pu filtrate losses. The precipitation occurred in an air-sparged ~13-cm diameter cylindrical glass tank at ambient temperature, solids were separated by suction filtration in a shallow pan, without washing, through glazed polypropylene felts, and drying and calcination occurred in Hastelloy boats (Delegard et al. 2002). Images from these steps are shown in Figure 3.21.

Plutonium(IV) oxalate precipitation also was used for Pu conversion operations in other U.S. facilities. At LASL, Pu(III) oxalate precipitation, rather than Pu(IV), generally was preferred (Christensen and Maraman 1968). Plutonium(IV) oxalate precipitation was only recommended to affect a gross separation from uranium because the chemically reducing conditions required for Pu(III) precipitation could coprecipitate uranium(IV) oxalate.

However, Pu(IV) oxalate precipitation was implemented in the 1980s in Pu recovery and purification from impure scrap at the Los Alamos National Laboratory

FIGURE 3.21 Pu(IV) oxalate precipitation (left), shallow pan filtration (top right), and hot-plate drying and initial thermal decomposition (bottom right) in solutions stabilization at Hanford, 2001–2002. (Delegard, C.H. et al., *Critical Mass Laboratory Solution Processing: Precipitation, Calcination, and Moisture Uptake*, Pacific Northwest National Laboratory, Richland, WA, PNNL-SA-56850, https://www.osti.gov/servlets/purl/15010137-OAVLyJ/native/, 2002.)

(LANL; the successor in name from LASL) (Christensen et al. 1988). The Pu(IV) oxalate process had three advantages: (1) separation from metallic (e.g., Fe, Si, Al) and anionic (e.g., chloride, sulfate) impurities; (2) concentration of valuable Pu to allow solution discard; and, (3) after calcination, formation of PuO_2 as an intermediate to metal and as a stable form for further processing. LANL performed Pu(IV) oxalate processing batch-wise but simultaneously added the Pu(IV) nitrate and 1 M oxalic acid streams to stirred and heated (~60°C) 1.5 M HNO_3 to enhance crystal growth. In an interesting modification, the LANL process valence-adjusted Pu(VI) to Pu(IV) with H_2O_2 during the precipitation and ripening of the oxalate particles, finding that the Fe(III) present in many scrap solutions that catalytically decomposed H_2O_2 did not do so, due to oxalate complexation, when oxalic acid was present. The Pu(IV) oxalate produced in this manner was medium-tan, coarse, and rapidly filtered and washed using 2.5 M HNO_3/0.1 M $H_2C_2O_4$ to produce a free-flowing PuO_2 product on calcination. These qualities were comparable to the best of the Pu(III) oxalates and provided better decontamination from Fe(III), Al(III), and U(VI). The Pu(IV) oxalate filtrates contained 0.2–0.5 g Pu/L, slightly more than the 0.1–0.5 g Pu/L found for Pu(III) oxalate.

Later, direct strike batch Pu(IV) oxalate precipitation processing was used at LANL to accomplish further plutonium polishing from U, Fe, and Al contaminants and provide better particle morphology for the desired PuO_2 product for a MOX fuel feed (Bluhm et al. 2005). The Pu(IV) solution was adjusted to 2.5 M HNO_3, and the precipitation conducted at ~65°C under continuous mixing with 0.7 M oxalic acid added slowly to the mixing slurry (Alwin et al. 2007). The Pu(IV) oxalate solids were settled in chilled suspension, vacuum filtered, and the cake washed with "dilute" oxalic acid in 2.0 M HNO_3 before being air-dried at reduced pressure.

Methods to produce PuO_2 particles for fabrication of $^{238}PuO_2$ items at the SRS were examined by Smith et al. (1976). Because of ^{238}Pu's high specific activity and resultant gassing in its solutions, only 2–8 g Pu/L concentrations were used. The HNO_3 concentrations ranged from 0.5 to 5 M. Tests included both Pu(III) and Pu(IV) precipitated by both direct and reverse strike with oxalic acid solution in batch mode at controlled temperatures. As noted in the Pu(III) oxalate discussion of this work, the calcined PuO_2 particle morphology and size distributions depended strongly on the mixing method, oxidation state, reactant concentrations, and precipitation temperature but not on the agitation rate or digestion times.

Batch Pu(IV) oxalate precipitation testing was performed at the Savannah River National Laboratory to identify optimum conditions for the second SRS H Area process line (i.e., the HB Line) (Crowder et al. 2012). Direct strike precipitation occurred by adding 0.9 M oxalic acid to ~56 g Pu(IV)/L solution adjusted to 1.5 M and 2.5 M HNO_3. The oxalic acid addition rate occurred over 25–48 minutes with the precipitation and digestion (5–30 minutes) conducted at 50°C to match expected HB Line precipitation conditions. The Pu(IV) oxalate slurry was filtered on a nominal 10 μm screen and the filter cake washed with 1.4 M HNO_3/0.1 M $H_2C_2O_4$ solution. Additional Pu(IV) oxalate precipitation tests were performed with the same Pu(IV) solution adjusted to 0.77–3.0 M HNO_3 and 5- to 30-minute digestion times.

Various methods to precipitate Pu from HNO_3 solution were surveyed at Rocky Flats as alternative means to convert Pu(IV) eluate from anion exchange purification to PuO_2 (Martella et al. 1984). Four Pu(IV) oxalate tests were conducted in this study with the stock feed being 4.09 M HNO_3 containing ~46 g Pu(IV)/L. The Pu(IV) oxalate was prepared by slow direct strike addition of 1.0 M oxalic acid followed by 30-minute digestion at 50°C. The final oxalate:Pu mole ratios ranged from 2.5:1 to 4:1. The precipitates were filtered and washed in 2 M HNO_3. Larger oxalate:Pu ratios gave larger particles with individual particles ranging from 2–7 μm spherical shapes at the lowest ratio, to 25–30 μm tetragonal particles at the highest ratio. The shapes became more geometric and had less relative size variation as the oxalate:Pu mole ratio increased. Filtrate concentrations were 0.28 g Pu/L at the lowest ratio and to 0.11 g Pu/L at the highest ratio with filtration rates lowest at the lowest tested oxalate:Pu mole ratio.

Batch Pu(IV) oxalate precipitation was used at the Lawrence Livermore National Laboratory to purify ^{238}Pu from ^{234}U in-grown by alpha decay (Mew et al. 2001). Weapons-grade Pu containing added uranium was used for preliminary testing. Alternatives in the Pu(IV) valence adjustment and precipitation steps, whether sequential or concurrent, at both 50°C–60°C, and with 0.05–0.15 excess oxalic acid and digestion times (one hour at 50°C–60°C versus 20 minutes after chilling to <10°C) were examined. In all cases, the precipitates were washed twice with 0.05 M $H_2C_2O_4$/2 M HNO_3 and decanted before being air-dried in preparation for calcination. Radiolytic decomposition of the ^{238}Pu(IV) oxalate precipitates was found to be significant. It also was found that the mixed 0.6 M H_2O_2/0.8 M $H_2C_2O_4$ precipitant decomposed with time and had to be freshly prepared to be effective. A DF from uranium of at least 31 was found in tests with weapons-grade Pu and ~520 in tests with the ^{238}Pu. The low ^{238}Pu recovery, only 78.5%, was attributed to radiolytic effects on the Pu(IV) stability.

Research into PuO_2 production for MOX applications took place at the Nuclear Materials and Equipment Corporation (NUMEC), in Apollo, Pennsylvania, in the early 1960s. Within these studies, Pu(IV) oxalate was synthesized by both batch and continuous methods. The process parameters were largely bounded by the Hanford production experience. For example, a Pu(IV) oxalate batch was produced in early 1961 by direct strike of 100 g Pu/L in 3 M HNO_3, at 55°C, with 1 M oxalic acid containing 0.8 M H_2O_2. The oxalic acid was added over 45 minutes followed by a 20-minute digestion. The slurry was separated using a 5 μm filter, and the cake washed with 2 M HNO_3/0.05 M $H_2C_2O_4$. The agitator tip speed was 3.3 ft/s (~100 cm/s), equivalent to ~380 RPM for a 5-cm diameter blade (NUMEC 1961). By the end of 1961, a continuous precipitation apparatus was operational using 100 g Pu/L and 1 M $H_2C_2O_4$/0.25 M H_2O_2 feed solutions with precipitations occurring at 28°C with a 20-minute precipitator residence time. Agitation was provided by a 2.5 in (6.35 cm) turbine paddle mixer run at 500 RPM, or ~166 cm/s tip speed. The cake was water-washed. In the report NUMEC P-90 (1961), it was found that a more dilute feed (20 and 12 g Pu/L in 3 M HNO_3) provided smaller solid particle agglomerates than a more concentrated feed in precipitations run at 35°C with 20–23 minutes residence time. Investigations in both continuous and batch modes were performed with a baseline feed of 100 g Pu/L in 3 M HNO_3 and 1 M $H_2C_2O_4$ containing 0.8 M H_2O_2 and with 35°C digestion temperature and 30 minutes time. The effects of Pu concentration, excess oxalic acid, and precipitator residence time, as well as cake-drying profiles and calcination temperature on PuO_2 qualities were studied (NUMEC P-100, -103, -105, and -106 (1962, 1963) and NUMEC 2389-2 and 2389-4 (1964)).

Internationally, the Pu(IV) oxalate precipitation process is favored over Pu(III) oxalate in commercial Pu nitrate conversion to PuO_2 because it provides better purification of Pu from Al, Fe, and U and provides better particle morphology (Cleveland 1967; Abraham et al. 2014). For these reasons, and despite the somewhat better Pu recovery (lower solubility) and filterability of Pu(III) oxalate, Pu(IV) oxalate precipitation has become the worldwide reference process, being implemented at UP1 (Marcoule, France), UP2, and UP3 (Cap La Hague, France), Windscale/Sellafield and THORP in the United Kingdom, Mayak in Russia, and Hanau in Germany (Abraham et al. 2014) as well as in India and China.

An overview with limited process detail of the use of both batch and continuous Pu(IV) oxalate precipitation for Pu finishing within the UP1 reprocessing plant at Marcoule, France, in the early 1960s is described by Jouannaud (1965). The batch process used 10 g Pu/L in 0.5 M HNO_3. No adjustment to higher HNO_3 was noted, and 1 M oxalic acid precipitant was used. A continuous precipitation process used 1 M oxalic acid and 6 g Pu/L in 1 M HNO_3 but with provision for HNO_3 addition to the precipitation vessel. Further detail on the conversion to PuO_2 and PuF_4 at UP1 also is provided in this reference.

Conversion to Pu(IV) oxalate and then to PuO_2 is being implemented for the Pu nitrate product from the reprocessing plants UP2 and UP3 at Cap La Hague (Grandjean et al. 2008). Precipitation conditions are not provided in detail, except that the operation is run at continuous conditions with a rotating filter, perhaps akin to the vacuum drum filter. This paper also describes work to improve and model the precipitation vessel design by using vortex mixing. In incidental remarks regarding ~1996 operations, it was stated that the La Hague PuO_2 production plant used "nitrous fumes" to adjust plutonium to the (IV) oxidation state (IAEA 1998). This account further states that the precipitation occurred with a slight oxalic acid excess, that the Pu(IV) oxalate crystals were aged about 15 minutes to increase particle size, but that aging time may vary depending on temperature and Pu and acid concentrations. The precipitate then overflowed the precipitation vessel to a rotating vacuum (drum) filter. The filter cake was removed by a scraper bar to deposit the cake into a drying furnace, while the residual oxalic acid and oxalate in the filtrate were destroyed by reflux in concentrated HNO_3. The cake was dried, converted to powder, and the solids advanced by an Archimedes screw calciner against a countercurrent air flow. The target 10 m^2/g specific surface area for the product oxide, accomplished in the calciner or in a subsequent furnace step, implies an ~600°C final calcination temperature.

The preparation of PuO_2 from Pu nitrate solution at the THORP plant in the United Kingdom, and in prior operations in another plant to produce military materials for which subsequent "fluorination" (hydrofluorination) and conversion to metal were performed, is described in a summary overview (IAEA 1998). The feed Pu nitrate was converted to Pu(IV) using H_2O_2 in several parallel vessels. A slight excess of oxalic acid (whether solution or solid is unstated) was used to precipitate Pu(IV) oxalate with 15 minutes of digestion to improve filtration. The slurry overflowed to a rotating vacuum filter, and the collected solids were scraped off into a drying furnace. The oxalic acid/oxalate destruction in the filtrates and the calcination were accomplished using techniques like those described above for the La Hague plant in France. Preparation of PuO_2 for dry-blended MOX fuel at Sellafield also evidently used Pu(IV) oxalate, but the process details were not elaborated (Edwards et al. 1991).

Preparation of Pu(IV) oxalate as part of PuO_2 production for MOX fuel at the ALKEM GmbH plant in Hanau, Germany, has been described (Kuhn and Koch 1973) with subsequent elaboration of the unique process features being presented in a patent (Kuhn and Koch 1977). The batch process used solid oxalic acid dihydrate ($H_2C_2O_4 \cdot 2H_2O$) crystals rather than $H_2C_2O_4$ solution to minimize filtrate volumes and subsequent handling, as well as to provide more process latitude for small batches within a nominal 5 kg Pu/day throughput. Valence was not adjusted for this process, which relied on the tendency toward Pu(IV) by radiolysis after "about 22 days" of storage. The effects of precipitation conditions—1.5–5 M HNO_3, digestion temperature between 40°C and 80°C, the solid oxalic acid addition rate, the stirring velocity and duration—on the product PuO_2 physical properties were investigated. However, detailed outcomes of these investigations were not published. The process patent (Kuhn and Koch 1977) describes how decontamination from ^{241}americium (^{241}Am), the beta decay product from ^{241}Pu, can be affected by applying the solid oxalic acid precipitating agent at a stoichiometric deficiency. Solid ammonium oxalate crystal, $(NH_4)_2C_2O_4$, also may be used instead of solid oxalic acid dihydrate to supply the necessary oxalate. The solid oxalate source was added over about one hour with stirring to the nominally 200 g Pu(IV) nitrate/L solution at 60°C–95°C (ideally 80°C) at quantities as much as 7% deficient from the stoichiometric 2:1 oxalate:Pu(IV) mole ratio. Digestion time was about one hour. The final HNO_3 concentration was less than 5.2 M, meaning that the starting HNO_3 concentration for 200 g Pu/L was <1.85 M. The slurry then was cooled to about 40°C and filtered to separate the solid Pu(IV) oxalate product. The DF of Pu from americium (Am) was 3 based on process experience (Kuhn and Koch 1973) but up to 15 was claimed in the patent (Kuhn and Koch 1977) with, though unreported, relegation of Pu to the filtrate in proportion to the stoichiometric deficiency in oxalate.

Industrial Pu(IV) oxalate processing dates from the 1949 beginnings at Mayak in the USSR/Russia (Sokhina et al. 2003). More recently, the preparation of PuO_2 at the Industrial Association Mayak in Russia is accomplished using Pu(IV) nitrate solutions at 7–24 g/L concentration and the PuO_2 produced from the Pu(IV) oxalate by calcination at 600°C–800°C (Bobylev et al. 2004). Studies of batch Pu(IV) oxalate precipitation methods to separate in-grown Am from Pu were performed in Russia (Pazukhin et al. 2003) based initially on information found in the patent of Kuhn and Koch (1977) in which solid oxalic acid was used as the precipitant. The Russian researchers found that stirring had to be adequate to keep the solid oxalic acid dihydrate suspended. Without adequate stirring, Pu(IV) oxalate deposited onto the oxalic acid solid and less effective decontamination of Pu from Am occurred. They also found that even under conditions reported to provide decontamination factors up to 15; i.e., oxalic acid at 93% of the full 2:1 $H_2C_2O_4$:Pu stoichiometry and regardless of direct or reverse strike, the DFs from Am were no more than 3 even with prolonged 25°C digestion. Decreased DFs were found at 90°C digestion temperature, in contradiction to the findings of Kuhn and Koch (1977). Pazukhin et al. (2003) performed further tests using diethyl oxalate as the precipitant. Like dimethyl oxalate (Abrahamson 1967), diethyl oxalate hydrolyzes in acid solution to slowly and homogenously generate oxalic acid, enhancing crystal growth and eliminating the large concentration gradients that occur especially in direct and reverse strike mixing. Whereas little to no decontamination of Pu from Am occurred in direct or reverse strike precipitation, Pu DFs from Am of 60 were found with diethyl oxalate at 108% stoichiometry (i.e., 2.16:1 $H_2C_2O_4$:Pu mole ratio). The DFs were ~13 at 127% stoichiometry and 3.6 at 150% when digested at 18–19 hours at 25°C. Extending digestion time to 60 hours at 127% stoichiometry (2.54:1 $H_2C_2O_4$:Pu mole ratio) decreased the DF to 1.9 by recrystallization and incorporation of the Am(III) oxalate into the Pu(IV) oxalate. Based on these findings, additional tests were performed using 12 g Pu/L in 3.4 M HNO_3 with 7.7 M diethyl oxalate as the precipitant and 20°C digestion. With these relative Pu and diethyl oxalate concentrations, little increase in total solution volume over that of the initial Pu solution was incurred. The DFs from Am(III) were ~2,000 at 101.3% stoichiometry but fell with increasing relative diethyl oxalate to ~11 at 128% stoichiometry, ~8 at 131.4% stoichiometry, and ~3 at 157.5% stoichiometry. However, losses of Pu to the filtrate were greater than would be expected by stoichiometry, indicating that the diethyl oxalate did not completely hydrolyze. Thus, ~29% of the Pu reported to the filtrate at 101.3% stoichiometry, ~13% at ~128% stoichiometry, ~11% at 131.4% stoichiometry, and ~2.2% at 157.5% stoichiometry.

Plutonium(IV) oxalate precipitation by direct strike at 50°C–60°C using slow addition of oxalic acid solution was studied at the Atomic Energy Establishment, Trombay, India. Ghosh-Mazumdar et al. (1965) found that using 2–3 M HNO_3 and 0.1 M oxalic acid in the final slurry obtained minimum solubility and coarse particles with good settling. No information in the technical literature was found on the preparation of PuO_2 at current Indian production facilities such as PREFRE at Tarapur and IGCAR at Kalpakkam.

Plutonium finishing at the planned non-military reprocessing pilot plant in China's Ganshu province was to be performed by way of a continuous precipitator with rotary vacuum filter (Wu and Jiang 1991). Although the authors did not elaborate upon the precipitation conditions and Pu oxidation state, the description appears to be similar to the continuous processes pioneered in Hanford Pu(IV) oxalate production. Development of a continuous Pu oxalate process from China, using Pu(III), Pu(IV), and mixed Pu(III)/Pu(IV) was described at the same conference (Sun et al. 1991). The process favored a draft-tube precipitator design (over a simple flow-through vessel) with 20–30 minutes residence time, 1.5–2.0 M HNO_3, variable (unspecified) Pu concentration, and 20°C–70°C digestion temperature. The Pu(III) process worked better in filtration, product recovery, and crystallization compared with similar Pu(IV)-only and mixed Pu(III)/Pu(IV) processes. Plutonium(III) also was favored because it was the predecessor material from the PUREX plant. Testing and imaging of the solids showed that particle size could be controlled by selection of precipitation conditions such as temperature, concentrations of HNO_3, Pu and oxalic acid, and the precipitator design.

3.5.6 Comparison of Major Plutonium Precipitation Conversion Processes

The major precipitation processes used for industrial Pu conversion from nitrate solution to solid forms are Pu(IV) peroxide, Pu(III) fluoride, Pu(III) oxalate, and Pu(IV) oxalate. All four alternatives have been conducted using batch processing. However, the last, Pu(IV) oxalate, also has been conducted through a continuous process using an overflow MSMPR design at Hanford and at other locations. For this reason, the continuous Pu(IV) oxalate process is currently the leading precipitation process for plutonium conversion. Nominal conditions for the precipitation processes are compared in Table 3.4.

The precipitation processes outlined in Table 3.4 also provide appreciable decontamination from impurities in feed reagents, from fission products, from process chemical impurities, and from metallic corrosion products. According to a

TABLE 3.4
Comparison of Conditions and Outcomes for Batch Pu(IV) Peroxide, Pu(III) Fluoride, Pu(III) Oxalate, and Pu(IV) Oxalate and Continuous Pu(IV) Oxalate Precipitation Processing

	Process				
Process Parameter	**Pu(IV) Peroxide**	**Pu(III) Fluoride**	**Pu(III) Oxalate**	**Pu(IV) Oxalate, batch**	**Pu(IV) Oxalate, continuous**
Formula	$Pu_2O_7 \cdot 0.5(HX) \cdot 6H_2O$	$PuF_3 \cdot xH_2O$	$Pu_2(C_2O_4)_3 \cdot 10H_2O$	$Pu(C_2O_4)_2 \cdot 6H_2O$	$Pu(C_2O_4)_2 \cdot 6H_2O$
Color	Olive-green	Lavender	Blue, teal	Tan, beige	Tan, beige
Feed [Pu], g/L	~50	~50	~5–100	~250	~150–200
Valence adjustment	H_2O_2	0.2 M NH_2SO_3H, 0.3 M $NH_2OH \cdot HNO_3$; ascorbic acid just before pptn.	0.05 M N_2H_4 + 0.03 M excess ascorbic acid	H_2O_2; almost instantaneous at 50°C	12% H_2O_2; almost instantaneous at 50°C
[HNO_3], M	~2–5	~4	~1	~2.5–3.5	~1.5–4.5
Precipitant	30%–50% (10–17 M) H_2O_2	4–28 M HF/0.5 M HNO_3	0.9–1 M $H_2C_2O_4$	0.9–1 M $H_2C_2O_4$	0.9–1 M $H_2C_2O_4$
Stoichiometric excess	~6–8× excess	~10×; free F^- controlled by HNO_3/HF ratio	0.2 M	0.1–0.2 M	0.1–0.2 M
Temperature, °C	~6–14	~25	~25	~55	~35–60
Strike direction	Direct	Direct or concurrent	Direct	Direct	Concurrent
Addition rate	2 mole H_2O_2/mole Pu-min	~9 mole F^-/mole Pu-min	30–45 min	45–90 min	8–60 min residence time in MSMPR
Digestion time	~30 min	~60 min	~30 min	~60 min	
Rinse composition	2% H_2O_2	0.8 M HF	0.5–1 M HNO_3/0.2 M $H_2C_2O_4$	2 M HNO_3/0.05 M $H_2C_2O_4$	2 M HNO_3/0.05 M $H_2C_2O_4$

(Continued)

TABLE 3.4 (*Continued*)
Comparison of Conditions and Outcomes for Batch Pu(IV) Peroxide, Pu(III) Fluoride, Pu(III) Oxalate, and Pu(IV) Oxalate and Continuous Pu(IV) Oxalate Precipitation Processing

	Process				
Process Parameter	**Pu(IV) Peroxide**	**Pu(III) Fluoride**	**Pu(III) Oxalate**	**Pu(IV) Oxalate, batch**	**Pu(IV) Oxalate, continuous**
Other conditions	Usually ~0.2 M H_2SO_4; Mainland et al. (1961) required no sulfate.	Two-steps/vessels—first for primary precipitation, second for crystal growth; HNO_3/HF ratio >4.	–	Chilling for digestion sometimes used for coagulation and to decrease solubility.	Process parameters varied according to feed solutions and to achieve desired product oxalate particle properties.
$[Pu]_{filtrate}$, g/L	0.04–0.2	0.01–0.1	<0.2	~0.1–0.2	~0.6–0.9
References	Leary et al. (1959), Mainland et al. (1961), Hagen and Miner (1980)	Orth (1963), Burney and Tober (1965)	Porter and Symonds (1965), Burney and Porter (1967)	Facer Jr. and Harmon (1954), Harmon and Reas (1957)	Lerch and Norman (1984), Rodgers (1986), Stubbs and Durnil (1991)

prominent 1967 review, the Pu(IV) peroxide precipitation provided the best decontamination followed by Pu(IV) oxalate; the relative efficacy of Pu(III) oxalate was not mentioned (pages 557–561 of Cleveland 1967). Plutonium(IV) peroxide precipitation was used at the SRS for Pu metal production in the 1950s (Mainland et al. 1961; Moore et al. 2000) but was supplanted by Pu(III) fluoride precipitation even though the fluoride process realized considerably less decontamination compared with peroxide and oxalate (whether Pu(III) or Pu(IV) is not stated; page 123 of Orth 1963). Winchester and Maraman (1958) conducted a single study for these four Pu precipitation processes comparing the trace element decontaminations from Fe and from common fission and activation product elements: Ce, Co, Mo, Ru, and Zr. These direct comparison values are shown in Table 3.5 with decontamination factors from various metallic element impurities for Pu(III) and (IV) oxalate, Pu(IV) peroxide, and Pu(III) fluoride precipitations. Besides accomplishing the desired chemical transformations, the ensuing calcination and, if undertaken, hydrofluorination steps also served to decontaminate the plutonium feed of volatile non-metallic constituents. Such constituents include nitrogen, from nitrate; sulfur, from sulfamic acid or hydroxylammonium sulfate from chemical reduction to Pu(III) (Mainland et al. 1961) and from the sulfate used in peroxide precipitation; carbon, from the oxalate precipitations; and chlorine from process impurities.

An alternative process for Pu nitrate solution conversion to PuO_2 is direct denitration, which is accomplished using a mechanical screw in a heated zone (screw calciner), with a fluidized bed, and by microwave heating. Although direct denitration provides no decontamination from impurities, it also does not create a side stream requiring processing as the precipitation processes do via their filtrates. Direct

TABLE 3.5
Decontamination Factors for Plutonium Nitrate Solutions by Precipitation

	Decontamination Factor			
Element	**Pu(IV) Peroxide**	**Pu(III) Fluoride**	**Pu(III) Oxalate**	**Pu(IV) Oxalate**
Al	>100	–	–	100
Ce	6	1.1	1	1
Co	30	8.6	47	>95
Cr	>700	–	–	100
Fe	50[a]	1.4	33	10
Mo	>140	1.1	>13	>15
Ni	>3000	–	–	100
Ru	>14	36	>38	12–33
Th	–	–	–	1
U	–	–	–	60
Zr	1	1.1	3.5	>44

Source: Taken from Table I of Winchester and Maraman 1958; Table 7.6 of Clark et al. (2006); Table 15.1 of Cleveland (1967); page 450 of Harmon et al. (1961).

[a] High iron concentrations cannot be tolerated as they catalyze H_2O_2 decomposition.

denitration using a screw calcination is described by Souply and Sperry (1977). A conceptual analysis of fluid bed direct denitration, largely based on experience in uranyl nitrate denitration, is provided by Souply and Neal (1977). Thermal microwave direct denitration conversion of Pu nitrate solutions has been implemented in 1:1 mole ratio mixture with uranium nitrate at the Plutonium Fuel Fabrication Facility and then at the Plutonium Conversion Development Facility at Tokai Mura, Japan (Oshima 1989), and is being implemented in MOX fuel fabrication at Japan's Rokkasho reprocessing plant (Collins 2015).

3.6 DECONTAMINATION OF PLUTONIUM PROCESS WASTE SOLUTIONS USING PRECIPITATION

The removal of fission products and of Pu from reprocessing and of plutonium from conversion process wastes has been accomplished using coprecipitation and scavenging. Processes to accomplish these separations are examined in this section.

3.6.1 Scavenging of Fission and Activation Products from Reprocessing Wastes Using Coprecipitation

Irradiated fuels reprocessing at Hanford and Mayak began the plutonium production industry. The early processes at both sites generated large waste volumes, both through chemically lavish coprecipitation methods (the Bismuth Phosphate Process and the Sodium Uranyl Acetate Process, respectively) and through the large quantities of irradiated uranium processed for weapons Pu in the cauldron of the Cold War. Both sites used underground tank storage for their reprocessing plant wastes but soon outgrew the tank capacities. As a result, both sites discharged waste solution to the environment. At Hanford, discharge of treated lower activity wastes occurred to the deep (~75–90 m above the water table) sand and gravel beds that underlie the site. The ion exchange retention of the trace radioactive elements, primarily Cs and strontium (Sr), on the clays associated with these soils was relied upon to decrease the residual activity reaching the groundwater, while the decades-long groundwater travel time to the Columbia River, about 15 km from the Hanford reprocessing areas, provided further margin to allow radioactive decay to occur (Burns and Stedwell 1957). At Mayak, solution wastes, initially even the high-level liquid (HLLW) reprocessing plant wastes, were discharged directly to nearby surface waters—Lake Karachay and the Techa River (Delegard et al. 2018).

To ameliorate the effects of the direct HLLW discharge to the environment at Mayak 1949–1951, the wastes first were pH-neutralized, forming oxyhydroxides of Fe, Cr, and Ni to coprecipitate polyvalent radionuclides. Methods implemented at Mayak by 1951 decontaminated the HLLW by carrying Cs using Ni ferrocyanide solids and Fe and Ni sulfides to coprecipitate radioactive Sr and rare earths (Peretroukhine and Delegard 2000; Delegard et al. 2018; page 52 of Cochran and Norris 1993). The settled Fe/Cr/Ni oxyhydroxides, Ni ferrocyanide, and Fe/Ni sulfides with their captured trace radioactive elements occupied less than 1% of the initial HLLW solution volume, sharply decreasing, though not eliminating, the radioactivity sent to the Lake Karachay and the Techa River.

In 1954–1958, Hanford's Scavenging Process used means very similar to those used at Mayak to decontaminate HLLW solutions of their contained fission and activation product radioactivity. By the time the Scavenging Process was implemented, three streams from the Bismuth Phosphate Process were available for treatment: (1) the Uranium Recovery Process TBP solvent extraction wastes, from the retrieval, acidification, and processing of Bismuth Phosphate Process Metal Wastes, (2) the wastes from the Uranium Recovery Process that already had been made alkaline and were stored in the Hanford waste tanks, and (3) the first cycle wastes still being produced in the Bismuth Phosphate Process (Figures 2-5, 2-6, and 2-7, respectively, of Cash et al. 1992).

Like Mayak, the Cs was removed from each of these Hanford waste streams using Ni ferrocyanide. The Ni ferrocyanide was prepared by the separate additions of sodium or potassium ferrocyanide ($Na_4Fe(CN)_6$ or $K_4Fe(CN)_6$) and nickel sulfate ($NiSO_4$) at about 0.005 M each, to the waste solutions, precipitating the resulting cesium nickel ferrocyanide compound. Selection of the ferrocyanide alternative likely did not arise from knowledge of its use in the USSR. Rather, it was based on a U.S. Mound Laboratory survey of alternatives for treating Bismuth Phosphate Process first and second cycle decontamination wastes (Lowe et al. 1951). As in the USSR, the Mound Laboratory also identified ferrous sulfide as a potent broad-spectrum solution decontamination agent, but it was not used at Hanford.

Radioactive Sr was removed from these three Hanford waste solutions by dilution with calcium and nonradioactive Sr nitrates and their precipitation as the corresponding sulfate and phosphate compounds using the sulfate and phosphate already present in the waste (Burns and Stedwell 1957; Cash et al. 1992). Radioactive cobalt also was scavenged from wastes present in underground tanks by the separate additions of up to 0.006 M each sodium sulfide (Na_2S) and nickel sulfate ($NiSO_4$) to form nickel sulfide (NiS) to carry the cobalt.

The precipitated $Cs_2NiFe(CN)_6$, $(Ca, Sr)SO_4$ and $(Ca, Sr)_3(PO_4)_2$, and NiS solids settled from these processes were relegated to the waste tanks, while the partially decontaminated supernatant solutions were discharged to underground soil columns in cribs (underground grid of waste solution distribution pipes). To improve the solid/liquid separation, cascades of underground waste tanks were used. The precipitate suspension formed at the first tank settled with the supernatant solution pumped to the second tank in the cascade and sometimes a third tank for more settling and decantation. The final supernatant solution from the cascade then was disposed to the soil. The volume sent to soil disposal thus freed the equivalent tank waste storage volume.

An alternative method to remove radioactive Cs from alkaline wastes was developed and process-tested in a single tank at SRS using sodium tetraphenyl borate (NaTPB or $NaB(C_6H_5)_4$), as a precipitation agent. The bulky TPB anion, $B(C_6H_5)_4^-$, precipitates with large monovalent cations such as potassium (the KTPB solubility product is ~5 × 10^{-8} moles^2/L^2) and Cs (the CsTPB solubility product is 1 × 10^{-10} moles^2/L^2), while the feed NaTPB reagent is relatively soluble in the Na-laden tank waste with the solubility product being 0.62 moles^2/L^2 (McCabe 1996). The process was demonstrated at SRS in 1983 and decontaminated 427,000 gallons (~1.6 million liters) of alkaline radioactive waste solution containing ~1 Ci ^{137}Cs/L. The product solution was ~3 nCi ^{137}Cs/g, or about 4 μCi/L (i.e., DF ≅ 250,000), leaving about 11,000 gallons of sludge containing the mixed KTPB and CsTPB precipitates (Walker et al. 1997). However, benzene release from decomposition of the TPB also was observed. Tank space ventilation and nitrogen gas purging kept vapors below flammable limits of benzene and radiolytically generated hydrogen. Later, laboratory and in-tank precipitation processing tests established that the TPB decomposition occurred primarily from the excess dissolved NaTPB, not the precipitated KTPB and CsTPB, and was catalyzed by the presence of transition metals (e.g., Fe, Mn) in the waste, likely as their low-solubility hydr(oxides), at a rate that increased with temperature (Walker et al. 1997). Analyses of the off-gases showed the decomposition products to be benzene (87%), phenol (C_6H_5OH; 7%), biphenyl ($(C_6H_5)_2$; 4.5%), and borate (BO_3^{3-}) (Walker et al. 1997).

A continuous TPB precipitation process to extract the Cs for immobilization has been outlined (Ponder et al. 2001) but not yet implemented. Cesium removal from acid wastes is impossible due to TPB hydrolysis.

3.6.2 Scavenging of Plutonium from Acidic Solutions by Alkaline Coprecipitation

Direct precipitation of Pu(IV) hydrated oxide ($PuO_2 \cdot xH_2O$) sometimes called plutonium hydroxide, or coprecipitation with other low-solubility metal (hydr)oxides can be done by simply making the solution alkaline using NaOH. This method was used at many sites including LASL to collect recoverable Pu and produce discardable solutions. The Pu(IV) hydrated oxide has very low solubility such that even in the presence of complexing agents such as citrate, $<10^{-3}$ g Pu/L ($<4 \times 10^{-6}$ M) concentrations can be attained (Section 8.4.5 of Christensen and Maraman 1968).

Later studies show that Pu in the (IV), (V), and (VI) oxidation states have solubilities in NaOH solution that increase both with increasing oxidation state and with increasing NaOH concentration as shown in Figure 3.22. The increase in solubility with increasing NaOH concentration suggests that the dissolved Pu species in all oxidation states are anionic hydroxide complexes. The divergence of the Pu(IV) solid phase solubility trace at greater than about 6 M NaOH likely relates to the oxidation state of the dissolved species. In the presence of air, the dissolved species present in contact with the Pu(IV) solid phase above 6 M NaOH likely is Pu(V). Plutonium(V) dissolved species is identified in 15 M NaOH by spectrophotometry, and Pu(V) is inferred less definitively by spectrophotometric methods at 10 M NaOH. In the presence of added N_2H_4, a strong chemical reductant, Pu(IV) dissolved species likely exists above 6 M NaOH. Under either condition, Pu(IV) dissolved species are likely below 6 M NaOH.

Although Pu(VI) has relatively high solubility compared with Pu(IV), radiolysis effects ultimately cause the Pu(VI) precipitate formed in NaOH solution to alter to Pu(IV) hydrated oxide and Pu concentration to drop correspondingly (Delegard 1985). Similar susceptibility to radiolysis would be expected for the Pu(V) precipitate with the Pu(V) subject to disproportionation to Pu(IV) hydrated oxide and Pu(VI) below about 8 M NaOH. Both Pu(VI) and Pu(V) precipitate as sodium salts of anionic hydroxide complexes. Plutonium(III) hydroxide ($Pu(OH)_3$) chemically

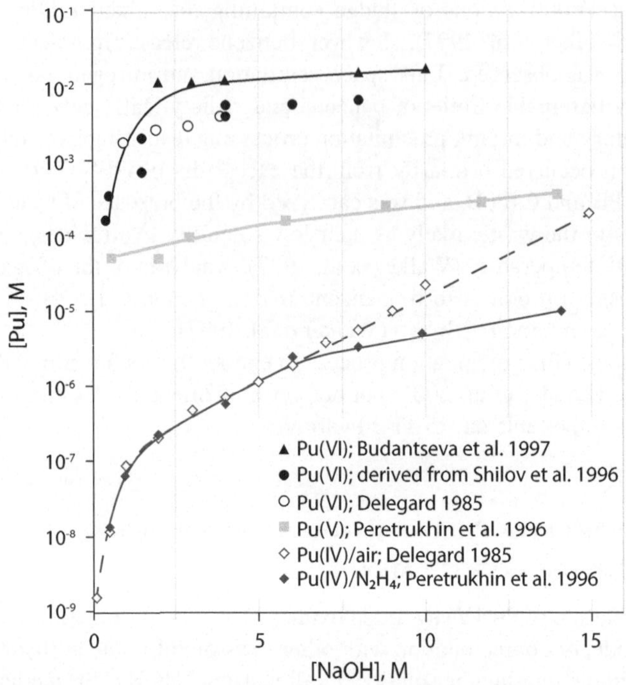

FIGURE 3.22 Plutonium(IV), (V), and (VI) solubilities in NaOH solution

reduces water and thus is unstable in NaOH solution in the absence of very strong reductants. The Pu product in this case also is Pu(IV) hydrated oxide (Section 5.1 of Delegard and Jones 2015 and associated references).

The Pu(IV) solid phase usually formed by treatment of acidic Pu-bearing solution with excess alkali, thus, is $PuO_2 \cdot xH_2O$ with nanometer-scale crystallites (Section 5.3.1 of Delegard and Jones 2015 and references therein); further properties of the Pu(IV) hydrated oxide formed by making Pu(IV) nitrate solutions alkaline are provided by Yusov et al. (2000a).

Plutonium recovery from acidic Pu-bearing process solutions and solution decontamination by precipitation and carrier precipitation methods were examined at LASL (Section 8.4 of Christensen and Maraman 1968). Among the methods used at LASL for recovering Pu were the Pu(IV) peroxide and Pu(III) and Pu(IV) oxalate processes already described in this chapter.

With carriers such as Fe(III) hydroxide, LASL experience showed that $<10^{-5}$ g Pu/L filtrate concentrations could be reached (Christensen and Maraman 1968). However, the same experience showed the Pu(IV) and Fe(III) hydroxide precipitates typically to be gelatinous, slow to filter, retaining large quantities of interstitial water, and difficult to re-dissolve making subsequent processing awkward.

A solution decontamination and Pu-scavenging method for acidic solutions using calcium oxalate carrier precipitation was implemented at LASL in the early 1950s

(Section 8.4.2 of Christensen and Maraman 1968). In this method, calcium nitrate ($Ca(NO_3)_2$) was added to the acidic waste solution, the acid concentration decreased to ~1 M by dilution or neutralization, and solid $H_2C_2O_4 \cdot 2H_2O$ added in sufficient quantity to leave 0.1 M excess oxalic acid in solution. Slow oxalic acid dissolution allowed near homogeneous precipitation conditions to be attained. A reducing agent such as $NH_2OH \cdot HNO_3$ was added to adjust to Pu(III). The alkalinity was increased to 0.1 M NaOH. The CaC_2O_4 precipitated, carrying the Pu and leaving <1 × 10^{-4} g Pu/L in solution. However, this method soon was abandoned because of the cumbersome but necessary oxalate destruction required before further Pu processing.

Despite the physical handling difficulties listed earlier for the Pu(IV)-Fe(III) hydroxide system, carrier precipitation of Pu from acidic solutions using Fe(III) hydroxide carrier was investigated further at LASL (Section 8.4.6 and Figure 50 in Christensen and Maraman 1968). In this method, the Pu was reduced to the (III) state using $NH_2OH \cdot HNO_3$. The acidic solution was made 0.03 M in Fe^{3+} and 0.02 M in Al^{3+} and then NaOH solution was added to precipitate the presumed Fe(III) hydroxide.[5] The aluminum was added as a pH indicator—its re-dissolution in excess alkali showed that sufficient NaOH (i.e., pH ~10) had been added. The filtrates arising from this treatment contained <1 × 10^{-5} g Pu/L.

An approach with similar chemical features as the Fe(III)-$NH_2OH \cdot HNO_3$ process described at LASL was investigated at Rocky Flats (Boyd et al. 1983). In this research, a simulated acidic Pu-bearing waste solution was amended by adding ferric nitrate and ferrous sulfate in a 2:1 Fe(III):Fe(II) mole ratio. The waste solution was adjusted to pH 10–12 by direct strike addition of NaOH solution and brought to volume to give 1.73 × 10^{-4} M Fe(III) and 8.64 × 10^{-5} M Fe(II) in the final suspension. The precipitate was allowed to digest for 10 minutes, and the supernatant solution passed through a steel wool matting between the poles of a 2,000 Gauss magnet, which captured the magnetic Fe_3O_4, ferrite, formed in precipitating the 2:1 Fe(III):Fe(II) mixture. The clarified supernatant solution from a simulated Rocky Flats waste contained only 3.4 × 10^{-9} g Pu/L (~1.4 × 10^{-11} M) from a feed containing 1 × 10^{-4} g Pu/L. Tests of the efficacy of this treatment for acidic Pu-spiked solutions in the presence of a variety of salts of individual anions (e.g., nitrate, sulfate, phosphate, fluoride, oxalate) showed decontamination to <1.4 × 10^{-8} g Pu/L (<6 × 10^{-11} M). Chloride and some organic agents (e.g., EDTA, acetate, commercial cleaner) increased the Pu concentration to ~3 × 10^{-8} g Pu/L.

Note that the solubility of $PuO_2 \cdot xH_2O$ at pH 10–12, as determined for ultrafiltered (~1.5–2 nm pore size; colloid-free) solution is ~4 × 10^{-11} M, while the Pu concentration for ultracentrifuged solution (i.e., colloids present) is ~3 × 10^{-9} M (Neck et al. 2007). These comparative studies attest to the stability and near-solute quality of the plutonium-bearing colloids.

In a series of tests focused on Hanford tank waste, the interactions of cationic and anionic tank waste constituents in the alkaline treatment of Pu-bearing HNO_3 solutions were investigated (Fedoseev et al. 1998; Krot et al. 1998a, 1998b). The experiments were performed to help understand Pu disposition in Hanford tank

[5] The $NH_2OH \cdot HNO_3$ also would have reduced at least some of the Fe(III) to Fe(II) (Barney 1971).

wastes, Pu's affinities with these constituents, the ways the constituents might alter the behavior of Pu, and means by which the tank waste solutions might become further decontaminated of Pu and other actinides.

The interactions of Pu(IV) with Al(III), Co(II), Cr(III), Fe(III), La(III), Ni(II), U(VI), and Zr(IV) metal ions and of Pu(VI) with the Al(III), Cr(III), and Fe(III) metal ions under alkaline coprecipitation conditions similar to those used in tank farm waste disposal at Hanford were examined in the first test series (Fedoseev et al. 1998). Although Pu is present only in trace amounts in actual process wastes, the tests used a wide range of generally higher Pu:metal mole ratios so that the Pu-metal ion solid phase interactions could be examined by X-ray diffraction (XRD) and other spectroscopic methods.

A second study's further investigations of the interactions of Pu(IV) and Am(III) with Al(III), Fe(III), and U(VI) at various mole ratios under alkaline coprecipitation conditions determined the leachability of Pu and Am from the product solid phases formed under alkaline conditions using alkaline carbonate and ethylenediamine tetraacetate (EDTA) solutions as potential complexation agents (Krot et al. 1998a).

A third study examined the potential influences of Hanford tank waste sodium salt anions on Pu(IV) nitrate disposition after its addition to alkaline solutions containing these salts (Krot et al. 1998b; Yusov et al. 2000b). Sodium nitrite, carbonate, sulfate, chromate, ferrocyanide, oxalate, glycolate, citrate, EDTA, and *N*-2-hydroxyethyl ethylenediamine triacetate (HEDTA) were tested as were sodium phosphate (Na_3PO_4) and sodium silicate (Na_2SiO_3).

The first and second studies (Fedoseev et al. 1998; Krot et al. 1998a) were directed to the interactions of Pu(IV) with Al(III), Co(III), Cr(III), Fe(III), La(III), Ni(II), U(VI), and Zr(IV) and Pu(VI) with Al(III), Cr(III), and Fe(III) under alkaline coprecipitation conditions. To determine the intimacy of mixing of Pu with the precipitating metal, the Pu-metal (hydr)oxide precipitates were dissolved in hydrochloric acid (Fedoseev et al. 1998). The dissolution rates showed whether the Pu was indeed coprecipitated with the metal ion. If the rates of the Pu and metal (hydr)oxides dissolution differed, separate Pu and metal phases formed. If the dissolution rates were the same, true coprecipitation occurred. In general, the $PuO_2 \cdot xH_2O$ is more difficult to dissolve so that if differential dissolution was observed, $PuO_2 \cdot xH_2O$ dissolution would be the one lagging. Most Pu:metal combinations, i.e., those with Fe(III), Co(III), Cr(III), La(III), Zr(IV), and U(VI), were true coprecipitates. Studies of the mixtures by spectroscopic methods also showed the (hydr)oxide structures were altered from those of the individual pure phases (Fedoseev et al. 1998). For the tests of Pu(IV) with Ni(II) and Al(III), however, no atomic-scale mixing was shown. Instead, separate $PuO_2 \cdot xH_2O$, $Ni(OH)_2$, and $Al(OH)_3$ phases were observed meaning simultaneous precipitation, not coprecipitation, occurred. The second study (Krot et al. 1998a) confirmed the findings of the first study and showed Pu(IV) interacted strongly under alkaline coprecipitation conditions with Fe(III) and with U(VI) but had no appreciable interaction with Al(III). Similar observations were made for SRS waste disposal conditions (Hobbs 1999). Plutonium(VI) showed no measurable association with Fe(III), Cr(III), and $Al(OH)_3$ (Fedoseev et al. 1998; Yusov et al. 2000c).

The product formed by introduction of Pu(IV) nitrate to the alkaline solutions of various sodium salts present in Hanford tank waste was $PuO_2 \cdot xH_2O$ for all the anions studied save phosphate and silicate (Krot et al. 1998b; Yusov et al. 2000b). Further studies of the $NaOH/NaNO_3$ system showed the well-washed product to be relatively pure $PuO_2 \cdot xH_2O$ containing <1 mole% occluded sodium and <0.03 mole% nitrate. Aluminate also is present in the Hanford tank waste solutions but it, too, showed no interaction with, or alteration of, the $PuO_2 \cdot xH_2O$ precipitated in the alkaline solution (Fedoseev et al. 1998; Yusov et al. 2000c). Of the anions investigated, only phosphate and silicate formed products other than the $PuO_2 \cdot xH_2O$ precipitated by hydroxide in the alkaline media (Krot et al. 1998b; Yusov et al. 2000b). The Pu(IV) interaction with silicate is stronger than that with phosphate. Neither the silicate nor the phosphate product seemed to be stoichiometric with the complement of silicate or phosphate diminishing with increasing NaOH concentration, thus indicating that strong alkali would metathesize the respective Pu(IV) silicate or phosphate. For example, the P:Pu mole ratio is ~0.15 for the well-washed solid product formed by reacting Pu(IV) nitrate solution with sodium phosphate in 0.2 M NaOH but decreases with increasing NaOH concentration. The Si:Pu mole ratio ranges from 1.4 to 1.8 in the well-washed solids formed by reacting Pu(IV) nitrate with 0.05 M Na_2SiO_3 solution in 0.2 M NaOH, but the mole ratio decreases to ~0.3 in 7 M NaOH with 0.05 M Na_2SiO_3. Freshly prepared $PuO_2 \cdot xH_2O$ treated with 0.05 M Na_2SiO_3 in 1 M NaOH at 120°C–140°C forms a product with 0.8 Si:Pu mole ratio, midway between that formed by precipitation in 0.2 and 7 M NaOH. This suggests that the Pu(IV) silicate material is not merely an artifact of a fast precipitation reaction but rather has demonstrable intrinsic stability. The plutonium silicate products examined from these limited tests were amorphous based on XRD.

3.6.3 Scavenging of Plutonium from Alkaline Solutions by Homogeneous Coprecipitation

The coprecipitation removal of Pu already dissolved in alkaline solution using low-solubility metal (hydr)oxide carriers is complicated by the ordinary means of their introduction; i.e., as acid solutions of these highly hydrolysable ions to the alkaline medium. The hydrolysis and precipitation of the carrier in this fashion only can occur at the limited acid-alkali interface and not throughout the Pu-bearing alkaline solution. The uptake that does occur is more by surface adsorption rather than by true coprecipitation.

To this end, chemists at the Frumkin Institute of Physical Chemistry and Electrochemistry of the Russian Academy of Sciences investigated methods to introduce the carrier as an alkali-soluble precursor so that homogeneous coprecipitation could take place and thus remove dissolved Pu. A variety of alkali-soluble precursors, such as nitroprusside $[Fe(CN)_5NO]^{2-}$ and ferrate $[FeO_4^{2-}]$, were used to achieve Pu(V) and Pu(VI) coprecipitation with Fe(III), Cr(III), Co(III), Co(II), and Mn(II) hydroxides and MnO_2 carriers (Krot et al. 1996; Bessonov et al. 1997; Peretrukhin et al. 1998). For the most part, the carriers were generated by alkaline hydrolysis, as was done with nitroprusside, or chemical reduction, as happens with ferrate by interaction with a chemical reductant. Both precursors generate $Fe(OH)_3$ homogeneously within the alkaline solution and thus offer much better possibility for Pu incorporation within the carrier. The Pu(V) and (VI) often reduce to Pu(IV) in the coprecipitation, particularly if a chemical reductant is used.

Uranium-based carriers triuranium octoxide (U_3O_8) and sodium uranate (Na_2UO_4) also coprecipitated Pu from alkaline solution (Krot et al. 1996; Peretrukhin et al. 1998). The U_3O_8 was generated by chemical reduction of solid sodium uranate (Na_2UO_4) using hydrazine and a catalyst and thus is not a homogeneous coprecipitation. The sodium diuranate used to capture Pu was produced from the soluble uranyl peroxide, $[UO_2(O_2)_3]^{4-}$, precursor upon its homogeneous decomposition by copper(II)-catalysis.

Organic complexants present in the Hanford tank waste such as EDTA and HEDTA decrease Pu coprecipitation by many of the tested carriers. However, introduction of chemical masking agents that bind with the complexants can thwart the deleterious effects (Bessonov et al. 1997; Gelis and Shilov 2000). Thus, calcium ion (Ca^{2+}) bound EDTA, Co(III) ion (Co^{3+}) was effective for HEDTA, and the effects of citrate and glycolate could be addressed by increasing NaOH concentration. Some representative findings for Pu(VI) are presented in Table 3.6. More detailed information on these studies is provided in papers in the journal *Radiochemistry* in 1998 and 1999 containing the phrase "Method of Appearing (or Arising) Reagents" in their titles.

TABLE 3.6
Decontamination of 0.5–4 M NaOH of Plutonium(VI) by the Method of Appearing Reagents

Precipitation Reaction	Rxn. Type[a]	Temp., °C	Carrier Conc., M	DF for Pu(VI) in NaOH 0.5 M	DF for Pu(VI) in NaOH 4 M
$MnO_4^- + NH_2OH \rightarrow Mn(OH)_2$	red'n.	25	0.01	430	200
$FeO_4^{2-} + N_2H_4 \rightarrow Fe(OH)_3$	red'n.	25	0.01	370	420
$[Fe(CN)_5NO]^{2-} \rightarrow Fe(OH)_3$	hyd.	80	0.01	2000	800
$[UO_2(O_2)_3]^{4-}$ + 0.005 M $Cu(II)_{catalyst} \rightarrow Na_2UO_4$	th. dec.	60	0.02	610	170
$[Co(NH_3)_6]^{3+} \rightarrow Co(OH)_3$	hyd.	70	0.024	600	1500
$[Co(NH_3)_5Cl]^{2+} \rightarrow Co(OH)_3$	hyd.	70	0.024	150	390
$[Co(NH_3)_5Cl]^{2+} + N_2H_4 \rightarrow Co(OH)_2$	red'n.	60	0.024	300	25
$Cr(O_2CCH_3)_3 \rightarrow Cr(OH)_3$	hyd.	80	0.01	260	300

[a] red'n. = reduction; hyd. = hydrolysis; th. dec. = thermal decomposition.

3.7 WASTE VOLUME REDUCTION AND FORMATION OF SOLIDS BY EVAPORATION AND CRYSTALLIZATION

Evaporative processing of alkaline tank wastes to gain waste tank storage space by disposal of the water condensates and decrease liquid quantities available to pass into the underlying soils in the event of tank leaks has occurred at the Hanford and Savannah River Sites using both radiolytic heating and externally provided heating. Vacuum evaporator-crystallizers, operated at about 40 Torr and 60°C and under MSMPR conditions, are the most recent process evolution at Hanford (Petrie et al. 1975).

The evaporative processes crystallize the sodium salts contained in the waste solutions, and the imposed alkaline conditions precipitate polyvalent metal ions as their (hydr)oxides, the most prominent of which are gibbsite ($Al(OH)_3$) boehmite (AlOOH), from gibbsite dehydration (see the following section), and hematite (Fe_2O_3). Sodium aluminosilicates such as cancrinite and sodalite, $NaAlSiO_4 \cdot xNaX \cdot yH_2O$, where X is a nitrate or other anion, also are present in both Hanford and Savannah River Site tank wastes. The salts largely are composed of $NaNO_3$ because the Pu and U recovery processes occurred in HNO_3, and the waste made alkaline with sodium carbonate (Na_2CO_3) and NaOH. Other solid salts crystallized in the sites' tank wastes include Na_2CO_3, sodium carbonate monohydrate ($Na_2CO_3 \cdot H_2O$) sodium fluoride phosphate, natrophosphate or ($Na_7F(PO_4)_2 \cdot 19H_2O$) and sodium diuranate ($Na_2U_2O_7$) (Hobbs and Coleman 1994; Wilmarth et al. 2000; Herting et al. 2015; Reynolds et al. 2013; Peterson et al. 2018).

While only the PUREX Process was used at SRS, Hanford also used the Bismuth Phosphate and REDOX Processes and thus has much more variable and complicated waste compositions. As a result, other salts also are crystallized in the Hanford tank wastes including sodium oxalate ($Na_2C_2O_4$); sodium aluminate ($NaAl(OH)_4$); and double salts such as sodium nitrate sulfate, darapskite or ($Na_3NO_3SO_4 \cdot H_2O$); sodium fluoride sulfate, kogarkoite or (Na_3FSO_4); sodium aluminum carbonate hydroxide, dawsonite or ($NaAlCO_3(OH)_2$) (Herting et al. 2015; Reynolds et al. 2013); and many other crystalline and non-crystalline (amorphous) solid phases (see Tables 2-1 and 5-1 of Harrington 2011; Peterson et al. 2018).

3.8 DISSOLUTION OF ALUMINUM SOLID PHASES FROM HANFORD AND SAVANNAH RIVER SITE TANK WASTES

Aluminum cladding was removed from irradiated fuel at Hanford and SRS by reaction with NaOH solution containing $NaNO_3$. The resulting waste, with aluminum dissolved as $NaAl(OH)_4$, was sent to underground tank storage. At Hanford, an additional aluminum complement went to tank storage, after being made alkaline with NaOH solution, from the aluminum nitrate used as a salting agent in the REDOX Process. With time, cooling, dilution by other wastes, and NaOH consumed by absorbed CO_2, aluminate precipitated as gibbsite: $NaAl(OH)_{4\ solution} \rightarrow Al(OH)_{3\ solid} + NaOH_{\ solution}$. At the Hanford Site, Al-bearing solid phases represent roughly 70% of the water-insoluble tank waste solids (Snow et al. 2009) and include not only gibbsite but also boehmite and sodium aluminosilicates.

Because the eventual path of the insoluble solids is vitrification and repository disposal, reducing the amount of aluminum in the solids would significantly reduce the glass volume and the associated production and lifecycle costs. Laboratory testing shows that >90% of the Al-bearing solids can be dissolved if adequate time, temperature, and hydroxide are applied, effectively reversing the gibbsite precipitation reaction. The amount of effort, as reflected by those three variables, depends on the existing aluminum form. Gibbsite dissolves readily, while boehmite is more recalcitrant and the sodium aluminosilicates impossible to dissolve in this manner. Most of the boehmite formed at the Hanford Site when REDOX Process waste tanks self-boiled due to heat from radioactive fission product decay. It was this self-boiling, and the deleterious effects it had on tank structural integrity, which led to the removal of radioactive Sr and Cs from the more concentrated PUREX Process wastes and REDOX Process waste solutions at Hanford (see Section 3.10 and Chapter 7).

Table 3.7 summarizes the aluminum distribution in the Hanford Site tank wastes. Aluminum phases at the SRS are similar but likely in different ratios due to process history. Table 3.7 shows that >60% of the Hanford tanks' aluminum can be readily diverted from the high-level waste glass by relatively simple phase separation to remove dissolved sodium aluminate and mild alkaline gibbsite dissolution.

Figure 3.23 shows how two Hanford Site waste solids, from cladding and from the REDOX Process, and SRS tank waste solids respond to alkaline leaching. The Hanford cladding waste contains gibbsite (Snow et al. 2009), while the REDOX waste that boiled during storage contains boehmite (Fiskum 2008). The SRS waste was also predominantly boehmite (Fondeur et al. 2004). Both Hanford waste tests were performed with 1 g of sludge solids in 100 mL of 3 M sodium hydroxide at 80°C. The gibbsite sample dissolved completely within 2–4 hours, whereas the boehmite sample required 170 hours. The SRS test conditions were slightly more aggressive at 3.5 M NaOH and 87°C but had threefold higher solids loading. Although the SRS Tank 12H material also was given nearly 170 hours, dissolution was incomplete, likely because of depletion in available NaOH due to the threefold higher solids loading compared with the Hanford tests.

The final aluminum leachate concentration in the two Hanford samples (Figure 3.23) is due to the difference in aluminum concentration in the starting sludge (Table 3.8). This difference is consistent with lower aluminum concentration

TABLE 3.7
Aluminum Distribution in Hanford Site Tank Wastes

Aluminum Source	Fraction Present
Already dissolved (sodium aluminate)	0.28
Readily dissolved (primarily gibbsite)	0.38
Difficult to dissolve (boehmite)	0.20
Intractable (primarily aluminosilicates)	0.06
Unassigned	0.08

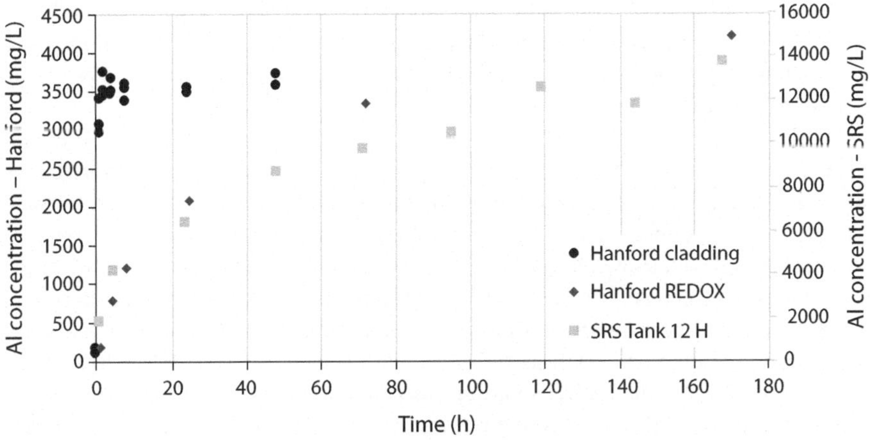

FIGURE 3.23 Gibbsite and boehmite dissolution rates in 3–3.5 M NaOH at 80°C–87°C.

TABLE 3.8
Compositions of Washed and Leached Sludge Samples

	Concentration, μg/g					
	Cladding (gibbsite)		REDOX (boehmite)		SRS Tank 12H (boehmite)	
Element	Before Leaching	After Leaching	Before Leaching	After Leaching	Before Leaching	After Dissolution
Al	332,500	38,000	388,250	83,675	359,148	123,300
Bi	2,310	33,000	<694	1,125	—	—
Ca	<7198	50,000	9,100	22,975	14,200	16,233
Cr	6,100	16,500	1,733	9,770	900	1,700
Fe	5,090	232,000	8,463	73,825	44,300	91,018
Mn	5,195	97,200	5,375	62,650	25,043	57,091
Na	37,000	38,000	55,000	76,500	12,150	20,339
P	4,480	13,800	<743	<664	2,710	2,189
Si	15,050	56,000	15,250	17,875	4,600	5,148
Sr	77	1,500	1,243	15,125	655	1,148
U	13,000	151,000	21,750	286,750	174,000	163,119

Source: Snow, L.A. et al., *Characterization and Leach Testing for PUREX Cladding Waste Sludge (Group 3) and REDOX Cladding Waste Sludge (Group 4) Actual Waste Sample Composites*, PNNL-18054, WTP-RPT-167, Pacific Northwest National Laboratory, Richland, WA, 2009; Fiskum, S.K. et al., *Characterization and Leach Testing for REDOX Sludge and S-Saltcake Actual Waste Sample Composites,* PNNL-17368, WTP-RPT-157, Pacific Northwest National Laboratory, Richland, WA, 2008; Fondeur, F.F. et al., *Aluminum Leaching of "Archived" Sludge from Tanks 8F, 11H, and 12H*, WSRC-TR-2004-00180, Westinghouse Savannah River Company, Aiken, SC, 2004.

Note: Elements not affected by the dissolution process are shaded gray; the non-shaded elements dissolved to some extent.

in gibbsite than in boehmite. While the gibbsite-rich cladding waste sample had a lower aluminum concentration than the pure aluminum phase (gibbsite), it also had a lower impurity concentration than the boehmite sample (REDOX Process waste), consistent with the complexity of the waste origins.

While aluminum leaching is ~99% complete for the Hanford samples, other element dissolutions were less complete. In Table 3.8, the elements not affected by the dissolution process are shaded gray; the non-shaded elements dissolved to some extent. Of the radionuclides, only Cs dissolved appreciably to >98%, as would be expected (results not shown here). However, these results do suggest that some of the cesium may have been retained in residual aluminosilicate matrices (e.g., cancrinite).

This alkaline leaching process has been deployed at the SRS and is planned for Hanford use. At the SRS, leaching in NaOH was utilized in Tank 51H in 2007 (Pike 2008, Krahn 2008). The NaOH solution was added to the tank, and the tank heated 60°C–65°C by operating the mixing pumps. Equilibrium was reached after 80 days with 65% of the aluminum dissolved. As in the bench scale tests described above, the dissolution extent was chemically limited by the NaOH provided. However, enough aluminum dissolved to significantly decrease the produced glass quantity.

Options for aluminum solids dissolution at the Hanford Site have been evaluated. Pilot scale demonstrations with a simulant prepared to represent an average Hanford feed demonstrated that 60% Al phase dissolution could be achieved in 16 hours at nominal 3 M NaOH and 100°C (Mahoney et al. 2010). This process was performed in a scale version of the planned Waste Treatment Plant pulse jet mixing vessels with heat supplied by direct steam addition which diluted the NaOH. All gibbsite and 30% of the boehmite dissolved. The aluminum dissolved also depends on the tank wastes blended, as other Hanford Site wastes are significantly lower in aluminum than the REDOX and cladding wastes.

Leach condition aggressiveness must be limited, however, to avoid undue leaching vessel corrosion. Stainless steel vessels lost chromium in tests with actual waste at 100°C and 3 M NaOH. Based on these results, leach temperatures should be <80°C if stainless steel vessels are to be used. SRS limited the leach temperature in carbon steel vessels to <70°C.

An additional concern is the potential to supersaturate the aqueous phase with respect to gibbsite, especially as the solutions cool or are diluted in subsequent processing. Tank waste solutions frequently are above the expected gibbsite solubility limit based on NaOH concentration (Figure 1 of Agnew and Johnston 2013). Simulant tests show that other dissolved salts (e.g., $NaNO_3$, $NaNO_2$) enhance gibbsite solubility. However, to date, these effects have not been adequately extended to account for the relatively high apparent gibbsite solubility in tank waste and to understand the bases of supersaturation.

3.9 CHROMIUM PHASE DISSOLUTION FROM HANFORD SITE TANK WASTES

As Table 3.8 shows, chromium concentrations in Hanford Site waste from the REDOX process and those associated with the S tank farm (which received REDOX wastes) are significantly higher than in SRS waste. The chromium arose from the chromate used to oxidize to Pu(VI) for its extraction in the REDOX Process.

TABLE 3.9
Behavior of Various Elements of S-Saltcake Solids during Dissolution with NaOH and Oxidation with $NaMnO_4$

	Concentration, µg/g or µCi/g		
Component	Before Treatment	After NaOH Leaching	After $NaMnO_4$ Oxidation
Al	187,000	113,500	82,500
Cr	92,850	168,500	28,425
Fe	14,700	26,650	35,125
Mn	4,680	8,540	214,000
Na	93,500	65,000	76,500
Si	15,500	31,500	24,200
U	6,575	12,650	12,700
$^{239+240}Pu$	0.923	2.06	2.45
^{90}Sr	266	599.5	749

Source: Fiskum, S.K. et al., *Characterization and Leach Testing for REDOX Sludge and S-Saltcake Actual Waste Sample Composites,* PNNL-17368, WTP-RPT-157, Pacific Northwest National Laboratory, Richland, WA, 2008.

Note: Components that are not dissolved are shaded gray.

Table 3.9 summarizes the composition of typical Hanford S tank farm solids after washing to remove water-soluble salts. Table 3.9 also shows the washed solids composition after leaching by 3 M NaOH at 100°C for 8 hours. The lines shaded in gray, including for Cr, indicate elements that did not dissolve in processing. Thus, large Cr concentrations remained in the heels.

Glass waste forms tolerate only low Cr due to spinel precipitation. Therefore, dissolution of low solubility Cr(III) phases such as $Cr(OH)_3$ to eliminate it from the high activity waste is desired. Chromium can be oxidatively dissolved from the Hanford waste solids by alkaline sodium permanganate ($NaMnO_4$) precipitating, in turn, manganese dioxide (MnO_2) by the reaction

$$Cr(OH)_{3\ solid} + MnO_4^- + OH^- \rightarrow CrO_4^{2-} + MnO_{2\ solid} + 2\ H_2O. \qquad (3.14)$$

Glass waste forms have a much higher tolerance for Mn than for Cr. Results of permanganate treatment show that this process also solubilizes other components, including aluminum, suggesting that the chromium solids likely contain other metals that are liberated with Cr dissolution (Table 3.9). The components that are not solubilized are shaded gray in the Table 3.9. The primary limitation of this process is the potential to oxidize and dissolve Pu, thus direct it away from its intended destination within the high-level solids stream. Above 0.25 M NaOH, the oxidative dissolution of Pu increases significantly, while at 0.25 M NaOH a nearly complete dissolution of the Cr phase can be achieved through the addition of a 25% stoichiometric excess of permanganate without undue Pu dissolution. This oxidative leach process was demonstrated at the bench scale with actual wastes and at the pilot scale with simulant and found to be effective (Mahoney et al. 2010) but has not yet been demonstrated beyond the lab scale with actual tank waste.

3.10 RECOVERY AND STORAGE OF RADIOSTRONTIUM AND RADIOCESIUM DERIVED FROM RADIOACTIVE WASTES BY THE WASTE FRACTIONIZATION PROCESS USING PRECIPITATION AND CRYSTALLIZATION

The ability to concentrate the alkaline Hanford tank wastes by evaporation of water, as discussed in an earlier section, first requires removal of heat-producing radioisotopes from the wastes. Without removal of the radiolytic heat sources, high temperatures in the waste can uncontrollably boil the wastes to dryness, particularly in settled sludge layers, and compromise the structural integrity of the mild-steel lined concrete tanks. At Hanford, the high specific heat loading was exacerbated by the decreasing relative waste volumes through the evolving Bismuth Phosphate, REDOX, and PUREX Processes.

Means to remove the Sr and Cs, primarily present as the isotopes ^{89,90}Sr and ^{134,137}Cs, were developed in the early 1960s and implemented in the Waste Fractionization Process (Buckingham 1967a–e). The Waste Fractionization Process consisted of several individual processes based on ion exchange (for ^{134,137}Cs, with a separate option for technetium as ^{99}Tc), solvent extraction (for ^{89,90}Sr, with separate options for Pu and Am), and precipitation (for ^{89,90}Sr, but also used to collect cerium, as ^{144}Ce, and promethium, as ^{147}Pm). Hanford Site waste streams treated by the Waste Fractionization Process(es) were: (1) dissolved acid wastes and Zirflex cladding wastes direct from the PUREX Process, containing both Sr and Cs, (2) stored alkaline solid sludge wastes from prior PUREX Process operations, containing mostly Sr, that were retrieved from the waste tanks and dissolved in acid, and (3) stored alkaline supernatant solution wastes from prior REDOX and PUREX Processes, containing mostly Cs.

A coprecipitation process was used to separate Sr from the acidic current and Zirflex wastes and the acidified sludges (pages 420–423 of Buckingham 1967b). In this process, the wastes first were treated with tartaric or hydroxyacetic acid to complex Fe(III) ions, thus retaining the iron in the higher pH solution for the next step. The wastes then were treated with sodium sulfate (Na_2SO_4) to attain about 1 M in sulfate, the pH adjusted to 0.7–2 using NaOH solution, and heated at 90°C for about 30 minutes to precipitate and digest the double salt $Na_2SO_4 \cdot Ln_2(SO_4)_3 \cdot 2H_2O$, where Ln represents the suite of fission product lanthanides present in the waste solutions. Lead(II) nitrate ($Pb(NO_3)_2$) then was added to the supernatant solution to reach 0.02 M, precipitating lead(II) sulfate ($PbSO_4$) and carrying the Sr by coprecipitation. This precipitate likewise was digested at 90°C for 30 minutes, the solutions decanted, and the precipitate cake washed in 1 M Na_2SO_4 adjusted to pH 1. This washing removed the dissolved Fe, Na, Al, Mg, Mn, Cr, Ni, Ca, Zr/Nb, Ru, and Cs present from the solid cake.

The cake with contained Sr and mixed Ln (including Ce and Pr) was metathesized by digesting in $NaOH/Na_2CO_3$ solution to form the corresponding low-solubility carbonates. This dissolved sulfate and lead, altered to soluble PbO_2^{2-} ion, which were removed by decantation and washing. The Sr- and Ln-bearing carbonate then was dissolved in acid for further processing to separate the Sr and Ln by precipitating the Ln using oxalic acid, in analogy with Pu(III) as described in Section 3.5.4. A similar precipitation sequence was used to treat Sr separated by solvent extraction processes (page 461 of Buckingham 1967b).

The Sr-bearing solution, which was nominally 0.35 M strontium nitrate, contained 0.09–0.17 M radiostrontium $^{89,90}Sr$, ~0.1 M HNO_3, and 3 M Na. It was made alkaline with NaOH and the Sr precipitated from solution as strontium fluoride (SrF_2) by addition of NaF solution (Jackson 1977). The SrF_2 cake was collected by filtration, water-washed, dried by self-radiolytic heating, and calcined to 1,100°C. The calcined cake was then chipped from the calcination boats, crushed to break up agglomerates, and the crushed powder rammed into Hastelloy C-276 capsules. The capsules were welded, decontaminated, leak-checked, and placed within welded secondary 316L stainless steel capsules for long-term storage in underwater pools. Details on these operations are provided by Jackson (1977).

Radiocesium was separated from PUREX acid waste raffinates and from Hanford tank waste solutions by ion exchange onto inorganic zeolite (pages 124–126 of Buckingham 1967a) and organic media (Barton et al. 1986). In both cases, the sorbed Cs was eluted using $[(NH_4)_2CO_3/NH_4OH]$ and the $(NH_4)_2CO_3/NH_4OH$ recovered and recycled after elution by distillation. In analogy with the process for Sr, the cesium carbonate eluate was treated with concentrated HCl to reach pH ~3–4, the solution heated to dryness to crystallize the cesium chloride, CsCl, and then taken to melting at ~640°C (Jackson 1977). The CsCl melt was vacuum transferred to 316L stainless steel capsules, capped and welded, and leak-checked before being loaded into a secondary 316L capsule, as was done for Sr.

3.11 RECOVERY OF PLUTONIUM PROCESS WASTE COMPONENTS USING PRECIPITATION/CRYSTALLIZATION

One of the primary disadvantages of plutonium separations processes by precipitation is the costly loss of process reagents to the radioactive waste streams. The cost is not primarily in the expense of the reagents but in the handling and disposal of these intrinsically nonradioactive materials containing the radioactive components. Efforts to recoup reagents from the Mayak process wastes arising from the Sodium Uranyl Acetate Process were undertaken in the USSR. Advanced lab demonstration of sodium nitrate ($NaNO_3$) recovery from Hanford tank wastes also has been done. In both cases, the recoveries used fractional crystallization as part of the recovery process.

The wastes arising from plutonium recovery at the Mayak reprocessing plant contained ~100 g $NaNO_3$/L (~1.2 M $NaNO_3$) and 60–80 g sodium acetate (NaO_2CCH_3)/L (~0.7–1 M NaO_2CCH_3). The radionuclide scavenging treatments using Fe and Cr hydroxides, Fe and Ni sulfides, and Ni ferrocyanide (see Section 3.6.1) first were applied to decrease the waste radioactivity. The wastes then were acidified using HNO_3, concentrated by fractional distillation of acetic acid (HO_2CCH_3) and the concentrated residue further volume-reduced to crystallize $NaNO_3$ from a solution slurry containing 1,100–1,150 g $NaNO_3$/L, ~13 M $NaNO_3$ (Cochran and Norris 1993).

Further details on processing wastes from the Sodium Uranyl Acetate Process have been published in Krapukhin et al. (1997). Thus, fractional crystallizations of sodium acetate trihydrate ($NaO_2CCH_3 \cdot 3H_2O$) and $NaNO_3$ were accomplished. The sodium acetate first was crystallized at ~20°C, and the solids separated from the liquors. The resulting $NaNO_3$-rich solutions, then, were concentrated to a 120°C boiling point and cooled to 70°C to effect $NaNO_3$ crystallization at a temperature where sodium acetate would not crystallize. The remaining acetate and nitrate denuded solutions were combined with fresh feed solution, and the process repeated to attain almost 100% recovery of both salts. At the same time, decontamination factors from radioactive contaminants of 30–50 were attained within these single crystallizations.

Process refinements became necessary to deal with the build-up of low concentration, but bothersome contaminants including not only radioactive constituents (ruthenium—Ru, Sr, and Cs) but also process impurities and reagents such as chromium present as chromate (used in Pu valence adjustment; Figure 3.7), aluminum, silicic acid ($SiO_2 \cdot xH_2O$), carbonate, and process surfactants. The chromate and carbonate were removed by precipitation using Ba and Ca, which also carried radiostrontium, and the Cs was carried by ferrocyanide with both processes occurring at neutral-to-alkaline conditions. It seems that aluminum and silicic acid likewise would combine to produce low-solubility aluminosilicate solids under these pH conditions. The treated partially decontaminated solution then was acidified, and the acetic acid removed by distillation.

The acid-side distillation purification of acetic acid could decompose excess ferrocyanide, generating poisonous hydrogen cyanide gas, HCN, and ferric ferrocyanide (Prussian blue), which would distribute, with its contained fission products, to the crystallized sodium nitrate in subsequent steps. Thus, Cs removal by use of preformed Ni and ferric ferrocyanide was done to avoid excess ferrocyanide carryover to the radioactive waste stream. Other means to address contaminants and improve $NaNO_3$ purity through recrystallization are outlined with description of actual methods and equipment to recover $NaNO_3$ and acetic acid/sodium acetate from Sodium Uranyl Acetate Process wastes (Krapukhin et al. 1997).

Laboratory studies at Hanford examined means to separate and recover $NaNO_3$ from Hanford tank wastes in a method called the Clean Salt Process (Herting 1993 and 1996). Included were reprocessing wastes from Bismuth Phosphate, REDOX, and PUREX Processes and the ancillary Hanford processes. Because processing largely occurred in HNO_3 media and the process wastes were made alkaline with NaOH and Na_2CO_3 prior to tank storage, $NaNO_3$ comprises 67% of the non-water complement in the Hanford tank wastes under Clean Salt acid-side conditions.

Tests of the Clean Salt Process from both the original, albeit diluted, alkaline wastes and these wastes after acidification were performed. The limited tests from the alkaline side were more difficult to control and achieve clean solid-liquid separation because of the accompanying precipitation of other sodium salts and the viscosity of the alkaline solutions. Waste acidification, to pH 2 using HNO_3, allowed recovery of the sodium contained in the alkaline-side sodium nitrite, carbonate, aluminate, and hydroxide by their conversions to the target $NaNO_3$ for crystallization. The acidified solution still contained insoluble solids such as aluminosilicates. The filtered solutions were evaporated until $NaNO_3$ crystallized and the filtrate separated for further treatment.[6] Meanwhile, trace radioactive elements (Cs, Sr, and others) were not incorporated into the $NaNO_3$ crystal matrix and thus were effectively removed, aside from that contained in interstitial or occluded solutions. The $NaNO_3$ crystals were washed, re-dissolved, and the crystallization process repeated (i.e., recrystallization) to effect further decontamination, as was demonstrated at Mayak. In a five-fold crystallization processing of an actual, and technically challenging, Hanford tank waste, a decontamination factor from radioactive constituents of 1.4×10^7 was achieved (Herting 1993).

3.12 CONCLUSIONS

Precipitation processes date from the earliest radiochemical separations used in recovering radium and polonium from uranium ores and tailings and in the identification and recovery of transuranium elements from irradiated uranium. Although precipitation processes have since been supplanted by other means in the recovery and initial purification of plutonium from irradiated uranium, further purification of Pu and its conversion from nitrate solution to technologically important pure compounds such as PuO_2 and plutonium metal still employ precipitation processes. Prominent among these is Pu(IV) oxalate precipitation, affording excellent decontamination and providing a useful precursor to PuO_2 and Pu metal. Precipitation processes also are used to remove contaminants from Pu process waste solutions, be they fission products or transuranic elements including Pu. Precipitation and dissolution of bulk chemical waste components also occur in stored tank waste. Dissolution of these bulk chemicals (e.g., $Al(OH)_3$ and $Cr(OH)_3$) is desired to avoid their costly treatment as high-level vitrified waste. Historically, bulk chemical recovery from the sodium uranyl acetate process wastes by crystallization was employed at Mayak to recycle $NaNO_3$ and sodium acetate/acetic acid and diminish waste quantities. Crystallization of $NaNO_3$ from Hanford tank wastes has been demonstrated on a laboratory scale and offers an opportunity to diminish the quantity of waste requiring expensive waste immobilization treatments.

---

[6] In a fully implemented process, the salt-depleted filtrate would be blended with fresh acidified waste and cycled for evaporation/crystallization treatment to form additional $NaNO_3$ crystals.

REFERENCES

Abraham, F., B. Arab-Chapelet, M. Rivenet, C. Tamain, and S. Grandjean. 2014. "Actinide Oxalates, Solid State Structures and Applications." *Coordination Chemistry Reviews* 266–267: 28–68. https://doi.org/10.1016/j.ccr.2013.08.036.

Abrahamson, S. G. 1967. "Precipitation of Plutonium-238 (IV) Oxalate from Homogeneous Solution." *Journal of Inorganic and Nuclear Chemistry* 29 (3): 842–844. http://dx.doi.org/10.1016/0022-1902(67)80345-2.

Adloff, J. P., and H. J. MacCordick. 1995. "The Dawn of Radiochemistry." *Radiochimica Acta* 70 (1): 13–22. https://doi.org/10.1524/ract.1995.7071.s1.13.

Agnew, S. F., and C. T. Johnston. 2013. *Aluminum Solubility in Complex Electrolytes*. Paper 13011 at WM2013 Conference, February 24–28, Phoenix, AZ. http://www.wmsym.org/archives/2013/papers/13011.pdf.

Alwin, J. L., F. Coriz, J. A. Danis, B. K. Bluhm, D. W. Wayne, D. W. Gray, K. B. Ramsey et al. 2007. "Plutonium Oxide Polishing for MOX Fuel Fabrication." *Journal of Alloys and Compounds* 444: 565–568. https://doi.org/10.1016/j.jallcom.2007.04.271.

Aquilano, D., and L. Pastero. 2013. "Anomalous Mixed Crystals: A Peculiar Case of Adsorption/Absorption." *Crystal Research and Technology* 48 (10): 819–839. https://doi.org/10.1002/crat.201200708.

Bagnall, K. W. 1957. "The Chemistry of Polonium." *Quarterly Reviews* 11 (1): 30–48. https://doi.org/10.1039/qr9571100030.

Baker, R. D. 1946. *Preparation of Plutonium Metal by the Bomb Method* (Los Alamos, New Mexico: Los Alamos Scientific Laboratory). LA-473. https://permalink.lanl.gov/object/tr?what=info:lanl-repo/lareport/LA-00473.

Baldwin, C. E., and J. D. Navratil. 1983. "Plutonium Process Chemistry at Rocky Flats." In *Plutonium Chemistry*, edited by W. T. Carnall and G. R. Choppin, 369–380. ACS Symposium Series. https://doi.org/10.1021/bk-1983-0216.ch024.

Barney, G. S., and C. H. Delegard. 1999. "Chemical Species of Plutonium in Hanford Site Radioactive Tank Wastes." In *Actinide Speciation in High Ionic Strength Media: Experimental and Modeling Approaches to Predicting Actinide Speciation and Migration in the Subsurface*, edited by D.T. Reed, S.B. Clark and L. Rao, 83–110. New York: Kluwer Academic/Plenum Publishers. https://doi.org/10.1007/978-1-4419-8690-0.

Barton, W. B., L. A. Gale, and M. E. Johnson. 1986. *Sixteen Years of Cesium Recovery Processing at Hanford's B Plant* (Rockwell Hanford Operations, Richland, WA). RHO-RE-SA-169. Also found in pages 1126–1136 of *Spectrum '86, Proceedings of the American Nuclear Society International Topical Meeting: Waste Management and Decontamination and Decommissioning*, edited by Pope, J. M., I. M. Leonard, and E. J. Mayer, American Nuclear Society. http://www.iaea.org/inis/collection/NCLCollectionStore/_Public/20/013/20013155.pdf.

Beede, R. L. 1956. *Continuous Plutonium(IV) Oxalate Precipitation, Filtration, and Calcination Process* (Richland, WA: General Electric). HW-45700. https://www.osti.gov/servlets/purl/7227473.

Bessonov, A. A., N. A. Budantseva, A. V. Gelis, M. V. Nikonov, and V. P. Shilov. 1997. *Investigation on the Coprecipitation of Transuranium Elements from Alkaline Solutions by the Method of Appearing Reagents. Study of the Effects of Waste Components on Decontamination from Np(IV) and Pu(IV)* (Richland, WA: Pacific Northwest National Laboratory). PNNL-11625. https://www.osti.gov/servlets/purl/554187.

Blaedel, W. J., and V. W. Meloche. 1963. "Contamination of Coprecipitates." In *Elementary Quantitative Analysis, Theory and Practice*. New York: Harper & Row Publishers.

Bluhm, E. A., K. D. Abney, S. Balkey, J. C. Brock, F. Coriz, J. T. Dyke, D. J. Garcia et al. 2005. "Plutonium Oxide Polishing for MOX Fuel Production." *Separation Science and Technology* 40 (1–3): 281–296. https://doi.org/10.1081/SS-200042232.

Bobylev, A. I., S. N. Elsukov, S. I. Rovnyi, I. V. Manakov, A. V. Kobyakov, and M. V. Pechenkina. 2004. "Fabrication of Fuel Pellets from Plutonium Dioxide". *Atomic Energy* 96 (1): 65–67. http://dx.doi.org/10.1023/b:aten.0000022817.53797.80.

Boyd, T. E., R. L. Kochen, J. D. Navratil, and M. Y. Price. 1983. "Actinide Aqueous Waste Treatment Studies Using Ferrites." *Radioactive Waste Management Nuclear Fuel Cycle* 4 (2): 195–209, http://inis.iaea.org/search/search.aspx?orig_q=RN:15014364.

Brown, P. M., and W. H. Reas. 1951. *Report of Invention*. HW-22434, Hanford Atomic Products Operation, Richland, WA. https://reading-room.labworks.org/Files/GetDocument.aspx?id=D3707309.

Brown, P. M., and W. H. Reas. 1952. *Preparation of Pu(III) Oxalate without Using Hydriodic Acid* (Richland, WA: General Electric). HW-25049. https://www.osti.gov/opennet/detail?osti-id=16431468.

Buckingham, J. S. 1967a. *Waste Management Technical Manual (Section 1 of 5)* (Richland, WA: Isochem Incorporated). ISO-100. https://pdw.hanford.gov/arpir/index.cfm/viewDoc?accession=0081113H.

Buckingham, J. S. 1967b. *Waste Management Technical Manual (Section 2 of 5)* (Richland, Washington: Isochem Incorporated). ISO-100. https://pdw.hanford.gov/arpir/index.cfm/viewDoc?accession=0081093H.

Buckingham, J. S. 1967c. *Waste Management Technical Manual (Section 3 of 5)* (Richland, Washington: Isochem Incorporated). ISO-100. https://pdw.hanford.gov/arpir/index.cfm/viewDoc?accession=0081092H.

Buckingham, J. S. 1967d. *Waste Management Technical Manual (Section 4 of 5)* (Richland, Washington: Isochem Incorporated). ISO-100. https://pdw.hanford.gov/arpir/index.cfm/viewDoc?accession=0081091H.

Buckingham, J. S. 1967e. *Waste Management Technical Manual (Section 5 of 5)* (Richland, Washington: Isochem Incorporated). ISO-100. https://pdw.hanford.gov/arpir/index.cfm/viewDoc?accession=0081090H.

Burney, G. A., and J. A. Porter. 1967. "Solubilities of Pu(III), Am(III), and Cm(III) Oxalates." *Inorganic and Nuclear Chemistry Letters* 3 (3): 79–85. https://doi.org/10.1016/0020-1650(67)80128-4.

Burney, G. A., and P. K. Smith. 1984. *Controlled PuO_2 Particle Size from Pu(III) Oxalate Precipitation* (Savannah River Laboratory, Aiken, SC: E.I. Du Pont de Nemours and Company). DP-1689. https://www.osti.gov/servlets/purl/6141411.

Burney, G. A., and F. W. Tober. 1965. "Precipitation of Plutonium Trifluoride." *Industrial & Engineering Chemistry Process Design and Development* 4 (1): 28–32. https://doi.org/10.1021/i260013a009.

Burns, R. E., and M. J. Stedwell. 1957. "Volume Reduction of Radioactive Waste by Carrier Precipitation." *Chemical Engineering Progress* 53 (2): 93–95.

Carritt, D. E., C. E. Hagen, and A. C. Wahl. 1946. *Plutonium Processes: D-Building Plutonium Purification* (Los Alamos, NM: Los Alamos Scientific Laboratory). https://fas.org/sgp/othergov/doe/lanl/lib-www/la-pubs/00350288.pdf.

Cash, R. J., G. T. Dukelow, and J. M. Atwood. 1992. *Quarterly Report on Defense Nuclear Facilities Safety Board Recommendation 90-7 for the Period Ending December 31, 1991* (Richland, WA: Westinghouse Hanford Company). WHC-EP-0474-3. http://pdw.hanford.gov/arpir/pdf.cfm?accession=E0019224.

Christensen, D. C., D. F. Bowersox, B. J. McKerley, and R. L. Nance. 1988. *Wastes from Plutonium Conversion and Scrap Recovery Operations* (Los Alamos, NM: Los Alamos National Laboratory). LA-11069-MS. http://www.osti.gov/scitech/biblio/5587648.

Christensen, E. L., and W. J. Maraman. 1968. *Plutonium Processing at the Los Alamos Scientific Laboratory* (Los Alamos, NM: Los Alamos Scientific Laboratory). LA-3542. https://www.osti.gov/servlets/purl/4794689.

Clark, D. L., S. S. Hecker, G. D. Jarvinen, and M. P. Neu. 2006. "Plutonium." In *The Chemistry of the Actinide and Transactinide Elements*, edited by L. R. Morss, N. M. Edelstein, J. Fuger and J. Katz. Dordrecht, the Netherlands: Springer Publishing.

Cleveland, J. M. 1967. "Plutonium Conversion Processes." In *Plutonium Handbook: A Guide to Technology*, edited by O.J. Wick. New York: Gordon and Breach, Science Publishers.

Cochran, T. B., and R. S. Norris. 1993. Russian/Soviet Nuclear Warhead Production. Washington, DC: Natural Resources Defense Council. https://fas.org/nuke/norris/nuc_09089301a_114.pdf.

Cochran, T. B., R. S. Norris, and O. A. Bukharin. 1995. *Making the Russian Bomb from Stalin to Yeltsin*. Boulder, CO: Natural Resources Defense Council, and Westview Press. https://fas.org/pubs/_docs/making_the_russian_bomb.pdf.

Collins, E. D. 2015. "Advanced Thermal Denitration Conversion Processes for Aqueous-Based Reprocessing and Recycling of Spent Nuclear Fuels." In *Reprocessing and Recycling of Spent Nuclear Fuel*, edited by Robin Taylor. Amsterdam, the Netherlands: Woodhead Publishing Series in Energy. https://doi.org/10.1016/B978-1-78242-212-9.00012-5.

Collins, P. E., and R. E. Isaacson. 1952. *Plant Process Evaluation Precipitation of Plutonium(IV) Oxalate* (Richland, WA: Hanford Atomic Products Operation). HW-24898. http://www.osti.gov/scitech/biblio/48239.

Crowder, M. L., R. A. Pierce, J. H. Scogin, W. E. Daniel, and W. D. King. 2012. *Small-Scale Testing of Plutonium(IV) Oxalate Precipitation and Calcination to Plutonium Oxide to Support the MOX Feed Mission* (Aiken, SC: Savannah River National Laboratory). SRNL-STI-2012-00338. http://sti.srs.gov/fulltext/SRNL-STI-2012-00338.pdf

Curie, P., and M. Curie. 1898. "Sur Une Substance Nouvelle Radio-Active, Contenue Dans La Pechblende." *Comptes Rendus de l'Academie des Sciences* 127 (26): 175–178, https://archive.org/details/surunesubstancen127curi/page/n7

Curie, P., M. Curie, and G. Bémont. 1898. "Sur Une Nouvelle Substance Fortement Radio-Active, Contenue Dans La Pechblende." *Comptes Rendus de l'Academie des Sciences* 127 (26): 1215–1217.

Delegard, C. H. 1985. "Solubility of $PuO_2 \cdot Xh_2o$ in Alkaline Hanford High-Level Waste Solution." 189th National Meeting of the American Chemical Society, Miami Beach, FL, April 28, 1985. http://inis.iaea.org/search/search.aspx?orig_q=RN:18050359.

Delegard, C. H. 2013. "Effects of Aging on $PuO_2 \cdot Xh_2o$ Particle Size in Alkaline Solution." *Radiochimica Acta* 101 (5): 313–322. https://doi.org/10.1524/ract.2013.2034.

Delegard, C. H., and S. A. Jones. 2015. *Chemical Disposition of Plutonium in Hanford Site Tank Wastes* (Richland, WA: Pacific Northwest National Laboratory). PNNL-23468, WTP-RPT-234. https://www.pnnl.gov/rpp-wtp/documents/WTP-RPT-234_rv1.pdf.

Delegard, C. H., V. F. Peretrukhin, and S. I. Rovny. 2018. "The Contributions of Radiochemistry to Mastering Atomic Energy for Weapons." *Bulletin for the History of Chemistry* 43 (2): 90–101. http://acshist.scs.illinois.edu/bulletin/v43-2/v43-2%20p90-101.pdf.

Delegard, C. H., S. I. Sinkov, B. K. McNamara, R. L. Sell, A. J. Schmidt, S. A. Jones, and G. S. Barney. 2002. *Critical Mass Laboratory Solution Processing: Precipitation, Calcination, and Moisture Uptake* (Richland, WA: Pacific Northwest National Laboratory). PNNL-SA-56850. https://www.osti.gov/servlets/purl/15010137-OAVLyJ/native/.

DOE. 2012. *Stabilization, Packaging, and Storage of Plutonium-Bearing Materials*. Washington, DC: United States Department of Energy. DOE-STD-3013-2012. https://www.standards.doe.gov/standards-documents/3000/3013-astd-2012/@@images/file.

Doerner, H. A., and W. M. Hoskins. 1925. "Co-Precipitation of Radium and Barium Sulfates." *Journal of the American Chemical Society* 47 (3): 662–675. https://doi.org/10.1021/ja01680a010.

DuPont. 1944. *Hanford Technical Manual Sections A, B and C*. HW-10475 ABC, Hanford Atomic Products Operation, Richland, WA. https://reading-room.labworks.org/Files/GetDocument.aspx?id=D197307786

DuPont. 1946a. *Memoranda for the File, Medical Department, Health Instrument Section, Personnel Meters Experience to August 1, 1945* (Richland, WA: E.I. Du Pont De Nemours & Company, Inc., Hanford Engineer Works). HAN-73214. https://reading-room.labworks.org/Files/GetDocument.aspx?id=D198144592.

DuPont. 1946b. *Memoranda for the File, S Department* (Richland, WA: E.I. Du Pont De Nemours & Company, Inc., Hanford Engineer Works). HAN-73214. http://reading-room.labworks.org/Files/GetDocument.aspx?id=D8766384.

Edwards, J., G. R. Chilton, J. R. Stanbridge, and W. Baxter. 1991. "The Conversion of Plutonium into Thermal Reactor MOX Fuel in the UK." Presented at *The Third International Conference on Nuclear Fuel Reprocessing and Waste Management (RECOD '91)*, Sendai, Japan, Atomic Energy Society of Japan and Japan Atomic Industrial Forum, April 15–18.

Facer, J., J. F., and K. M. Harmon. 1954. *Precipitation of Plutonium(IV) Oxalate* (Richland, WA: Hanford Atomic Products Operation). HW-31186 (Del.). https://www.osti.gov/servlets/purl/4634035.

Fedoseev, A. M., N. N. Krot, N. A. Budantseva, A. A. Bessonov, M. V. Nikonov, M. S. Grigoriev, A. Y. Garnov, V. P. Perminov, and L. N. Astafurova. 1998. *Interaction of Pu(IV, VI) Hydroxides/Oxides with Metal Hydroxides/Oxides in Alkaline Media* (Richland, WA: Pacific Northwest National Laboratory). PNNL-11900. http://www.osti.gov/scitech/biblio/665966.

Fiskum, S. K., E. C. Buck, R. C. Daniel, K. Draper, M. K. Edwards, T. L. Hubler, L. K. Jagoda et al. 2008. *Characterization and Leach Testing for REDOX Sludge and S-Saltcake Actual Waste Sample Composites* (Richland, WA: Pacific Northwest National Laboratory). PNNL-17368, WTP-RPT-157. https://www.pnnl.gov/main/publications/external/technical_reports/PNNL-17368.pdf.

Fondeur, F. F., D. T. Hobbs, and S. D. Fink. 2004. *Aluminum Leaching of "Archived" Sludge from Tanks 8F, 11H, and 12H* (Aiken, SC: Westinghouse Savannah River Company). WSRC-TR-2004-00180. https://digital.library.unt.edu/ark:/67531/metadc782154/m2/1/high_res_d/828685.pdf.

Fröman, N. 1996. "Marie and Pierre Curie and the Discovery of Polonium and Radium." Royal Swedish Academy of Sciences, Stockholm, Sweden, February 28, 1996. https://www.nobelprize.org/prizes/themes/marie-and-pierre-curie-and-the-discovery-of-polonium-and-radium.

Fugate, G. A., and J. D. Navratil. 2006. "Plutonium Peroxide in Process Chemistry: Oxidation State, Coordination Chemistry, and Solubility." In *Separations for the Nuclear Fuel Cycle in the 21st Century*, edited by G. J. Lumetta, K. L. Nash, S. B. Clark and J. I. Friese, In ACS Symposium Series, 167–181. Washington, DC: American Chemical Society. https://doi.org/10.1021/bk-2006-0933.ch011.

Gelis, A. V., and V. P. Shilov. 2000. "Coprecipitation of Transuranium Elements from Alkaline Solutions by the Method of Arising Agents: XIX. Masking of the Interference of Complexones in Plutonium Recovery." *Radiochemistry* 42 (4): 390–392.

Gelman, A. D., A. I. Moskvin, L. M. Zaitsev, and M. P. Mefodeva. 1962. *Complex Compounds of Transuranium Elements*. Translated by C.N. Turton and T.I. Turton. New York: Consultants Bureau.

Gerber, M. S. 1997. *History and Stabilization of the Plutonium Finishing Plant (PFP) Complex, Hanford Site* (Richland, WA: Fluor Daniel Hanford, Inc.). HNF-EP-0924. http://www.osti.gov/scitech/biblio/325360.

Gergen, J. B., N. L. Brown, D. E. Carritt, C. E. Hagen, W. W. Lowe, D. F. Mastick, E. Maxwell, and A. C. Wahl. 1946. *Chemical Research Building D Plutonium Purification* (Los Alamo, NM: Los Alamos Scientific Laboratory). LA-404. https://fas.org/sgp/othergov/doe/lanl/lib-www/la-pubs/00350133.pdf.

Ghosh-Mazumdar, A. S., M. N. Namboodiri, and H. D. Sharma. 1965. "Chemistry of Separation of Plutonium from Irradiated Fuel." Proceedings of the Third International Conference on the Peaceful Uses of Atomic Energy, Geneva, Switzerland, 31 August–9 September 1964.

Gibson, W. M. 1961. *The Radiochemistry of Lead* (Washington, DC: National Academy of Sciences). NAS-NS-3040. https://library.lanl.gov/cgi-bin/getfile?rc000004.pdf.

Gosling, F. G. 2010. The Manhattan Project: Making the Atomic Bomb. Washington, DC: U.S. Department of Energy. https://www.energy.gov/sites/prod/files/Manhattan_Project_2010.pdf

Grandjean, S., B. Arab-Chapelet, M. Bertrand, and G. Leturcq. 2008. Plutonium Conversion and Uranium-Plutonium Co-Conversion to Oxide for LWR Fuel Applications. Paris, France: Commissariat à l'Énergie Atomique et aux Énergies Alternatives. CEA-R-6181. https://inis.iaea.org/collection/NCLCollectionStore/_Public/40/013/40013983.pdf.

Grebenshchikova, V. I. 1972. "Coprecipitation Processes in Radiochemistry." *Soviet Atomic Energy* 33 (2): 723–727. https://doi.org/10.1007/BF01261842.

Greintz, R. M., and D. H. Neal. 1978. *Plutonium(IV) Oxalate Precipitation and Calcination Process for Plutonium Nitrate to Oxide Conversion* (Golden, CO: Rockwell International). RFP-2602. http://sti.srs.gov/fulltext/RFP-2602.pdf

Hagen, P. G., and F. J. Miner. 1980. "Plutonium Peroxide Precipitation: Review and Current Research." In *ACS Symposium Series*, edited by J.D. Navratil and W.W. Schulz, 51–67. Washington, DC: American Chemical Society. https://doi.org/10.1021/bk-1980-0117.ch005.

Harmon, K. M., B. F. Judson, W. L. Lyon, R. A. Pugh, and R. C. Smith. 1961. "Plutonium Reconversions." In *Reactor Handbook*, edited by S. M. Stoller and R. B. Richard. New York: Interscience Publishers.

Harmon, K. M., and W. H. Reas. 1957. *Conversion Chemistry of Plutonium Nitrate* (Richland, WA: General Electric, Hanford Atomic Products Operation). HW-49597-A. http://www.osti.gov/scitech/biblio/4364697.

Harmon, K. M., and W. H. Reas. 1958. "Conversion Chemistry of Plutonium Nitrate." In *Progress in Nuclear Energy, Series III, Process Chemistry*, edited by F. R. Bruce, J. M. Fletcher and H. H. Hyman. New York: Pergamon Press.

Harrington, S. J. 2011. *Hanford Waste Mineralogy Reference Report* (Richland, Washington: Washington River Proetection Solutions LLC). RPP-RPT-46618, Rev. 2.

Hawkins, H. T. 2013. *History of the Russian Nuclear Weapon Program* (Los Alamos, NM: Los Alamos National Laboratory). LA-UR-13-28910. https://fas.org/nuke/guide/russia/lanl-history.pdf.

Henderson, L. M., and F. C. Kracek. 1927. "The Fractional Precipitation of Barium and Radium Chromates." *Journal of the American Chemical Society* 49 (3): 738–749. https://doi.org/10.1021/ja01402a017.

Herting, D. L. 1993. *Clean Salt Process Applied to Double Shell Slurry (Tank 101-SY)* (Richland, WA: Westinghouse Hanford Company). WHC-SD-WM-DTR-029. https://pdw.hanford.gov/arpir/pdf.cfm?accession=E0032758.

Herting, D. L. 1996. *Clean Salt Process Final Report* (Richland, WA: Westinghouse Hanford Company). WHC-EP-0915. https://www.osti.gov/biblio/325064.

Herting, D. L., G. A. Cooke, J. S. Page, and J. L. Valerio. 2015. *Hanford Tank Waste Particle Atlas* (Richland, WA: Washington River Protection Solutions LLC). LAB-RPT-15-0005 RO. https://www.osti.gov/servlets/purl/1423895.

Hobbs, D. T. 1999. "Precipitation of Uranium and Plutonium from Alkaline Salt Solutions." *Nuclear Technology* 128 (1): 103–112. https://doi.org/10.13182/NT128-103.

Hobbs, D. T., and C. J. Coleman. 1994. "Characterization of High-Level Waste Saltcake Stored at the Savannah River Site." 208th National American Chemical Society Meeting, Washington, DC, August 21–26, 1994. https://www.osti.gov/servlets/purl/10110952.

Hommaa, S., J.-I. Ishii, T. Kikuchi, T. Chikazawa, A. Shibata, T. Koyama, J. Koga, and S. Matsumoto. 2008. "Flowsheet Study of U-Pu Co-Crystallization Reprocessing System." *Journal of Nuclear Science and Technology* 45 (6): 510–517. https://doi.org/10.1080/18811248.2008.9711874.

Hyde, E. K. 1954. "Radiochemical Separation of the Actinide Elements." In *The Actinide Elements*, edited by G. T. Seaborg and J. J. Katz, In National Nuclear Energy. New York: McGraw-Hill Book Company.

Hyde, E. K., and G. T. Seaborg. 1956. *The Transuranium Elements* (Berkeley, CA: University of California). UCRL-3312. https://escholarship.org/uc/item/56h2r51f.

IAEA. 1998. *Safe Handling and Storage of Plutonium* (Vienna, Austria: International Atomic Energy Agency). ISSN 1020–6450, no. 9. http://www-pub.iaea.org/MTCD/Publications/PDF/P061_scr.pdf.

IAEA. 1999. *Status and Trends in Spent Fuel Reprocessing. IAEA-TECDOC-1103.* Vienna, Austria: International Atomic Energy Agency, 1999.

IAEA. 2012. *Experiences and Trends of Manufacturing Technology of Advanced Nuclear Fuels, IAEA-TECDOC-1686.* Vienna, Austria: International Atomic Energy Agency, 2012.

Il'enko, E. I., and N. A. Abramova. 1999. "Development of the First Soviet Radiochemical Process for Recovery of Plutonium from Irradiated Uranium." *Radiochemistry* 41 (2): 101–107.

IUPAC. 2014. *Compendium of Chemical Terminology Gold Book.* 2.3.3 ed. Research Triangle Park, NC: International Union of Pure and Applied Chemistry. https://goldbook.iupac.org/.

Jackson, R. R. 1977. "Hanford Waste Encapsulation: Strontium and Cesium." *Nuclear Technology* 32 (1). 10–15. https://doi.org/10.13182/NT77-A31732.

Jenkins, I. L., F. H. Moore, and M. J. Waterman. 1965. "X-Ray Powder Crystallographic Data on Plutonium and Other Oxalates.1. Oxalates of Plutonium(3) and Plutonium(6) and Their Isomorphs." *Journal of Inorganic & Nuclear Chemistry* 27 (1): 77–80. https://doi.org/10.1016/0022-1902(65)80193-2.

Jones, M. M. 1953. *A Study of Plutonium Trifluoride Precipitated from Aqueous Solution* (Richland, WA: General Electric, Hanford Atomic Products Operation). HW-30384. https://www.osti.gov/servlets/purl/4369891.

Jouannaud, C. 1965. "Expérience De Six Années De Fonctionnement De L'usine De Retraitement De Marcoule." Proceedings of the Third International Conference on the Peaceful Uses of Atomic Energy, Geneva, Switzerland, 31 August–9 September 1964.

Kathren, R. L., J. B. Gough, and G. T. Benefiel. 1993. *The Plutonium Story: The Journals of Professor Glenn T. Seaborg 1939–1946.* Columbus, OH: Battelle Press.

Kennedy, J. W., G. T. Seaborg, E. Segre, and A. C. Wahl. 1946. "Properties of 94(239)." *Physical Review* 70 (7–8): 555–556. https://doi.org/10.1103/PhysRev.70.555.

Kennedy, K. K., and W. L. Lyon. 1950. *A Plutonium Peroxide Process for the Purification Building* (Richland, WA: Hanford Atomic Products Operation). HW-18092. https://reading-room.labworks.org/Files/GetDocument.aspx?id=D8427733.

Kerr, W. B. 1951. *Final Report – Production Test 231-10 – Supplement B – Process Evaluation-Production of Concentrated Plutonium Nitrate Solution in Building 231 – Transfer to and Processing in Building 234-5* (Richland, WA: General Electric). HW-22333. http://www.osti.gov/scitech/biblio/86307.

Kerr, W. B. 1952. *Final Report – Production Test 234-5 – Plant Process Evaluation Precipitation of Plutonium IV Oxalate* (Richland, WA: General Electric). HW-26450. http://www.osti.gov/scitech/biblio/85028.

Kirby, H. W., and M. L. Salutsky. 1964. *The Radiochemistry of Radium* (Washington, DC: National Academy of Sciences). NAS-NS-3057. https://www.osti.gov/biblio/4560824.

Krahn, S. L. 2008. EM Engineering & Technology Roadmap and Major Technology Demonstrations. Washington, DC: DOE, Office of Environmental Management. https://www.energy.gov/sites/prod/files/em/KrahnRoadmapwtihSTI.pdf.

Krapukhin, V. B., E. P. Krasavina, and A. K. Pikaev. 1997. *Crystallization of Sodium Nitrate from Radioactive Waste* (Richland, WA: Pacific Northwest National Laboratory). PNNL-11616. https://inis.iaea.org/collection/NCLCollectionStore/_Public/30/024/30024569.pdf.

Krot, N. N., V. P. Shilov, A. A. Bessonov, N. A. Budantseva, I. A. Charushnikova, V. P. Perminov, and L. N. Astafurova. 1996. *Investigation on the Coprecipitation of Transuranium Elements from Alkaline Solutions by the Method of Appearing Reagents* (Richland, WA: Westinghouse Hanford Company). WHC-EP-0898. http://www.iaea.org/inis/collection/NCLCollectionStore/_Public/30/047/30047157.pdf.

Krot, N. N., V. P. Shilov, A. M. Fedoseev, A. B. Yusov, A. A. Bessonov, N. A. Budantseva, S. I. Nikitenko et al. 1998a. *Alkali Treatment of Acidic Solution from Hanford K Basin Sludge Dissolution* (Richland, WA: Pacific Northwest National Laboratory). PNNL-11901. http://www.osti.gov/scitech/biblio/2507.

Krot, N. N., V. P. Shilov, A. B. Yusov, I. G. Tananaev, M. S. Grigoriev, A. Y. Garnov, V. P. Perminov, and L. N. Astafurova. 1998b. *Plutonium(IV) Precipitates Formed in Alkaline Media in the Presence of Various Anions* (Richland, WA: Pacific Northwest National Laboratory). PNNL-119010. http://www.osti.gov/bridge/purl.cover.jsp?purl=/665911-NwS0BW/webviewable/.

Kruglov, A. *The History of the Soviet Atomic Industry*. London, UK: Taylor & Francis Group, 2002.

Kuhn, K.-D., and K.-H. Koch. 1973. "Description of a Conversion Facility for Plutonium Nitrate." *Kerntechnik* 15 (6): 257–263.

Kuhn, K.-D., and K.-H. Koch. 1977. Method of Producing PuO_2 by Calcination of Pu Oxalate Produced by Discontinuous Precipitation from Solutions Contaminated with Americium. Patent Application US4298579A.

Leary, J. A., A. N. Morgan, and W. J. Maraman. 1959. "Plutonium Peroxide Precipitation." *Industrial and Engineering Chemistry* 51 (1): 27–31. https://doi.org/10.1021/ie50589a029.

Lehmkuhl, G. D. 1977. *Alternate Fuel Cycle Technologies – Nitrate-to-Oxide Conversion Project Progress Report – January–June 1977* (Golden, CO: Rockwell International). RFP-2591. http://www.osti.gov/scitech/biblio/7211272.

Lerch, R. E., and R. E. Norman. 1984. "Nuclear-Fuel Conversion and Fabrication Chemistry." *Radiochimica Acta* 36 (1–2): 75–88. https://doi.org/10.1524/ract.1984.36.12.75.

Lowe, C. S., R. L. Bates, L. L. Bentz, G. W. Frink, M. McEwan, R. W. Moshier, E. Orban et al. 1951. *Decontamination of Process Waste Solutions Containing Fission Products by Adsorption and Coprecipitation Methods* (Miamisburg, OH: Mound Laboratory). MLM-621. https://www.osti.gov/scitech/biblio/1369640-decontamination-process-waste-solutions-containing-fission-products-adsorption-co-precipitation-methods.

Luerkens, D. W. 1983. *An Improved Plutonium Trifluoride Precipitation Flowsheet* (Aiken, SC: Savannah River Laboratory). DPST-83-1065. https://digital.library.unt.edu/ark:/67531/metadc722439/m2/1/high_res_d/782818.pdf.

Maffei, H. P. 1957. *Continuous Task I – II* (Richland, WA: General Electric, Hanford Atomic Products Operation). HW-51006. https://www.osti.gov/opennet/detail?osti-id=16441914.

Mahoney, L. A., S. D. Rassat, P. W. Eslinger, R. L. Aaberg, P. M. Aker, E. C. Golovich, B. D. Hanson et al. 2010. *EFRT M-12 Issue Resolution: Caustic-Leach Rate Constants from PEP and Laboratory-Scale Tests* (Richland, WA: Pacific Northwest National Laboratory). PNNL-18502, Rev 1, WTP-RPT-186, Rev 1. https://www.pnnl.gov/main/publications/external/technical_reports/PNNL-18502rev1.pdf.

Mainland, E. W. 1961. "Production of Plutonium Metal." *Industrial and Engineering Chemistry* 53 (9): 685–694. https://doi.org/10.1021/ie50621a019.

Mandleberg, C. J., K. E. Francis, and R. Smith. 1961. "475. The Solubility of Plutonium Trifluoride, Plutonium Tetrafluoride, and Plutonium(IV) Oxalate in Nitric Acid Mixtures." *Journal of the Chemical Society (Resumed)* (0): 2464–2468. https://doi.org/10.1039/JR9610002464.

Marsh, S. F., and T. D. Gallegos. 1987. *Chemical Treatment of Plutonium with Hydrogen Peroxide before Nitrate Anion Exchange Processing* (Los Alamos, NM: Los Alamos National Laboratory). LA-109047. http://www.iaea.org/inis/collection/NCLCollectionStore/_Public/19/001/19001141.pdf.

Martella, L. L., M. T. Saba, and G. K. Campbell. 1984. *Laboratory-Scale Evaluations of Alternative Plutonium Precipitation Methods* (Golden, CO: Rockwell International). RFP-3589. http://www.osti.gov/scitech/biblio/5318991.

McCabe, D. J. 1996. *Cesium, Potassium, and Sodium Tetraphenylborate Solubility in Salt Solution* (Aiken, SC: Westinghouse Savannah River Company). WSRC-TR-96-0384. https://www.osti.gov/servlets/purl/626456.

McMillan, E., and P. H. Abelson. 1940. "Radioactive Element 93". *Physical Review* 57 (12): 1185–1186. https://journals.aps.org/pr/pdf/10.1103/PhysRev.57.1185.2.

Mew, D. A., O. H. Krikorian, K. E. Dodson, and J. A. Schmitz. 2001. *Purification of ^{238}PU Oxide Using the Pu Oxalate Process* (Livermore, CA: Lawrence Livermore National Laboratory). UCRL-ID-146750. http://www.osti.gov/scitech/biblio/15005300.

Moore, E. N., D. A. Orth, W. B. Summer, and J. A. Purcell. 2000. *Development of Pu-239 Processes and Facilities* (Aiken, SC: Westinghouse Savannah River Site). WSRC-MS-2000-00061. https://sti.srs.gov/fulltext/WSRC-MS-2000-00061.pdf.

Mudge, L. K. 1959. *Influence of Temperature on the Continuous Precipitation of Plutonium(IV) Oxalate* (Richland, WA: Hanford Atomic Products Operation). https://reading-room.labworks.org/Files/GetDocument.aspx?id=D8368290.

Muscatello, A. C., and M. E. Killion. 1990. *Chloride Anion Exchange Coprocessing for Recovery of Plutonium from Pyrochemical Residues and Cs_2PuCl_6 Filtrate* (Golden, CO: EG & G Rocky Flats). RFP-4325. https://www.osti.gov/servlets/purl/6303236.

Myers, M. N. 1956. *Reduction of Plutonium(VI) with Hydrogen Peroxide* (Richland, WA: General Electric, Hanford Atomic Products Operation). HW-44987. https://www.osti.gov/opennet/detail?osti-id=16440409.

Neck, V., M. Altmaier, A. Seibert, J. I. Yun, C. M. Marquardt, and T. Fanghänel. 2007. "Solubility and Redox Reactions of Pu(IV) Hydrous Oxide: Evidence for the Formation of PuO_{2+X}(S, Hyd)." *Radiochimica Acta* 95 (4): 193–207. https://doi.org/10.1524/ract.2007.95.4.193.

NUMEC. 1961. *Progress Reports. NUMEC P-60* https://www.osti.gov/servlets/purl/4840029/ *(1 Jan–31 Mar 1961), NUMEC P-70,* https://www.osti.gov/servlets/purl/4837198/ *(1 Apr–30 Jun 1961), NUMEC P-80* https://www.osti.gov/servlets/purl/4807381/ *(1 Jul–30 Sep 1961), NUMEC P-90* https://www.osti.gov/servlets/purl/4797591/ *(1 Oct–31 Dec 1961)* (Apollo, PA: Nuclear Materials and Equipment Corporation).

NUMEC. 1962. *Progress Reports. NUMEC P-100* https://www.osti.gov/servlets/purl/4815636/ *(1 Jan–31 Mar 1962), NUMEC P-103* https://www.osti.gov/servlets/purl/4747237/ *(1 Oct–31 Dec 1962)* (Apollo, PA: Nuclear Materials and Equipment Corporation).

NUMEC. 1963. *Progress Reports. NUMEC P-105* https://www.osti.gov/servlets/purl/4630019/ *(1 Apr–30 Jun 1963), NUMEC P-106* https://www.osti.gov/servlets/purl/4040704/ *(1 Jul–30 Sep 1963)* (Apollo, PA: Nuclear Materials and Equipment Corporation).

NUMEC. 1964. *Progress Reports. NUMEC 2389-4* https://www.osti.gov/servlets/purl/4636160/ *(1 Jan–31 Mar 1964); NUMEC 2389-2* https://www.osti.gov/servlets/purl/4079185/ *(Jan 1964)* (Apollo, PA: Nuclear Materials and Equipment Corporation).

Oshima, H. 1989. "Development of Microwave Heating Method for Co-Conversion of Plutonium-Uranium Nitrate to MOX Powder". *Journal of Nuclear Science and Technology* 26 (1): 161–166. http://dx.doi.org/10.1080/18811248.1989.9734282.

Pazukhin, E. M., B. Y. Galkin, A. P. Krinitsyn, Y. A. Pokhitonov, and V. I. Ryazantsev. 2003. "Separation of Am(III) and Pu(IV) at Their Joint Oxalate Precipitation." *Radiochemistry* 45 (2): 159–166. https://doi.org/10.1023/A:1023885310014.

Peretroukhine, V. F., and C. H. Delegard. 2000. "Some Comparisons of Plutonium-Bearing Radwaste Management in the USA and Russia". In *Environmental Challenges of Nuclear Disarmament*, edited by T. E. Baca and T. Florkowski, Kluwer Academic Publishers, Dordrecht, Netherlands. Presented at "Environmental Challenges of Nuclear Disarmament", NATO Advanced Research Workshop, November 1998, Crakow, Poland. https://link.springer.com/chapter/10.1007/978-94-011-4104-8_2.

Peretrukhin, V. F., V. I. Silin, A. V. Kareta, A. V. Gelis, V. P. Shilov, K. E. German, E. V. Firsova, A. G. Maslennikov, and V. E. Trushina. 1998. *Purification of Alkaline Solutions and Wastes from Actinides and Technetium by Coprecipitation with Some Carriers Using the Method of Appearing Reagents* (Richland, WA: Pacific Northwest National Laboratory). PNNL-11988. https://inis.iaea.org/collection/NCLCollectionStore/_Public/30/013/30013957.pdf.

Peterson, R. A., E. C. Buck, J. Chun, R. C. Daniel, D. L. Herting, E. S. Ilton, G. J. Lumetta, and S. B. Clark. 2018. "Review of the Scientific Understanding of Radioactive Waste at the U.S. DOE Hanford Site". *Environmental Science & Technology* 52: 381–396. https://doi.org/10.1021/acs.est.7b04077.

Petrie, J. C., R. I. Donovan, R. E. Van der Cook, and D. C. Christensen. 1975. *Processing of Radioactive Waste Solutions in a Vacuum Evaporator-Crystallizer* (Richland, WA: Atlantic Richfield Hanford Company). ARH-SA-240. https://www.osti.gov/servlets/purl/4140194.

Pike, J. A. 2008. *Review of SMR Project Process Development since CD-0* (Aiken, SC: Washington Savannah River Company). LWO-LWE-2008-00117.

Pittman, F. K., W. A. Brown, D. Duffy, G. B. Fleming, W. B. Gibson, B. L. Kelchner, R. Schaffer, and D. L. Wolfe. 1944. *Recovery and Purification of Plutonium* (Los Alamos, NM: Los Alamos Scientific Laboratory). LA-141. https://permalink.lanl.gov/object/tr?what=info:lanl-repo/lareport/LA-00141.

Ponder, S. M., R. Helkowski, and T. E. Mallouk. 2001. "Continuous-Flow Process for the Separation of Cesium from Complex Waste Mixtures." *Industrial & Engineering Chemistry Research* 40 (15): 3384–3389. https://doi.org/10.1021/ie001058p.

Porter, J. A., and A. E. Symonds Jr. 1965. *Precipitation of Plutonium(III) Oxalate and Calcination to Plutonium Dioxide* (Aiken, SC: Savannah River laboratory). DP-981. https://www.osti.gov/biblio/4649172.

Reed, T., and A. Kramish. 1996. "Trinity at Dubna." *Physics Today* 49 (11): 30–35. https://doi.org/10.1063/1.881546.

Reynolds, J. G., G. A. Cooke, D. L. Herting, and R. W. Warrant. 2013. "Salt Mineralogy of Hanford High-Level Nuclear Waste Staged for Treatment." *Industrial & Engineering Chemistry Research* 52 (29): 9741–9751. https://doi.org/10.1021/ie400822r.

Rodgers, L. H. 1986. *1985 Remote Mechanical C (RMC) Line Campaign Report* (Richland, WA: Rockwell Hanford Operations). RHO-RE-EV-85.

Roepenack, H., V. W. Schneider, and W. G. Druckenbrodt. 1984. "Experience with the Ammonium Uranyl Plutonyl Carbonate Coconversion Process for Mixed-Oxide Fuel Fabrication." *American Ceramic Society Bulletin* 63 (8): 1051–1053.

Ryabev L. D. (Ed.). 2000. "USSR Atomic Bomb Project: Documents and Materials. " Volume II. Book 2. "Atomic Bomb. 1945–1954" (in Russian). Page 503. Fizmatlit. Moscow, Russia. "Атомный проект СССР: документы и материалы." Т. 2. Кн. 2. http://elib.biblioatom.ru/text/atomny-proekt-sssr_t2_kn2_2000/go,502/

Ryan, J. L., G. H. Bryan, M. C. Burt, and D. A. Costanzo. 1985. "Preparation of Acid Standards for and Determination of Free Acid in Concentrated Plutonium Uranium Solutions." *Analytical Chemistry* 57 (7): 1423–1427. https://doi.org/10.1021/ac00284a055.

Salutsky, M. L., J. G. Stites, and A. W. Martin. 1953. "Fractionation of Barium-Radium Mixtures as Chromates – by Precipitation from Homogeneous Solution." *Analytical Chemistry* 25 (11): 1677–1681. https://doi.org/10.1021/ac60083a028.

Sarsfield, M. J. 2015. "The Coprecipitation and Conversion of Mixed Actinide Oxalates for Aqueous-Based Reprocessing of Spent Nuclear Fuels." In *Reprocessing and Recycling of Spent Nuclear Fuel*, edited by Robin Taylor, 325–351. Amsterdam, the Netherlands: Woodhead Publishing Series in Energy: Number 79, Elsevier. https://doi.org/10.1016/C2013-0-16483-5.

Schlyter, K., and L. G. Sillen. 1950. "On the Coprecipitation of Tetrapositive Ions with Lanthanum Fluoride." *Acta Chemica Scandinavica* 4 (8): 1323–1325. https://doi.org/10.3891/acta.chem.scand.04-1323.

Schneider, K. J. 1951. *Flow Sheets and Flow Diagrams of Precipitation Separation Processes* (Richland, WA: Hanford Atomic Products Operation). HW-23043. https://www.osti.gov/opennet/detail?osti-id=16430616.

Seaborg, G. T. 1958. *The Transuranium Elements* (New Haven, CT: Yale University Press).

Seaborg, G. T., E. M. McMillan, J. W. Kennedy, and A. C. Wahl. 1946a. "Radioactive Element-94 from Deuterons on Uranium." *Physical Review* 69 (7–8): 366–367. https://doi.org/10.1103/PhysRev.69.366.2.

Seaborg, G. T., A. C. Wahl, and J. W. Kennedy. 1946b. "Radioactive Element-94 from Deuterons on Uranium." *Physical Review* 69 (7–8): 367. https://doi.org/10.1103/PhysRev.69.367.

Silver, G. 2010. *Plutonium Oxalates as Sources of Plutonium Dioxide* (Los Alamos, NM: Los Alamos National Laboratory). LA-UR-10-02596. https://permalink.lanl.gov/object/tr?what=info:lanl-repo/lareport/LA-UR-10-02596.

Smith, P. K., G. A. Burney, D. T. Rankin, D. F. Bickford, and R. D. Sison Jr. 1976. "Effect of Oxalate Precipitation on PuO_2 Microstructures." Sixth International Materials Symposium, Berkeley, California, August 24–27, 1976. http://www.osti.gov/bridge/product.biblio.jsp?osti_id=7234786.

Smyth, H. D. 1945. "Atomic Energy for Military Purposes." *Reviews of Modern Physics* 17 (4): 351–471. https://doi.org/10.1103/RevModPhys.17.351.

Snow, L. A., E. C. Buck, A. J. Casella, J. V. Crum, R. C. Daniel, K. E. Draper, M. K. Edwards et al. 2009. *Characterization and Leach Testing for PUREX Cladding Waste Sludge (Group 3) and REDOX Cladding Waste Sludge (Group 4) Actual Waste Sample Composites* (Richland, WA: Pacific Northwest National Laboratory). PNNL-18054, WTP-RPT-167. https://www.pnnl.gov/main/publications/external/technical_reports/PNNL-18054.pdf.

Sokhina, L. P. 2000. *Pages of History and Radiochemical Plant Production Association "Mayak"*. Ozersk, Russia: Printing House Mayak. http://elib.biblioatom.ru/text/sohina_stranitsy-istorii-zavoda-mayak_2000/?bookhl=

Sokhina, L. P., Y. I. Kolotinsky, and G. V. Khalturin. 2003. *Plutonium in Girl's Hands*. Ekaterinburg, Russia: South Ural Institute of Biophysics. http://elib.biblioatom.ru/text/sohina_plutoniy-v-devichih-rukah_2003/go,0/?bookhl=

Souply, K. R., and D. H. Neal. 1977. *Fluid Bed Direct Denitration Process for Plutonium Nitrate to Oxide Conversion*. RFP-2600 (Golden, CO: Rocky Flats Plant). https://www.osti.gov/servlets/purl/7305433

Souply, K. R., and W. E. Sperry. 1977. *Screw Calciner Mechanical Direct Denitration Process for Plutonium Nitrate to Oxide Conversion*. RFP-2601 (Golden, CO: Rocky Flats Plant). https://digital.library.unt.edu/ark:/67531/metadc1069289/m2/1/high_res_d/5254490.pdf

Stoll, M. E., D. J. Spengler, and M. L. Vil. 2012. "Processing and Purification of Heat Source $^{238}PuO_2$ (HS-PuO_2) at Los Alamos National Laboratory." *Nuclear and Emerging Technologies for Space 2012*, The Woodlands, TX, March 21–23, 2012. http://www.lpi.usra.edu/meetings/nets2012/pdf/3038.pdf.

Strutt, R. J. 1904. "The Treatment of Pitchblende Residues on a Large Scale." In *The Becquerel Rays and the Properties of Radium*, 37–41 and Appendix C. London, UK: Arnold, Edward.

Stubbs, A. M., and J. F. Durnil. 1991. *Process Flowsheet Document, Remote Mechanical C (RMC) Line, Plutonium Finishing Plant* (Richland, WA: Westinghouse Hanford Company). PFD-Z-190-00002, Rev C02. http://pdw.hanford.gov/arpir/index.cfm/viewDoc?accession=D196240593.

Sun, Z. H., B. K. Han, and H. Z. Hu. 1991. "Continuous Precipitation Process of Pu(III) Oxalate and Its Equipment." *The Third International Conference on Nuclear Fuel Reprocessing and Waste Management (RECOD '91)*, Atomic Energy Society of Japan and Japan Atomic Industrial Forum, Sendai, Japan, April 15–18.

Vertés, A., and I. Kiss. 1987. "The Chemistry of Ultra-Low Concentrations." In *Nuclear Chemistry*. Amsterdam, the Netherlands: Elsevier.

Volk, V. I., T. A. Boytsova, S. N. Veselov, K. N. Dvoeglazov, and A. Y. Shadrin. 2018. "Crystallization Purification Full Scale Test of U–Pu Mixture." *Journal of Radioanalytical and Nuclear Chemistry* 318: 2501–2504. https://doi.org/10.1007/s10967-018-6336-9.

Walker, D. D., M. J. Barnes, C. L. Crawford, R. A. Peterson, R. F. Swingle, and S. D. Fink. 1997. *In-Tank Precipitation with Tetraphenylborate: Recent Process and Research Results* (Aiken, SC: Westinghouse Savannah River Company). https://www.osti.gov/src/fulltext/565164.

Walker, D. D., M. J. Barnes, C. L. Crawford, R. A. Peterson, R. F. Swingle, and S. D. Fink. 1998. "In-Tank Precipitation with Tetraphenylborate: Recent Process and Research Results." In *Science and Technology for Disposal of Radioactive Tank Wastes*, edited by W.W. Schulz and N. L. Lombardo, 219–230. Boston, MA: Springer.

White, G. D. 1977. *RMA Oxide Line Flowsheet for Production of Plutonium Dioxide* (Richland, WA: Atlantic Richfield Hanford Company). ARH-F-105.

Wilmarth, W. R., C. J. Coleman, J. C. Hart, and W. T. Boyce. 2000. *Characterization of Samples from the 242-16H Evaporator Wall* (Aiken, SC: Westinghouse Savannah River Company). WSRC-TR-2000-00089. https://www.osti.gov/biblio/754457-tuS4vA/webviewable/.

Winchester, R. S., and W. J. Maraman. 1958. "Aqueous Decontamination of Plutonium from Fission Product Elements." *Second United Nations International Conference on the Peaceful Uses of Atomic Energy*, Geneva, Switzerland, September 1–13, 1958.

Work, J. B. 1950. *Conversion from Oxalate to Peroxide Cycle in 234 Building* (Richland, WA: Hanford Atomic Products Operation). HW-18948. https://reading-room.labworks.org/Files/GetDocument.aspx?id=D8480411.

Wu, Z.-Y., and Y.-Q. Jiang. 1991. "Some Aspects of a Civil Reprocessing Pilot Plant." *3rd International Conference on Nuclear Fuel Reprocessing and Waste Management (RECOD '91)*, Sendai, Japan, April 15–18, 1991.

Yarbro, S. L., S. B. Schreiber, S. L. Dunn, and C. W. Mills. 1992. "Homogeneous Oxalate Precipitation of Pu(III)." In *Transuranium Elements: A Half Century*, edited by L. R. Morss and J. Fuger. Washington, DC: American Chemical Society.

Yusov, A. B., N. A. Budantseva, A. V. Anan'ev, and A. M. Fedoseev. 2000c. "Coprecipitation of Aluminum with Hydroxides of Tetra-, Penta-, and Hexavalent Actinides." *Radiochemistry* 42 (5): 413–416, https://inis.iaea.org/search/search.aspx?orig_q=RN:32022468.

Yusov, A. B., A. Y. Garnov, V. P. Shilov, I. G. Tananaev, M. S. Grigoriev, and N. N. Krot. 2000a. "Plutonium(IV) Precipitation from Alkaline Solutions: I. Effect of Precipitation and Coagulation Conditions on Properties of Hydrated Plutonium Dioxide $PuO_2 \cdot xH_2o$." *Radiochemistry* 42 (2): 151–156.

Yusov, A. B., A. Y. Garnov, V. P. Shilov, I. G. Tananaev, M. S. Grigoriev, and N. N. Krot. 2000b. "Plutonium(IV) Precipitation from Alkaline Solutions: II. Effect of Anions on Composition and Properties of Hydrated Plutonium Dioxide $PuO_2 \cdot xH_2o$." *Radiochemistry* 42 (2): 157–160.

4 Solvent Extraction in the Nuclear Fuel Cycle

Gabriel B. Hall, Susan E. Asmussen, and Amanda J. Casella

CONTENTS

4.1 INTRODUCTION TO SOLVENT EXTRACTION

The present discussion will be limited to solvent extraction as it pertains to the reprocessing of used nuclear fuel. Solvent extraction is the chemical purification and isolation of a specific element or group of elements, by the use of mutually immiscible solvents. Solvent extraction is not used to provide isotopic purification, although isotope effects have been observed. The quest to successfully isolate plutonium and recover uranium from irradiated nuclear fuel (i.e., reprocessing) resulted in the bulk of the mid-twentieth century's nuclear fuel cycle solvent extraction developments. Thus, it was also necessary in reprocessing of irradiated fuels to remove high-activity fission products such as Cesium (Cs), Strontium (Sr), and rare earths. During the latter half of the twentieth century and the early 2000s, much of the research focus shifted from simple reprocessing to remediation of legacy nuclear waste and recycling of actinides within expanded nuclear fuel cycle flow sheets. This included both the separation of the minor actinides (Neptunium (Np), Americium (Am), Curium (Cm)) from used nuclear fuel and the removal of major dose contributors (e.g., Cs, Sr) and long-lived environmentally mobile nuclides (Technitium-99, ^{99}Tc) from legacy nuclear waste.

Isolation of plutonium and recovery of uranium was initiated in the 1940s to use plutonium in nuclear weapons and recover the uranium for additional plutonium production. Today, plutonium/uranium recovery is also focused around the production of mixed oxide fuels used for additional civilian power production. The recovery of plutonium from commercial nuclear fuel also has the benefit of reducing the radiotoxic lifetime of the residual materials to be disposed in a geological repository. Separation of the minor actinides is motivated by further reducing the radiotoxic lifetime of the used fuel residues and reducing the heat load to increase the effective capacity of the geological repository (Poinssot, Boullis, and Bourg 2015). The minor actinides, after separation from the lanthanide elements with large neutron absorption cross sections, then may be transmuted to short-lived or stable isotopes in a nuclear reactor or accelerator.

To perform solvent extraction on an industrial scale, several chemical and engineering challenges need to be overcome. The preeminent chemical challenge is to extract a single element (or combination of elements) while leaving the remaining elements behind. The efficiency of these processes are measured by the distribution ratio (D) and separation factor (SF) calculated according to Equations 4.1 and 4.2, where: D_M is the distribution ratio for a given metal; $[M]_{org}$ is the organic phase metal concentration; $[M]_{Aq}$ is the aqueous phase metal concentration; $SF_{M1/M2}$ is the SF between two metals; and D_{M1} and D_{M2} are the distribution ratios of two different metals. Because the SF is a ratio of D-values, and the D-values are fundamentally thermodynamic values, the SF is driven by the difference in thermodynamic favorability for extraction of each metal. Chemical separations can be complicated by the high radiation fields that degrade solvents and ligands, the marked chemical similarity of some of the elements involved (e.g., the trivalent actinides and lanthanides), and the complex redox behavior of the lighter actinides (especially Np and Plutonium (Pu)).

$$D_M = \frac{[M]_{Org}}{[M]_{Aq}} \tag{4.1}$$

$$SF_{M_1/M_2} = \frac{D_{M1}}{D_{M2}} \tag{4.2}$$

The engineering challenges are much more diverse and include implementing remote processes/handling of material, minimizing and treating secondary waste, maintaining of criticality safety, minimizing the dose, mitigating corrosion processes, ensuring complete phase separation, eliminating third phase formation, addressing solid entrainments (crud), designing reliable and rugged equipment, accommodating production volume, and modeling proper flowsheets, among other challenges. As augmentations to the engineered system are made, the chemistry often needs to be revised, and the same is true in the reverse direction. For example, the radiolytic stability of an extractant is determined both by the inherent cross section for absorption as well as the exposure to radiation. Selection of an extraction process and equipment that allows for fast extraction kinetics with fast solvent cycling will minimize dose and thus simplify the engineering surrounding solvent washing and replacement. However, if a process achieves an exceptional SF but has slow chemical kinetics, then equipment with longer residence times and additional solvent conditioning steps must be engineered into the flowsheet.

The engineering challenges surrounding solvent extraction do not cease once successful extraction has occurred. Secondary nuclear waste that is generated during the process must be managed and properly disposed. This has led to the "CHON principle": using extractants and solvents comprised solely of carbon, hydrogen, oxygen, and nitrogen, which generate water, CO_2, and gaseous nitrogen compounds upon oxidation or incineration, thus leaving no condensable waste residue. This greatly simplifies waste handling but requires chemical research to identify, characterize, and prepare suitable extractants, diluents, phase modifiers, complexants, and oxidation/reduction reagents.

4.2 PROCESSES: BASIC CHEMISTRY

Table 4.1 and the text below provide a brief overview of chemical processes used for nuclear fuel cycle separations, while Figures 4.1 and 4.2 provide the chemical structure of some commonly used extractants. The extractants are generally categorized as either neutral, which require a counterion to charge-balance the extracted metal ion during phase transfer, or acidic, which transfer a proton to the aqueous phase as the metal is transferred to the organic phase. Figure 4.3 shows some reagents that are commonly used as complexants and/or reductants in the aqueous phase.

The discussion is broken into three sections depending on the desired objective: (1) recovery of uranium and plutonium, (2) recovery of minor actinides, and

TABLE 4.1
Selected Solvent Extraction Processes and the Desired Separation

Process Name	Desired Separation
REDOX	Pu, U
PUREX	Pu, U
COEX	Mixed Pu/U, U
CoDCon	Mixed Pu/U, U
Advanced PUREX	Mixed Pu/U/Np, U
GANEX	U, Np, Pu, Am, Cm
UREX	Tc, U
UNEX	Cs, Sr, An, Ln
TRUEX	Pu, U, Am, Cm
BUTEX	Pu, U
TALSPEAK	Post-PUREX An, Ln
Advanced TALSPEAK	Post-PUREX An, Ln
Reverse TALSPEAK	Post-PUREX An, Ln
ALSEP	Post-PUREX An, Ln
DIAMEX	Post-PUREX An, Ln
DIAMEX-SANEX	Am, Cm from DIAMEX raffinate
POR	An
ChCoDiC	Transplutonium and rare earth elements
CSEX	Cs
SREX	Sr
CSSX	Cs
SRTALK	Tc, Cs, Sr

Abbreviations: An, actinide; Ln, Lanthanide.

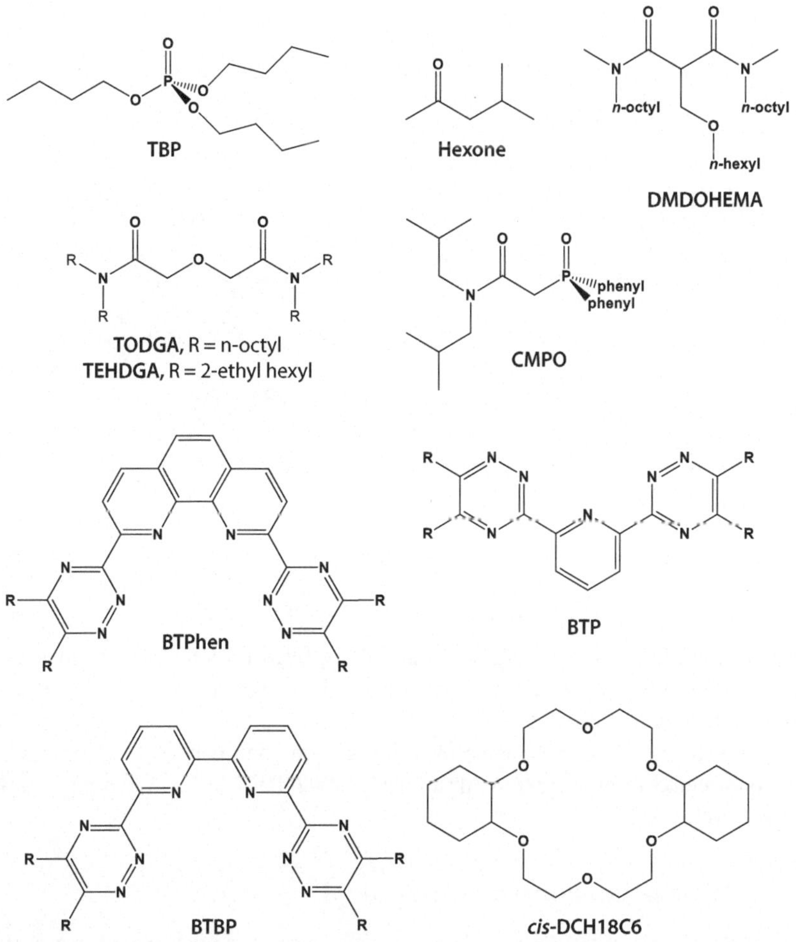

FIGURE 4.1 Neutral extractants commonly used for solvent extraction in the nuclear fuel cycle.

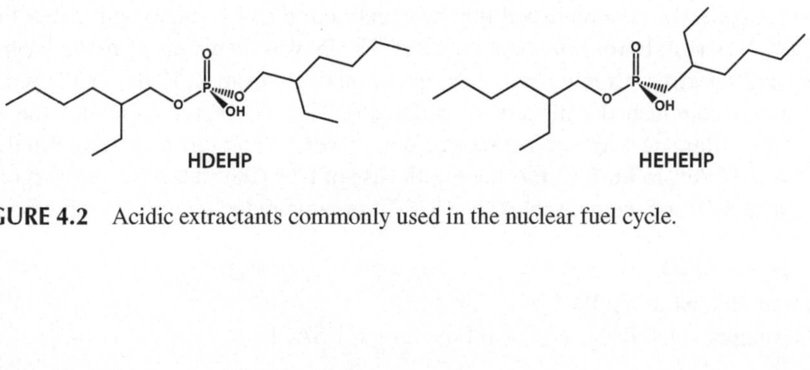

FIGURE 4.2 Acidic extractants commonly used in the nuclear fuel cycle.

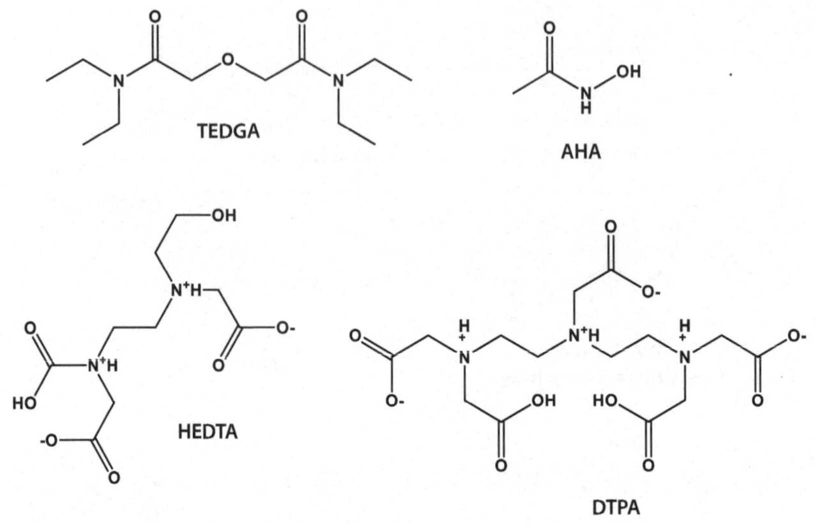

FIGURE 4.3 Aqueous phase complexants used in nuclear fuel cycle solvent extraction schemes.

(3) separation of high-dose and problematic long-lived fission products such as Cs, Sr, and Technetium (Tc). The process chemistry developed around solvent extraction for the nuclear fuel cycle is too diverse of a subject to be covered in a single book chapter. For a more comprehensive overview of the chemistry of each process, the reader is directed to other texts with a heavier chemistry focus (Nash and Lumetta 2011; R. Taylor 2015).

4.2.1 Uranium and Plutonium Separations

Purification of Pu and recapture of U for recycle from used nuclear fuel in a continuous operation with high throughput were the driving forces for developing solvent extraction during the middle of the twentieth century (Paulenova 2011). Due to the hard Lewis acid nature of the actinides, typical extractants are hard Lewis base in nature with the typical ligating atom being oxygen. This includes phosphine oxides, ethers, diglycolamides, ketones, phosphonates, and organic phosphates. Extraction schemes generally accomplished this by coextracting UO_2^{2+} along with either Pu^{IV} or $Pu^{IV/VI}$ (Herbst, Baron, and Nilsson 2011). The Pu was then reduced to Pu^{III}, which promoted selective stripping and thus separation of Pu from U. Early on, the reduction was accomplished with ferrous sulfamate. The Fe^{II} served to reduce the Pu, while the sulfamate scavenged nitrous acid to prevent oxidation of Fe^{II} to Fe^{III} and re-oxidation of Pu^{III} to Pu^{IV}. Unfortunately, the use of Iron (Fe) adds significant secondary waste; U^{IV} or decomposable (i.e., CHON) organic reductants are now preferred.

4.2.1.1 REDOX

Desired Separation: Pu, U
Developed: Oak Ridge National Laboratory, USA, 1949

The REDOX process was the first industrial-scale solvent extraction process applied to uranium and plutonium recovery from irradiated fuel in the United States (O.F. Hill and Cooper 1958). Methyl isobutyl ketone (MIBK, or hexone) was used with no diluent to extract uranium and plutonium (Woodfield 1951). The first step was to oxidize Pu^{IV} to Pu^{VI} with sodium dichromate, $Na_2Cr_2O_7$, to make it extractable with U^{VI}, into methyl isobutyl ketone (O.F. Hill and Cooper 1958). Due to the instability of hexone in the presence of concentrated nitric acid, $Al(NO_3)_3$ was used as a salting agent. As a further protection of the MIBK but mainly to improve separation from fission products, partial neutralization of the dissolved irradiated fuel uranyl nitrate feed stream and of the $Al(NO_3)_3$ scrub solution was done, making both aqueous phases "acid deficient" (Delegard and Jones 2015); i.e., containing partially hydrolyzed species such as $UO_2(OH)_x(NO_3)_{2-x}$ and $Al(OH)_y(NO_3)_{3-y}$. $Uranium^{VI}$ and Pu^{VI} were co-extracted into the hexone solvent. After extraction, the plutonium D-value was altered by reduction of Pu^{VI} to Pu^{III} with ferrous sulfamate, thus stripping the Pu into the aqueous phase. Uranium was stripped by contacting the U-loaded solvent with water. The REDOX process proved to have a much higher throughput rate than previously used precipitation-based methods. However, due to safety concerns related to the low flash point of methyl butyl ketone, the once-through use of large amounts of $Al(NO_3)_3$ and $Na_2Cr_2O_7$, and the higher throughput and improved chemical efficiencies of the newer PUREX process, the REDOX process, which only was used at the Hanford Site from 1952 to 1967, was ultimately replaced by PUREX at Hanford and became the worldwide industry standard (Gerber 1996).

4.2.1.2 BUTEX

Desired Separation: Pu, U
Developed: United Kingdom and Canada, 1940s–1950s

The name BUTEX is derived from the extractant used in the process, dibutylcarbitol (butex). Unfortunately, due to disadvantages of the BUTEX process (i.e., the BUTEX extractant had a high viscosity, undesirable complexes formed with uranyl nitrate, among others), the BUTEX process was never adopted for broad use (Gray et al. 2015). Using dibutylcarbitol allowed the simultaneous recovery of both U and Pu from nitric acid solutions. The loaded BUTEX solvent was treated with ferrous sulfamate to reduce the Pu to Pu^{III}, allowing facile striping of Pu^{III} along with some U (Gray et al. 2015). The bulk of the uranium was stripped using a dilute nitric acid solution. The uranium remaining in the plutonium stream was then removed with a 20% tri-n-butyl phosphate (TBP) in kerosene solution (Ursu 1985; Gray et al. 2015).

4.2.1.3 PUREX

Desired Separation: Pu, U
Developed: Oak Ridge National Laboratory, USA, 1949

The Plutonium Uranium Reduction EXtraction process (PUREX) was developed in the late 1940s and utilizes TBP dissolved in aliphatic hydrocarbon as an

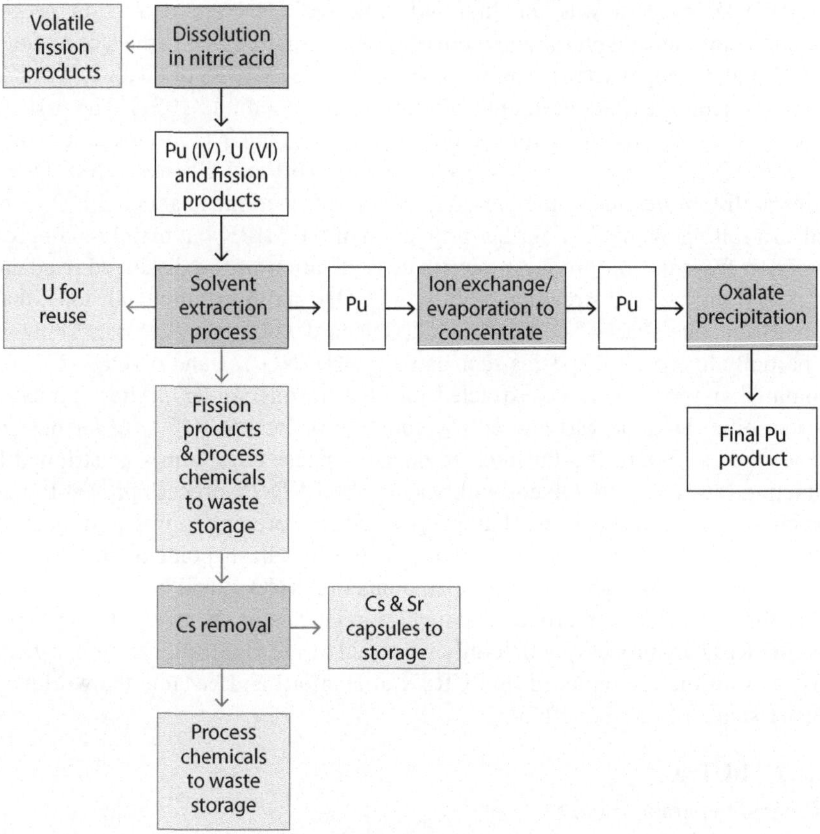

FIGURE 4.4 The PUREX process as implemented at the Hanford Site.

extractant (Lanham and Runion 1949). To date, it has been implemented in some variation by the United States, Russia, France, the United Kingdom, Germany, India, Belgium, Italy, and Japan. A simplified version of the original application of PUREX can be seen in Figure 4.4, with a more detailed view of the process as implemented at the Thermal Oxide Reprocessing Plant (THORP) in the UK seen in Figure 4.5.

After fuel rods have been dissolved in aqueous nitric acid solution, the organic phase quantitatively extracts the plutonium (as Pu^{IV}) and uranium. The plutonium is then selectively stripped from the organic phase by valence adjustment from Pu^{IV} to Pu^{III}, which lowers the D-value. Valence adjustment is typically accomplished by ferrous sulfamate (Fe^{II}), hydroxylamine nitrate, or U^{IV}. The use of Fe^{II} is disadvantageous due to the addition of iron to the nuclear waste (McKibben, Chostner, and Orebaugh 1983). Tetravalent uranium is advantageous due to being a stronger reducing agent than either hydroxylamine nitrate or Fe^{II}, and the inherent presence of U in the process stream prevents creation of additional waste. The disadvantage of U^{IV}

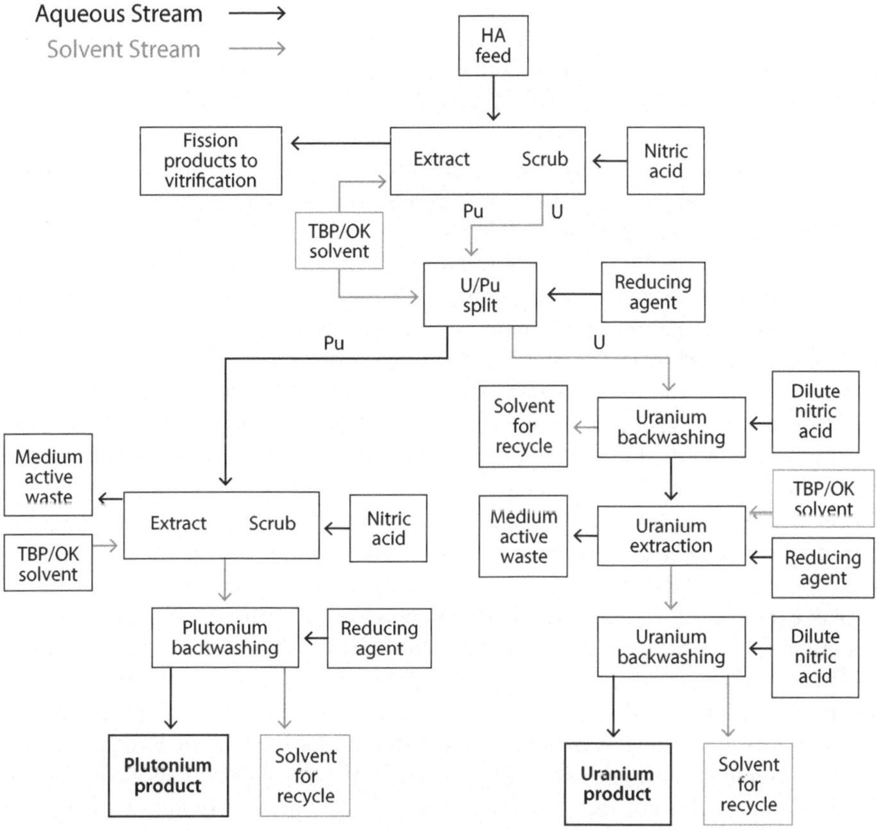

FIGURE 4.5 THORP (UK) implementation of the PUREX process.

is that it has a tendency toward oxidation to UO_2^{2+} in nitric acid solutions; thus, U^{IV} must be generated and stabilized. This is typically achieved either through electrochemical means or reduction by H_2 with a platinum (Pt) catalyst. Regardless of the selected reducing agent, a nitrous acid scavenger such as $N_2H_5^+$ is necessary to prevent oxidation of Pu^{III} and/or U^{IV}. The uranium can then be selectively stripped with dilute nitric acid, leaving a barren organic phase that can be reutilized after regeneration. During the stripping steps, the concentration of nitric acid in the aqueous phase needs to be carefully controlled as the D-value of both Pu and U depend heavily on nitric acid concentration (Tachimori et al. 1983).

While the PUREX process has been broadly adopted for reprocessing and is the worldwide industry standard, processing complications do exist that must be worked through including interfering species and third phase formation. The most common species that interfere with PUREX are Zr, Niobium (Nb), Np, and Tc (HAPO 1955). Neptunium is unextractable as Np^V but extracts as either Np^{IV} or Np^{VI}. While U^{IV} is a sufficiently strong reductant to reduce Np to the IV oxidation state, the kinetics of

$Np^{V} \rightarrow Np^{IV}$ reduction are slow while the reduction of $Np^{VI} \rightarrow Np^{V}$ is fast. This has the potential to result in mixed Np oxidation states, and as a consequence of the unextractable nature of Np^{V}, the routing of Np in PUREX flowsheets can become split between product streams (R.J. Taylor et al. 2013). Additionally, coextraction of Np with U and Pu can affect the Pu oxidation state, thus complicating the routing of Pu. Recent focus on routing Np to a stream other than waste has resulted in the Advanced PUREX process discussed later in Section 4.2.1.6. Tc exists in nitric acid solution as TcO_4^- and, as an anionic species, can coextract as an ion pair with either the positively charged U or Pu. Thus, similar to Np, Tc can become split between product streams without careful control (Dileep et al. 2008). Tc is problematic not only due to contamination of pure Pu and U, but also due to Tc acting as a catalyst for the decomposition of $N_2H_5^+$ and the oxidation of U^{IV}, which is necessary to stabilize Pu in the lower oxidation states. Zirconium (Zr) and Nb are reasonably extractable at high acid concentrations, but relatively unextractable at low acid concentrations due to the formation of hydrolysis and polymeric compounds. However, this leads to the possible formation of third phase and crud due to these species. Under high acid, if Zr and Nb are extracted into TBP and stripped into the final products, they can be cleaned from the final product by adsorption of the Zr and Nb onto silica gel (HAPO 1955).

An example of the historic PUREX process as implemented at Hanford, Washington, can be seen in Figure 4.4. This simple scheme shows the recapture of U for further use in a reactor, the separation of Pu for processing into nuclear weapons, and the remainder of the fission, activation, and corrosion products disposed of as waste. The Savannah River Site (SRS) modified the PUREX flowsheet with two separate designs. One design followed the same basic flowsheet process as Hanford but did not recapture the U, which was instead routed with the fission products. A second SRS flowsheet called the H-area modified (HM) process was capable of processing enriched uranium fuel. Uranium, Pu, and Np were harvested in three separate streams, where each stream underwent further purification through multistage countercurrent solvent extraction in mixer-settlers. Using 7.5 vol % of TBP in an aliphatic hydrocarbon mixture, fuel containing 1.1%–94% U^{235} could be reprocessed (Shook 1988). The HM PUREX suffers from the same issues as the PUREX process, with solvent degradation proving problematic. A more detailed flowsheet of PUREX as implemented in the United Kingdom at THORP can be seen in Figure 4.5. The PUREX plant at THORP was designed to produce PuO_2 and U_3O_8 products that could then be processed into mixed oxide (MOX) fuel as needed.

The PUREX process has formed the basis of many subsequent extraction schemes designed to more readily produce MOX fuel or more efficiently manage nuclear waste. Each country has taken a slightly different approach depending on their individual long-term nuclear vision, but all extraction schemes retain the TBP extractant in hydrocarbon (generally kerosene) diluent used in PUREX. The resulting processes include Advanced PUREX, UREX, NPEX, TRUEX, COEX, and CoDCon. An example flowsheet of the PUREX process as implemented at THORP in the United Kingdom is shown in Figure 4.5 (Denniss and Phillips 1990).

4.2.1.4 COEX

Desired Separation: Combined Pu and U product for increased proliferation resistance during recycling; bulk U
Developed: France, 2000s

Coextraction, or COEX™, is a French process developed in a collaboration between the Commissariat à l'énergie atomique et aux énergies alternatives (CEA) and Areva. COEX™ is a modification of the PUREX flowsheet designed to provide increased criticality safety and weapons proliferation resistance by ensuring that no pure plutonium is separated from used nuclear fuel at any point in the process (Drain et al. 2008). Two product streams are created, one containing pure uranium and the second containing a mixture of uranium and plutonium by allowing bleed over of U^{IV} into the plutonium product. The two streams can subsequently be blended to achieve the desired ratio of plutonium to uranium needed given the isotopic ratios present in the fuel being recycled. The design of the COEX™ process is modular, which allows for individual adaptation depending on a given host country's nuclear policies. For example, with slight modifications to this flowsheet, Np could be routed as its own separate waste, become part of the pure uranium stream, or become part of the plutonium/uranium product.

A major advantage of COEX™ is that it is relatively mature and well-tested; this means that there is a possible economic advantage in implementing this process over competing recycling technologies and a once-through fuel cycle (Peters et al. 2006).

4.2.1.5 CoDCon

Desired Separation: Combined Pu and U product; bulk U
Developed: The United States, 2016

CoDeContamination (CoDCon) is a laboratory-scale investigation sponsored by the U.S. Department of Energy's Office of Nuclear Energy looking into a modified PUREX flowsheet to obtain a mixed U/Pu product (Lumetta et al. 2018). The effort is not specifically directed at applying such a scheme industrially at this time; rather, the goal is to determine the uncertainty with which a controlled ratio of U:Pu can be achieved in the mixed U and Pu product (Lumetta et al. 2018; Lumetta et al. 2019). The practical differentiation between CoDCon and other flowsheets designed to produce a mixed U/Pu product is that the U/Pu ratio is controlled to a specific target value by adjusting one particular flowsheet variable, specifically the flowrate of the barren TBP during the Pu stripping step. The adjustments to the flowsheet parameter are made based on real-time online monitoring of the U/Pu ratio in the product stream, which is based on optical spectroscopic techniques coupled with chemometric modeling.

4.2.1.6 Advanced PUREX

Desired Separation: Np routing in the PUREX process
Developed: United Kingdom, 1990s

Advanced PUREX is an evolution of the PUREX process developed in the United Kingdom and focuses on precisely controlling the routing of Np (R.J. Taylor, Denniss, and Wallwork 1997). During the first extraction cycle in the original PUREX processes, the Np is distributed between the aqueous raffinate, the U product, and the Pu product. While isolation of Np from the U and Pu streams was historically pursued (Regalbuto 2011), more recent interest in closed fuel cycles seeks to recover Np as a Np/U/Pu mixed product to be burnt as MOX fuel (Lumetta and Nash 2011; R. Taylor 2015). Routing Np with the Pu also has the added benefit of decreasing the attractiveness level of the Pu for nuclear weapons. However, routing of Np is complicated due to the complex and difficult to manage $Np^{IV/V/VI}$ redox chemistry. While $Np^{IV/VI}$ is readily extracted, Np^{V} favors the aqueous phase.

The splitting of Np during a PUREX-type flowsheet can occur during either the loading or stripping steps. Although low concentrations of HNO_2 can catalyze the oxidation of Np^{V}, higher concentrations reduce Np^{VI} to Np^{V}. Consequently, reaching equilibrium between Np^{V} and Np^{VI} is slow. Slow chemical kinetics are a design challenge if short residence time contacting equipment, advantageous for preventing solvent degradation, is used. Experimental evidence shows that raising the concentration of HNO_3 in the feed and the temperature of the feed during extraction increases the steady state concentration of Np^{VI}, thus facilitating complete extraction during the loading of TBP (R.J. Taylor et al. 2013). The current reductant of choice for Pu stripping, U^{IV}, is sufficient to reduce Np^{VI} to Np^{IV}; however, while the initial reduction of Np^{VI} is fast, the reduction of Np^{V} is slow due to the slow chemical step of breaking the neptunyl bonds in the $O = Np^{V} = O$ moiety. Thus, to route Np to the product, careful consideration must be given to reaction kinetics and contact time.

4.2.1.7 GANEX

Desired Separation: Removal of bulk uranium from high-level waste; transuranics as a group

Developed: CEA, Marcoule, France, 2000s

The Group ActiNide Extraction (GANEX) process reprocesses all transuranic elements (Np, Pu, Am, and Cm) together, without separation (Aneheim 2012). As opposed to other separations processes that recover minor actinides (such as the DIAMEX-SANEX process; see Section 4.2.2.6), GANEX looks to recover the other minor actinides with Pu recovery in conjunction with U (Brown et al. 2012). The extraction process is designed to process thermal MOX and fast reactor fuels and has the capability of handling 5–20 wt % concentrations of heavy metals. The GANEX process first removes uranium from the dissolved fuel (GANEX-1, an initial process) leaving the transuranic elements and lanthanides in the aqueous raffinate. This is followed by a second extraction cycle in which Np, Pu, Am, and Cm (transuranics), as well as any remaining U, are removed together; a final lanthanide strip comes immediately after (Aneheim 2012; R. Taylor et al. 2016). Using GANEX in place of processes such as PUREX is beneficial in that there are fewer extraction steps (Aneheim 2012). The ultimate goal of the GANEX process is to function as a once-through cycle with the simultaneous extraction of actinides, which would benefit both waste management and reprocessing safety (R. Taylor et al. 2016).

To begin the GANEX process, GANEX-1 begins by removing the bulk U content. In this process, GANEX-1 extracts U with N, N-di-(ethyl-2-hexyl)isobutyramide (DEHiBA), which has a high capacity for U (Miguirditchian et al. 2008). For the second step, the extractant N, N,N',N'-tetraoctyl diglycolamide (TODGA) was chosen because of its extractant ability exhibited in other separations processes (Ansari et al. 2012). However, due to the high capacity of Pu required in the extraction, TODGA alone proved inadequate due to its low Pu capacity (R. Taylor et al. 2016). Through laboratory testing, a mixture of 0.2 M TODGA in combination with 0.5 M N, N'-dimethyl-N, N'-octyl-hexylethoxy-malonamide (DMDOHEMA) was eventually selected; the presence of TODGA could not be avoided. This TODGA/DMDOHEMA combination solvent provided reasonable loading capacity (acceptable at ≤6 M HNO_3), physical properties, and ease of stripping extracted metals (R. Taylor et al. 2016). Laboratory-scale results measured and confirmed that the solvent also features efficient distribution ratios of actinides and lanthanides in the organic phase in the GANEX flowsheet (R. Taylor et al. 2016).

4.2.1.8 UREX

Desired Separation: Tc, U, Cs, Sr
Developed: USA, 2000s

As part of the effort to enhance the proliferation resistance of the PUREX process, the URanium EXtraction (UREX) process has been heavily researched in the USA, France, and United Kingdom. While each country is looking at similar methodologies, they are examining different pathways due to differing political climates (Laidler et al. 2001). The goal of the UREX process is U and Tc extraction. The evolution of the UREX process illustrates the modularity of flowsheet development and design. With time, the process has evolved to include the separation of various other species, resulting in the UREX+ processes (UREX+1, UREX+1a, UREX+1b, UREX+2, UREX+2a, UREX+3, UREX+3a, UREX+4, and UREX+4a) (Regalbuto 2011). Originally, PUREX was developed and used to separate plutonium from used fuel, with uranium also captured as a product. While similar to the PUREX process, the UREX process instead prevents plutonium extraction using the addition of acetohydroxamic acid (AHA), which can complex and reduce Pu, reducing its extraction into the organic phase; Pu would instead follow the remaining transuranics (Pereira, Vandegrift, and Swanson 2002). Low acidity also allowed for extraction of Tc with U (Pereira, Vandegrift, and Swanson 2002). This process focused on handling all minor actinides and fission products for reuse as fuel and for waste management purposes (Vandegrift et al. 2004).

Studies of the UREX process began in the USA in 2003 and fell under the Advanced Fuel Cycle Initiative (AFCI) and the Global Nuclear Energy Partnership (GNEP) (Vandegrift et al. 2004). Under the U.S. Department of Energy's Office of Nuclear Energy, this initiative focused on researching technology for treating and processing used nuclear fuel (Pereira et al. 2007) with the goal of showing that used fuel could be effectively separated solely by aqueous processing in preparation for final disposition (Regalbuto 2011). The UREX+ process was designed to generate separate U and Tc product streams, a lanthanide stream, and a mixed fission product stream (Regalbuto 2011).

While the UREX+ process has multiple iterations, all effectively capture U, Tc, and Cs/Sr as the first three product streams (Regalbuto 2011). Subsequent product streams vary depending on the process iteration. For example, the UREX+1a included both a Cs/Sr product stream and a Pu, Np, Am, and Cm product stream; the UREX+3 (investigated 2003–2007) separated Cs and Sr, Am and Cm, and Pu and Np (Regalbuto 2011). Due to the number of process iterations, this text will focus on only the UREX1+a process, completed near the end of the AFCI/GNEP initiative.

The UREX+1a flowsheet consists of: recovery of Tc and U (UREX), ion exchange to separate U and Tc, recovery of Cs and Sr using chlorinated cobalt dicarbollide (CCD) for the extraction of Cs and polytethylene glycol (PEG) for the synergistic extraction of Sr in a phenyltrifluoromethyl sulfone diluent (Law et al. 2005), separation of transition metal fission products from TRU elements and lanthanide fission products by the TRUEX process, and TALSPEAK (see Section 4.2.2.1) for recovery of rare earth fission products and Pu, Np, Am, and Cm (see Section 4.2.2.1) (Pereira et al. 2007). Following dissolution of the fuel, U and Tc are captured using the UREX process, but in this process, Pu^{IV} and Np^{V} (from reduced Np^{VI}) extraction is prevented using a complexant/reductant (AHA) in the scrub feed (Regalbuto 2011). Two approaches were pursued in the separation of U and Tc. In the first instance, dilute nitric acid stripped the loaded solvent of U and Tc before Tc was isolated from the U/Tc strip by anion exchange. Technetium was stripped into a higher concentrated nitric acid, while the U was stripped into dilute nitric acid (Pereira, Vandegrift, and Swanson 2002). The second approach stripped U and Tc together from the loaded solvent with dilute nitric acid (Regalbuto 2011). Following this, Tc is separated by ion exchange. Results indicated that Tc recovery by ion exchange provided better recovery. The feed into the CCD-PEG flowsheet is the raffinate from the UREX flowsheet. The CCD-PEG in phenyltrifluoromethyl sulfone diluent extracts Cs and Sr, while nitric acid scrubs transuranic and lanthanide elements from the solvent. Ammonium carbonate salt and a complexing agent strip alkali and alkaline-earth cations. Once the aqueous stream of CCD-PEG in diluent was adjusted by concentrated nitric acid, the aqueous stream becomes feed to the TRUEX process, where the solvent feed carbamoylphosphine oxide (CMPO) and TBP diluted in *n*-dodecane extracts non-rare earth fission products (Schroeder, Attrep, and Marrero 2001). The product of the TRUEX flowsheet was subsequently used as the feed for purification with Cyanex 301 (a dialkyldithiophosphinic acid, bis(2,4,4-trimethylpentyl)dithiophosphinic acid) or a TALSPEAK-based flowsheet, where the Pu/Np are separated (Pereira et al. 2007). This process was to separate trivalent actinides from the TRUEX product. The solvent is scrubbed with the acid form of the weak complexant and stripped using an ammonium salt complexant to ultimately obtain minor actinide fission products.

4.2.1.9 UNEX

Desired Separation: Cs, Sr, Actinides, Lanthanides

Developed: Idaho National Lab, USA, and the Khlopin Radium Institute, Russia, 1990s–2000s

The UNiversal EXtraction process, UNEX, was developed by scientists at the Khlopin Radium Institute (RI) and Idaho National Laboratory (INL). The goal of the UNEX process was to simplify radionuclide separations by recovering all species of interest in one extraction cycle as opposed to using various flowsheet operations (Herbst et al. 2002; Romanovskiy et al. 2011), resulting in significant cost savings (Luther et al. 2006). UNEX uses CCD, PEG, and diphenyl-N-N-dibutylcarbamoyl-phosphinoxide, which are all commercial reagents. Research of the process and chemistry quickly revealed that concentrations of the extractants in the mixture can affect the radionuclide extraction (Romanovskiy et al. 2011). Thus, effort was taken to determine the optimal composition of extractant, thus maximizing the extraction while maintaining solvent stability (Romanovskiy et al. 2011).

The UNEX flowsheet extracts Cs, Sr, actinides, and lanthanides. Cesium is extracted with a liquid–liquid phase cation exchange that uses CCD dissolved in PEG (Dozol, Dozol, and Macias 2000; Luther et al. 2006). The CCD acts to stabilize metal cation extraction by acting as a counterion in the process. When in strong acidic media, the CCD-PEG complex increases recovery of Sr. When using CCD-PEG in combination with bifunctional neutral organophosphorus compounds (BNOPC), high extraction ability is illustrated, specifically with diphosphine dioxides or with CMPOs. The CMPO compounds are generally preferred due to the complex synthesis of diphosphine dioxides, their poor solubility in polar diluent, and the success of CMPO compounds in the TRUEX (see Section 4.2.1.10) process (Romanovskiy et al. 2011). Ultimately, through various iterations and trial processes, UNEX's high chemical and radiation resistance, increased safety during operating conditions due to fire and explosion resistance, and reduced volume of high-level waste (HLW) proved its success (Romanovskiy et al. 2011).

4.2.1.10 TRUEX

Desired Separation: Pu, U, Am, and Cm from used fuel waste to reduce the high alpha activity

Developed: Argonne National Laboratory, 1995

The TRansUranium Extraction (TRUEX) process utilizes carbamoylmethylphosphine oxide to remove all actinides (non-selectively) from chloride and nitric acid waste (Horwitz and Schulz 1990). TRUEX acts much like the PUREX process, with the addition of a second compound for extraction. Having the second agent could remove transuranic elements and reduce the TRU activity of the waste. Specifically, CMPO is the TRUEX solvent and TBP is used as a phase modifier; CMPO and TBP are diluted in *n*-dodecane (Regalbuto 2011). The CMPO is selective for trivalent actinides and lanthanides. Through this separation, curium, americium, and lanthanide elements are extracted. Three scrub sections occur in the following order: (1) oxalic acid is used to remove impurities, (2) oxalic acid is scrubbed using nitric acid of reasonable concentration, and (3) acid concentration is lowered by diluting the third scrub with less concentrated nitric acid for stripping (Regalbuto 2011). As part of a campaign for developing UREX+1a, TRUEX was demonstrated using irradiated nuclear fuel and shown to be a viable option (Pereira et al. 2007).

4.2.2 Minor Actinide Recovery Processes

The separation of the lanthanides from the minor actinides is possibly the most challenging goal of solvent extraction for the nuclear fuel cycle. The minor actinides and the lanthanides both exist as trivalent cations in aqueous solution and have similar ionic radii, causing them to have similar chemical behavior. The separation is accomplished by leveraging the slightly harder character of the lanthanide 4*f* valence orbitals (i.e., hard Lewis acid) as compared to the slightly softer actinide 5*f* orbitals.

4.2.2.1 TALSPEAK

Desired Separation: Separation of actinides from lanthanides from a PUREX raffinate

Developed: Oak Ridge National Laboratory, USA, 1960s

The Trivalent Actinide Lanthanide Separation with Phosphorus—Reagent Extraction from Aqueous Komplexes, TALSPEAK, was developed in the 1960s at Oak Ridge National Laboratory to separate the lanthanides from the trivalent minor actinides after the PUREX process (Weaver and Kappelmann 1964, 1968). TALSPEAK has been widely studied and has well-understood thermodynamics and mechanism for isolating the minor actinides from the lanthanides (Martell, Smith, and Motekaitis 2004; Nilsson and Nash 2007; Nash 2015). Consequently, TALSPEAK has provided inspiration for many subsequent separation methods, including Advanced TALSPEAK and Reverse TALSPEAK, among others.

TALSPEAK utilizes three principle components: an acidic phosphorus extractant, di-(ethyl-2-hexyl)-phosphoric acid (HDEHP), to extract the lanthanides into the organic phase; a weakly coordinating buffer to control aqueous phase pH; and a stronger complexing actinide holdback reagent to prevent extraction of the minor actinides. The original formulation of TALSPEAK utilized lactate as the buffer and diethylenetriaminepentaacetic acid (DTPA) as the holdback reagent, though alternatives for both the buffer and holdback reagent have been extensively studied. The separation occurs due to a very slight difference in affinity of DTPA for the trivalent actinides versus the lanthanides. Thus, the lanthanides are extracted into the organic phase, while the minor actinides are left in the aqueous phase. The lanthanides can then be stripped from the organic phase for subsequent concentration and processing of the metals and recycling of the solvent. While this process provides adequate separation, several issues arise from the engineering standpoint. Namely, a large change in the aqueous phase is necessary; the PUREX raffinate that would be a typical feed for this process is >1 M HNO_3, but TALSPEAK operates in a pH range from 3.0 to 4.0. The second major engineering disadvantage to this process is its high sensitivity to pH; small changes in aqueous phase pH cause large changes in extraction D-values (Figure 4.6), which further complicates the engineering process (Nilsson and Nash 2007). Additionally, phase transfer kinetics can be slow, requiring long contact times and consequently higher radiolytic degradation of the solvent. However, for minor actinide separation flowsheets, TALSPEAK has a relatively high technology readiness level as it has been successfully demonstrated with irradiated nuclear fuel (Pereira et al. 2007).

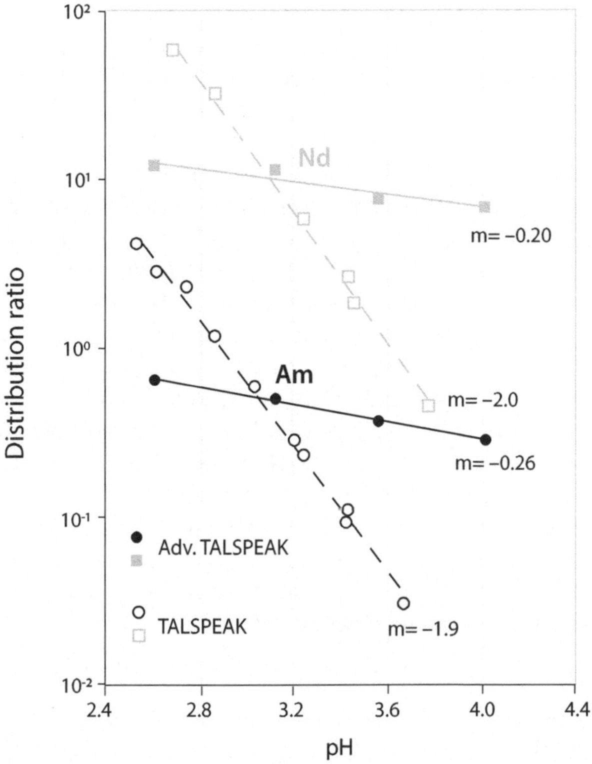

FIGURE 4.6 pH dependence of the TALSPEAK and Advanced TALSPEAK systems for the separation of Am and Neodymium (Nd). (Lumetta, G.J., et al. *Solvent Extr. Ion Exc.*, 33, 211–233, 2015.)

4.2.2.2 Advanced TALSPEAK

Desired Separation: Separation of actinides and lanthanides from a PUREX raffinate

Developed: USA, 2010s

While studying the extraction mechanism of TALSPEAK, and more specifically the role of extracted lactate, a weaker extractant, HEH[EHP], was used instead of HDEHP to suppress lactate extraction (Grimes et al. 2012). Observations from this experiment led to the separation factors of the lanthanides and minor actinides being tested with HEH[EHP] (Braley et al. 2012; Braley, Grimes, and Nash 2012). Due to HEH[EHP] being a weaker extractant than HDEHP, it is necessary to use a weaker complexing holdback reagent in the aqueous phase. Several holdback reagents have been tested, with HEDTA being used in the current version. The main advantage to this system is that it provides much more consistent and predictable D-values in the pH range of 2.0–4.5 (Figure 4.6) (Braley, Grimes, and Nash 2012). While this is still a relatively immature solvent extraction scheme, it has been tested with both 1-cm and 2-cm centrifugal contactors with very promising results (Lumetta et al. 2017).

4.2.2.3 Reverse TALSPEAK

Desired Separation: Actinides and lanthanides in a PUREX raffinate
Developed: Oak Ridge National Laboratory, USA, 1970s

Reverse TALSPEAK is an early modification (1979) of the TALSPEAK flowsheet (Svantesson et al. 1979). The same extractant (HDEHP) and actinide holdback reagent (DTPA) used in TALSPEAK are employed, although the actinide holdback reagent is omitted during the loading step to coextract the lanthanides and actinides simultaneously. Separation is then achieved by selectively stripping the minor actinides into an aqueous phase containing a holdback reagent. There are two main advantages to a Reverse TALSPEAK flowsheet over TALSPEAK. The first is that the initial aqueous feed that results from a PUREX or UREX process does not need a dramatic adjustment in acidity. Previous studies have dropped the nitric acid concentration to 0.1 M HNO_3 through extraction of HNO_3 with TBP, creating a pathway for more facile pH control in subsequent steps (Nilsson and Nash 2007). The narrow pH requirements of TALSPEAK can then be maintained in the stripping solution rather than the feed solution, thus simplifying the process from an engineering standpoint. The second advantage is that potential problematic elements such as zirconium or molybdenum have already been removed prior to lanthanide/minor actinide separation, and thus are less likely to affect the fragile separation step.

4.2.2.4 ALSEP

Desired Separation: Am/Cm from lanthanides from an organic TBP raffinate, such as PUREX
Developed: U.S. Department of Energy (DOE), January 2013

The Actinide Lanthanide SEparation Process (ALSEP) is a relatively new process that utilizes components of both Advanced TALSPEAK and Reverse TALSPEAK (Gelis and Lumetta 2014). The ALSEP process is designed to reduce the number of steps and directly utilize the aqueous raffinate of a PUREX-type process, with the aqueous phase consisting of ~3 M HNO_3. The process utilizes a dual extractant system consisting of the phosphonic acidic extractant HEH[EHP] and a neutral diglycolamide extractant (T2EDGA). The flowsheet operates in a Reverse TALSPEAK-type fashion, where the lanthanides and minor actinides are coextracted and then selectively stripped. The intended design is for the neutral diglycolamide to extract the metals when the aqueous phase is high in nitric acid and consequently nitrate. The phosphonic acid is designed to retain the lanthanides when the aqueous phase acidity is decreased in the presence of the polyaminocarboxylate actinide stripping agent. In practice, it has been observed that the HEH[EHP] and diglycolamide act synergistically, increasing the organic phase loading over either individual extractant at all likely aqueous phase acidities (Holfeltz et al. in preparation). This process has been tested with simulants, as well as with a test batch of dissolved irradiated fuel. There were intermittent observations of crud formation during complex simulant testing, something that was completely absent during fuel testing. An important note for these tests is that the purity of HEH[EHP] has a large impact on the minor actinide stripping D-value and consequently a large impact on the separation factor (Holfeltz et al. 2018).

4.2.2.5 DIAMEX

Desired Separation: Coextraction of trivalent actinides and lanthanides from PUREX raffinate

Developed: Commissariat à l'énergie atomique et aux énergies alternatives (CEA), France, 1993

The DIAMide EXtraction (DIAMEX) process was developed to reduce the volume of secondary waste from the high-activity raffinate that resulted from the PUREX process (Heres et al. 2008). This arose as an alternative to the TRUEX process and focused on applying the CHON principle (Pereira, Vandegrift, and Swanson 2002). Like TRUEX, the DIAMEX process was designed to coextract transuranic elements and lanthanides from the high-activity PUREX raffinate.

DIAMEX uses a malonamide to coextract the actinides and lanthanides from the raffinate that remains from the PUREX process (Madic et al. 2002). The extractant, DMDOHEMA, showed the most promising results in the extraction flowsheet (Madic et al. 2002; C. Hill 2011). Testing of the flowsheet using DMDOHEMA with high-activity raffinate demonstrated an An^{III} and Ln^{III} recovery of >99.9%, proving the effectiveness of the separation (C. Hill 2011). While comparable, the TRUEX process does have an advantage over the DIAMEX process because it efficiently extracts species at lower nitric acid concentrations; typical raffinate streams into the flowsheet, such as raffinate from UREX, have low nitric acid concentrations.

4.2.2.6 DIAMEX-SANEX

Desired Separation: Am and Cm from DIAMEX raffinate

Developed: Various European research centers, including ITU (Germany), *Jülich* (Germany), CEA (France), and Chalmers University (Sweden), early 2000

The DIAMEX-Selective ActiNide Extraction (SANEX) process works to separate lanthanides (III) and actinides (III) from PUREX raffinate in one cycle (Heres et al. 2008). Research performed by the CEA Marcoule made this possible by using the acidic malonamide DMDOHEMA in the DIAMEX process. After extracting Ln^{III} and An^{III} from the original raffinate, An^{III} can be selectively isolated via back extraction (Heres et al. 2008).

Following the initial coextraction of the lanthanides and actinides by DMDOHEMA in the solvent hydrogenated tetrapropylene (HTP), an aqueous poly-aminocarboxylate reagent strips trivalent actinides at a pH greater than 2 (Heres et al. 2008). At this pH, DMDOHEMA cannot effectively keep actinides in the organic phase. While additives can counteract this measure, the impact on the final waste disposition is detrimental. Thus, HDEHP was added in the flow sheet; this extractant was removed from the process following the lanthanide strip step (Heres et al. 2008).

Studies advancing the flowsheet have focused on alternative molecules to HDEHP that were designed and optimized specifically for the separation of the actinides from the lanthanides. Nitrogen-based 6,6′-bis(5,5,8,8-tetramethyl-5,6,7,8-tetrahydro-benzo[1,2,4]triazin-3-yl)-[2,2′] bipyridine ($CyMe_4$-BTBP) has exhibited the most favorable results in terms of hydrolytic and radiolytic stability, extraction and back extraction ability, extraction properties, and kinetics (Magnusson et al. 2009). The solvent mixture was composed of 0.015 M $CyMe_4$-BTBP and 0.25 M DMDOHEMA in octanol (Magnusson et al. 2009). Use of this solvent in laboratory-scale centrifugal contactors recovered Am and Cm at ≥99.9%, with <0.1% of the trivalent lanthanides remaining. The process had low flow rates and slow kinetics, which could limit the separation and loading of the organic phase in cases of waste with high minor actinide content. Further research is needed to overcome these issues with this solvent (Magnusson et al. 2009).

4.2.2.7 POR

Desired Separation: U, Np, Pu, transplutonium elements
Developed: Idaho National Laboratory, USA, and Khlopin Radium Institute, Russia, 1990s

Due to their high solubility in organic diluents, phosphine oxides have an excellent extraction ability with a high actinide recovery rate. The name POR (Phosphine Oxide different-Radical) is derived from the reference to the phosphine oxides used (Romanovskiy 2002). This process was developed to provide actinide recovery (U, Np, Pu, and transplutonium elements (TPEs)) and recovery of rare earth elements (REEs) as well as Cs and Sr and utilized iso-amyldialkylphosphine oxide as an ab extractant (Romanovskiy et al. 2001). Evaluation of flowsheet testing results from a high-level waste POR extraction showed that the actinides, TPEs, and REEs were extracted in combination, followed by a stripping process that sequestered the TPEs from the REEs. Recovery from the real waste, which includes U, Np, Pu, Am, and Tc at various concentrations, was successful, with the lowest recovery being Np at 99.71% and the highest recovery being >99.99% for Pu and Am (Romanovskiy 2002). The POR process can be paired with the Chlorinated Cobalt DICarbollide (ChCoDiC) process (discussed below) to afford further separation for Cs and Sr.

4.2.2.8 ChCoDiC (with and without PEG)

Desired Separation: Trans-plutonium elements and REEs
Developed: Russia, Czech Republic, 1990s

The ChCoDiC process extracts Cs from waste solutions that are acidic (Romanovskiy 2002). PEG can also be added to enable Sr extraction. ChCoDiC strives to separate TPEs and REEs at lower acidities. The process was successfully used at the Mayak Production Facility in Russia for waste treatment until the facility was no longer in operation (Romanovskiy 2002).

Czech scientists studied the process and produced results that showed it was possible to have successful extraction of TPE and REEs from an acidic solution. A solution of 0.3 M ChCoDiC was generated using the polar nitroaromatic solvent m-nitrobenzotrifluoride (MNBTF) with PEG (6% vol). However, safety concerns about using nitrobenzene derivatives have resulted in research focused on identifying an acceptable non-nitroaromatic diluent for the removal of Cs and Sr from acidic waste. The tested batch contacts and flowsheets were processed with and without the addition of PEG (Todd, Law, and Herbst 2001). Without PEG addition, the process achieved 99.998% separation for Cs. Alternatively, with PEG addition, removal of Cs was 99.3%, while Sr removal reached 96.2% (Todd, Law, and Herbst 2001).

4.2.3 Other Applications of Solvent Extraction—Molecular Recognition Applications

4.2.3.1 CSEX

Desired Separation: Cs
Developed: USA, 1990s

Molecular recognition allows for high selectivity in the separation of fission products from aqueous streams using macrocyclic compounds and is an important development for separations that has emerged in recent decades (Collins, Del Cul, and Moyer 2011). Crown ethers initially showed the greatest promise for Cs extraction (CSEX) using molecular recognition. However, more recently, calixarenes have replaced crown ethers as the extractant of choice, as Cs extraction by crown ethers is difficult due to weak extraction strength and only moderate selectivity (Collins, Del Cul, and Moyer 2011). Calixarenes have a dramatically increased binding strength and selectivity for Cs in comparison to crown ether compounds (Collins, Del Cul, and Moyer 2011). Structurally, these compounds resemble crown ethers with the addition of aromatic rings that protrude above and below the crown cavity in a vase-like structure.

The initial CSEX process solvent utilized bis-4,4'(5')[(1-hydroxy-2-ethylhexyl) benzo-18-crown-6 in either undiluted TBP or TBP-paraffinic hydrocarbon mixtures for the extraction. Provided the aqueous phase was at or below approximately 4 M HNO_3, the recovery of Cs was adequate and efficient. However, with the advent of calixarenes, the process was improved to increase the selectivity and binding ability of the extractant. In the CSEX process, bis(*tert*-octylbenzo-crown-6)calix[4]arene (BOBCalixC6) with modifier 1-(1,1,2,2-tetrafluoroethoxy)-3-(4-*tert*-octylphenoxy)-2-propanol (Cs-3) in Isopar® L acts as the solvent (Bonnesen et al. 1998; Delmau 1999).

4.2.3.2 SREX

Desired Separation: Sr
Developed: Argonne National Laboratory, USA, late 1980s–early 1990s

The StRontium Extraction (SREX) process was developed to extract Sr from acidic waste streams by using a crown ether extractant. Initial SREX studies extracted $Sr(NO_3)_2$ from a nitric acid solution using the crown ether *cis*-dicyclohexano-18-crown-6 (*cis*-DCH18C6) (Horwitz, Dietz, and Fisher 1991; Collins, Del Cul, and Moyer 2011). The results of further research encouraged the use of bis(*tert*-butylcyclohexano)-18-crown-6 (DtBuCH18C6), a crown ether with higher lipophilicity, in 1-octanol or Isopar® L modified with TBP (Collins, Del Cul, and Moyer 2011). A dilute HNO_3 strip followed the extraction. Developed at Argonne National Laboratory, the first SREX tests studied extraction on a synthetic dissolved sludge waste solution (Horwitz, Dietz, and Fisher 1991). Barium, Sr, and Tc were successfully removed using the process (Horwitz, Dietz, and Fisher 1991). It is worth noting, however, that the system was not selective for Barium (Ba) over Sr, meaning that the Sr product stream contained Ba (Horwitz, Dietz, and Fisher 1991). Following the SREX process, the TRUEX process was expected to follow, which is why actinide separation was not also investigated in the studies (Horwitz, Dietz, and Fisher 1991).

Following the successful results from individual CSEX and SREX processes (results from INL demonstrated 99.9997% Sr removal from the waste (Law, Herbst, and Todd 2002)), the two processes were combined to allow for simultaneous Sr and Cs extraction. CSEX-SREX uses a solution of dibenzo-18-crown-6 (with 1-hydroxy-2-ethylhexyl substituents on each benzo group) mixed with DtBuC18C6 in Isopar® L (a synthetic isoparaffinic hydrocarbon solvent) and modified with TBP and lauryl nitrile. This combined process was run using an INL waste simulant in a countercurrent test (Collins, Del Cul, and Moyer 2011).

4.2.3.3 CSSX

Desired Separation: Cs
Developed: Oak Ridge National Laboratory, USA, late 1990s–early 2000s

The Caustic-Side Solvent Extraction (CSSX) process uses calixarenes to remove Cs from highly alkaline sodium nitrate reprocessing wastes, like those found in the USA at SRS and the Hanford site (Delmau et al. 2010). CSSX uses 0.007 M calix[4]arene-bis(tert-octylbenzocrown-6) (BOBCalixC6), 0.75 M 1-(2,2,3,3,-tetrafluoropropoxy)-3-(4-sec-butylphenoxy)-2-propanol (Cs-7SB modifier), and 0.003 M tri-n-octylamine in Isopar® L (Delmau et al. 2010; Collins, Del Cul, and Moyer 2011; C. Hill 2011). Using this methodology, an almost pure $CsNO_3$ product stream was recovered following a scrub and strip of the loaded solvent with 0.05 M HNO_3 and 0.001 M HNO_3 (Delmau et al. 2010). Figure 4.7 illustrates the flow sheet for the CSSX operation (Leonard 2010). Studies were also performed to identify alternative CSSX solvents to determine if extraction efficiency could be improved. Using the extractant BEHBCalixC6 at 0.020 M, a stripping method with a scrub of 0.1 M NaOH, and finally a strip with 0.010 M boric acid, the Cs distribution ratios were tripled with few impacts on the flowsheet design (Delmau et al. 2010).

The CSSX process has been in operation at the SRS since 2008 at the Modular CSSX Unit (MCU) (Collins, Del Cul, and Moyer 2011). The MCU is a pilot facility that has treated in excess of seven million gallons of tank waste from low-level SRS salt waste (Collins, Del Cul, and Moyer 2011). The CSSX system is slated for implementation at the SRS Salt Waste Processing Facility (SWFP) (Collins, Del Cul, and Moyer 2011). While the infrastructure was complete as of 2016, the current focus is on plant systems and workforce training to prepare for the start of operations.

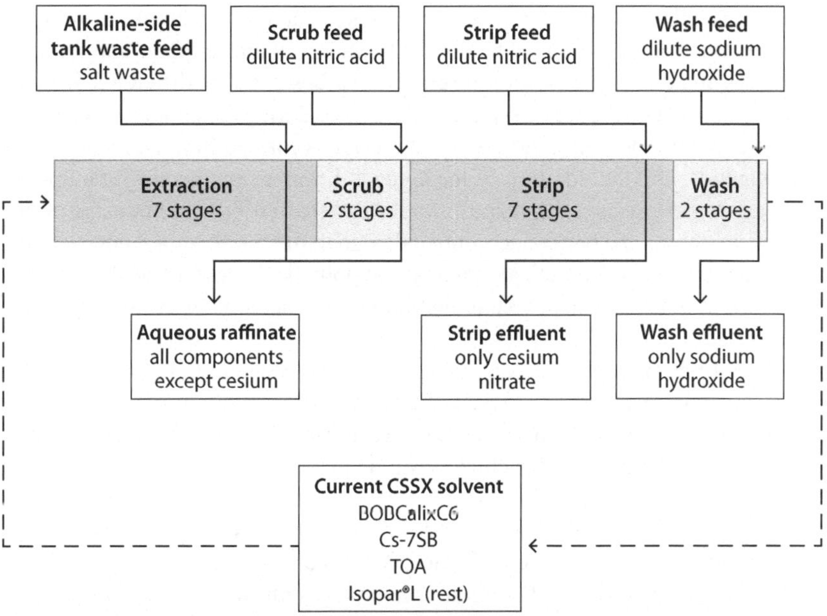

FIGURE 4.7 The CSSX flow sheet. (Leonard, R. A. *Initial Flowsheet for the Next Generation CSSX Process in the MCU. Report for FY 2010.* (Argonne National Laboratory), 2010. https://publications.anl.gov/anlpubs/2012/11/72864.pdf.)

4.2.3.4 SRTALK

Desired Separation: Tc, Cs, Sr
Developed: Oak Ridge National Laboratory, USA, 1990s

The SRTALK extraction process, invented at Oak Ridge National Laboratory, was initially focused on the extraction of Tc, specifically in the form of pertechnetate, for used nuclear fuel wastes in the USA. With development, the extraction has been expanded to also include Cs and Sr extraction. As with SREX, CSEX, and CSSX, the extraction looks at alkaline-side removal and recovery of the species of interest. Using a crown ether extractant, bis-4,4′(5′)[(*tert*-butyl)cyclohexane-18-crown-6, in a 1:1 mixture of TBP and Isopar® L, the extraction separates and concentrates technetium to reduce the waste for downstream operations (Leonard et al. 1997). The process was developed for alkaline wastes, meaning there is no requirement to acidify the used fuel solution prior to extraction (Leonard et al. 1997). Successful results indicated that the extraction would work adequately on larger, industrial-scale sizes and equipment (Leonard et al. 1997).

4.3 EQUIPMENT

The equipment used in solvent extraction processes is designed to operate continuously, allowing for high throughput. The contacting equipment receives two immiscible liquids as input streams, thoroughly mixes the two immiscible liquids to form an emulsion, thus allowing phase transfer of metal ions, and finally breaks the emulsion to allow the immiscible liquids to be directed away from one another. The separation of the phases and disengagement of the emulsion relies upon the difference in density of the two phases. The degree of separation achieved during the contact is determined by multiple factors including the size of the droplets formed in the emulsion, phase transfer kinetics, and the duration of the contact. From an engineering standpoint, it is necessary to design a system that allows for the greatest possible transfer of metal ions between phases. To better accomplish this goal, the metric of a theoretical stage has been developed. A theoretical stage is a hypothetical segment of the extraction equipment that reaches a steady state distribution of metal between the organic and aqueous phases.

The three types of contacting equipment that have been employed for the nuclear fuel cycle are columns (pulsed and packed), mixer settlers, and centrifugal contactors. Mixer settlers and centrifugal contactors are intrinsically single-stage processes, whereas pulsed columns ideally achieve a single theoretical stage between each plate and packed columns have an intermediate density of theoretical separation stages. Ultimately, the number of theoretical stages for columns is determined by the height of the column and of the theoretical stage. Multiple stages achieve higher SF than can otherwise be achieved with a single stage that is allowed to reach equilibrium. Each piece of equipment has inherent advantages and disadvantages depending on the specific application. It is common to have multiple types of extracting equipment operate alongside one another in the same plant. Table 4.2 and the following paragraphs explain each method's advantages for PUREX-type flowsheets and the basic mechanisms by which they operate.

TABLE 4.2
A Comparison of Separation Equipment

	Centrifugal Annular Contactor	**Pulse Column**	**Packed Column**	**Mixer Settler**
Separation stages	One	Ideally, one per plate	Intermediate	One
Residence time	Short	Long	Long	Very Long
Equipment cost	Moderate	Low	Low	Moderate
Reliability	Acceptable	High	High	Acceptable
Third phase tolerance	Low	High	High	Low
Emulsion tolerance	Low	High	High	Medium
Crud tolerance	Low	High	High	Medium
Throughput	High	High	High	Low
Start-up time	Fast	Intermediate	Intermediate	Low
Floor print	Moderate	Low (but tall)	Low (but tall)	High
Scalability	Readily Scalable	Moderate	Moderate	Moderate

Source: Law, J. D., and T. A. Todd, *Liquid-Liquid Extraction Equipment.* (Idaho National Laboratory), 2008. INL/CON-08-15151. https://inldigitallibrary.inl.gov/sites/sti/sti/4156452.pdf.

4.3.1 Columns

Column separations are a mature technology with respect to nuclear reprocessing. Columns have an advantage over annular contactors and mixer settlers because of their few moving parts within the plant's limited access high radiation zones. Because of this, column separation equipment has been a favorite of commercial reprocessing plants and has been employed by all countries that have performed nuclear reprocessing. Two types of columns are employed in the nuclear fuel cycle: packed columns and pulsed columns, which can be viewed on the left-hand side and right-hand side of Figure 4.8, respectively.

4.3.1.1 Packed

Packed columns are the simplest both in terms of operation and design. Long column tubes are packed with a material, such as Raschig Rings, to create a long, torturous pathway for the dispersed phase to pass through. Typically, the dispersed phase is designed to be the phase that is loaded with metal to be transferred into the bulk phase. The dense phase (typically aqueous) is fed into the top of the column, while the less dense phase (typically organic) is fed into the bottom of the column. Due to the continuous contact between the phases throughout the column, the number of stages is determined by the length of the column. With a theoretical stage height of several feet, packed columns are typically very tall. The simplicity of the packed columns is advantageous, as there is little that can cause upset conditions that are difficult to remediate in a high-dose environment.

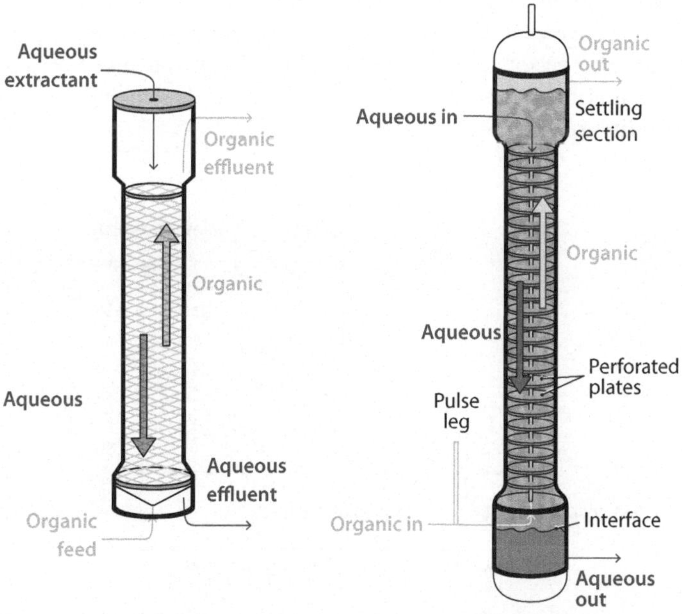

FIGURE 4.8 Schematics of a packed columns (left) and pulsed column (right).

4.3.1.2 Pulsed

Pulsed columns operate by applying mechanical energy to "pulse" liquid in the column up and down. As the fluid passes through plates in the column, it is forced into small droplets forming an emulsion. Flow in a pulsed column is countercurrent, like the flow in a packed column. The most common way to obtain the pulse in the system is to inject pressurized air into the system, which in turn pushes liquid into the column. Pulsing acts to push the solution up and down in the column, increasing the turbulence and kinetics of extraction. Perforated plates in the column act to generate droplets as the dispersed phase is pulsed through the plates. Plate design varies between countries and systems (Law and Todd 2008). While only slightly more complex than packed columns, the pulsing action reduces droplet size, thus increasing contact area and mass transfer rate making pulsed columns more efficient in achieving extraction.

4.3.2 Mixer-Settlers

As the name implies, mixer-settlers contain two compartments for separations (Mayo 2018). Two liquid phases are mixed in the smaller mixer section of the equipment (see left of Figure 4.9). With the use of an impeller driven by a motor or gearbox, the two phases are agitated, and mass transfer occurs (Arm and Phillips 2011). Once mixing is complete, the liquids settle and separate by gravity in the larger settler compartment. A barrier, or weir, separates the mixer and settler compartments (Mayo 2018). One mixer-settler unit provides one stage of extraction (Law and Todd 2008); thus, multiple mixer-settlers are set up in series to achieve desired separation.

In general, mixer-settlers are beneficial for processes that have slow kinetics requiring a long residence time. This, of course, is a decided disadvantage for processes in a high-dose environment as solvent degradation is increased. Additionally, mixer-settlers have a relatively large footprint with a comparatively low throughput, which must be taken into account for large throughput industrial processes. Designing mixer-settlers in critically safe geometries can also be challenging, thus restricting their applicability to fissile materials.

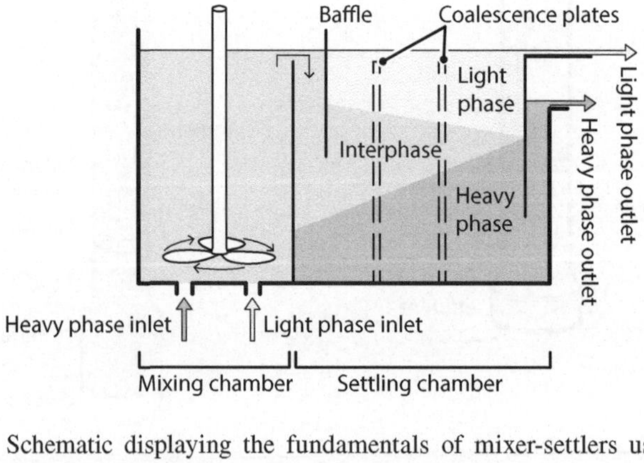

FIGURE 4.9 Schematic displaying the fundamentals of mixer-settlers used for solvent extraction, with the mixing chamber on the left and the settling chamber on the right.

4.3.3 Centrifugal Contactors

Centrifugal contactors for radiochemical solvent extraction were originally developed at Savanah River National Laboratory; subsequent improvements at Argonne National Laboratory produced the annular centrifugal contactors that are widely employed today (Leonard et al. 1980). The basic design of an annular centrifugal contactor can be seen in Figure 4.10. Generally, contactors are arranged as "banks" of several contactors with countercurrent flow. The aqueous and organic phases flow into opposite sides of each individual contactor. Both phases move to the bottom of the contactor in the space between the central rotor and the contactor housing where the two phases mix. After mixing in the contactor's bottom, both phases are drawn into the hollow cavity of the central rotor. As the liquid rises through the rotor, the emulsion is broken, and the phases are separated according to density.

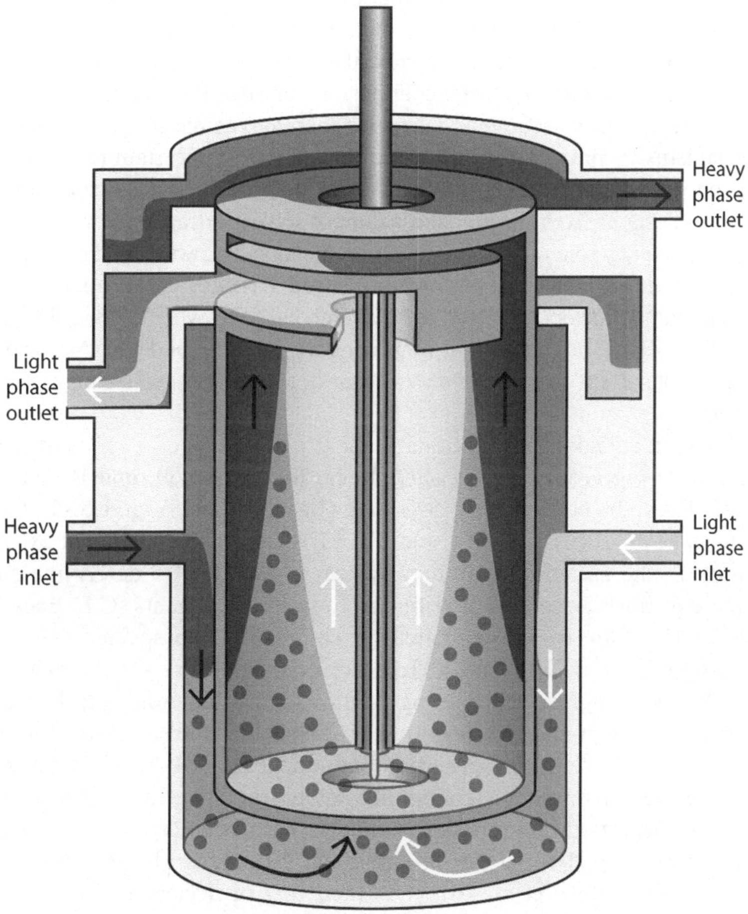

FIGURE 4.10 Cross section of a centrifugal annular contactor. The heavy phase is depicted as dark shading, while the light phase is depicted as light shading.

Several advantages exist for the centrifugal contactors over columns and mixer-settlers, including short residence times, compact size, and scalability. These advantages, particularly the scalability, make annular centrifugal contactors ideal for developing process chemistry at a laboratory scale, which can be readily scaled up to a process scale. The two main disadvantages of centrifugal contactors are (1) the comparatively large number of moving parts by comparison to column separation methods, and (2) that contactors are a much newer technology with less process experience than either columns or mixer-settlers. Contactors are employed extensively at national laboratories and in academia to test flowsheet designs, but commercial reprocessing has been slow to adopt contactors. Even so, utilization is likely to increase in the future.

4.4 SOLVENT EXTRACTION MODELING

Research on novel solvent extraction processes begins at the bench scale by determining kinetic and thermodynamic distribution ratios with batch contacts for select analytes. These bench processes are then scaled up to an industrial-scale continuous flow process utilizing mixer-settlers, columns, or centrifugal contactors. Modeling is generally used to either accurately determine thermodynamic parameters and reaction mechanisms by fitting laboratory-measured data or to ascertain (with predictive models) appropriate flow rates, number of stages, and ideal input concentrations during flowsheet design. Given the modular nature of solvent extraction flowsheet design, modeling plays a key role in coupling the aforementioned flowsheets. For example, if a solvent extraction scheme will perform Advanced TALSPEAK to isolate the minor actinides, input parameters will need to be altered based on whether the prior process was UREX, PUREX, or COEX. Thus, utilizing a predictive model is a powerful tool for the design of solvent extraction flowsheets prior to experimental testing at the laboratory scale.

C.F. Baes, Jr. at Oak Ridge National Laboratory developed a series of thermodynamic modeling programs: SXLSQ (Baes, Jr., McDowell, and Bryan 1987), SXLSQA (Baes, Jr. et al. 1990), SXLSQI (Baes, Jr. 1998), and SXFIT (Baes 2001; Moyer et al. 2001). SXLSQI and SXFIT are used by the solvent extraction community in the national laboratory system in the USA to determine solvent extraction equilibria based on distribution ratio measurements (C.F. Baes 2001; Zalupski 2011). This is accomplished by calculating the species and solvent activity coefficients. SXFIT and SXLQI are designed more for chemists rather than the chemical engineers designing the flowsheets, but crossover does exist.

The majority of predictive modeling programs for solvent extraction in the nuclear fuel cycle have been focused around modeling TBP-based systems, with many of the predictive models containing databases of thermodynamic data for common constituents. Many predictive models have been created, with the earliest models developed for pulse columns and utilized in the PUREX process in the early 1960s (Law et al. 2011). As modeling of nuclear fuel cycle solvent extraction developed, nations not only developed computer code independent of one another, but so did the different laboratories within the DOE national laboratory system. A DOE program named Nuclear Energy Advanced Modeling and Simulation (NEAMS) program was focused on creating comprehensive models for the entire nuclear fuel cycle, including fuel fabrication, neutronics, fluid dynamics, and solvent extraction (Bradley 2013). To address the solvent extraction component, the NEAMS program decided to update Argonne's Model for Universal Solvent Extraction (AMUSE), which was originally developed as an expansion of the TRUEX model to include additional processes of interest such as PUREX and UREX. The initial AMUSE code assumed a series of well-mixed stages and calculated the distribution of each component based on the thermodynamic parameters. A literature survey at the time determined there were three existing models that had readily available code and were of particular note with respect to the updated AMUSE, SEPHIS, PUBG, and MIXSET (Law et al. 2011). In 2012, AMUSE was updated to include time-dependent calculations to allow for mass transfer and kinetics (Frey, Krebs, and Pereira 2012). While solvent extraction processes are theoretically run as a continuous operation, nuclear material inventory restrictions and upset conditions, such as start-up and shutdown, disturb the steady state and must be considered to accurately model material accountability. AMUSE is valuable for modeling the distribution of components that exhibit slow extraction kinetics in equipment with short residence times. The chemical speciation and mass balances are handled in two separate subprograms, SASPE and SASSE. Currently, AMUSE is the only predictive solvent extraction modeling program still supported in the USA.

Two codes are commonly used in Europe: the PAREX code (Bisson et al. 2016), which is used to model the PUREX process at La Hague, and the SX Process, a computer program developed in Germany (Magnusson, Geist, and Malmbeck 2013). Unfortunately, PAREX appears to have limited availability outside of the CEA. The SX Process focuses on short residence time centrifugal contactors and, similar to the AMUSE code, calculates mass transfer based on D-values and transfer kinetics (Magnusson and Malmbeck 2012). The SX Process has been extensively tested against experimental SANEX flowsheets with good agreement (Magnusson et al. 2013; Magnusson, Geist, and Malmbeck 2013).

4.5 PROCESSING COMPLICATIONS

Ideally, the aqueous and organic phases will separate cleanly and efficiently. Numerous instances from early on in nuclear fuel separations have demonstrated that this is not always the case (HAPO 1955). Common occurrences observed in these operations are buildup of interfacial crud, formation of stable emulsions, foaming, and third phase formation (i.e., phase splitting and inversion). These occurrences can hinder process flow, reduce separation factors, and jeopardize both criticality and fire safety. Figure 4.11 illustrates various examples of separations, including a clean separation, crud formation at an interface, third phase formation, and stable emulsion formation.

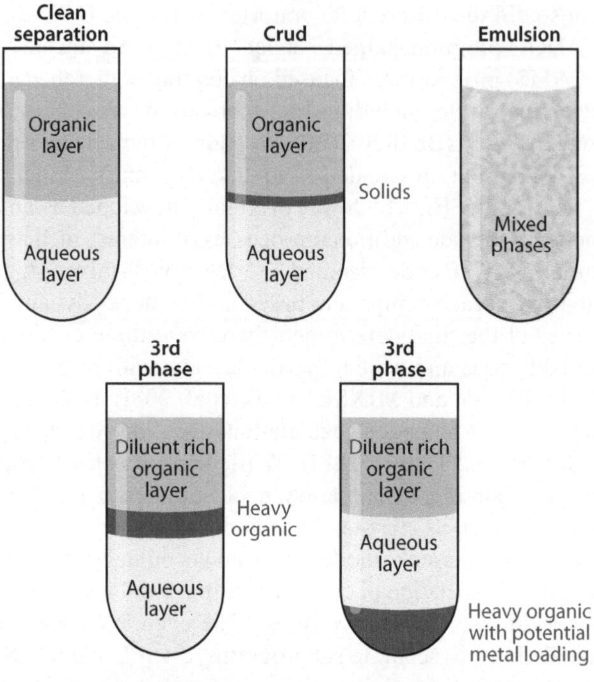

FIGURE 4.11 Examples of separations in an aqueous-organic environment.

4.5.1 Interfacial Crud and Stable Emulsions

The International Union of Pure and Applied Chemistry (IUPAC) (Rice, Irving, and Leonard 1993) defines crud that is related to solvent extraction as:

A deposit or emulsion at the interface between two partially settled phases. Notes:

1. *The phenomenon of crud formation arises from many causes, and this definition does not imply any single one.*
2. *Other terms—some unprintable—have been used, but crud is the generally accepted term.*

This definition of crud should not be confused with another common term in the nuclear fuel industry, Chalk River Unidentified Deposits (CRUD), which are occasionally found in reactors during irradiation (Hurst 1997; Wellock 2015). CRUD can consist of solid, solvent-based organic materials and/or inorganic materials that aggregate at the aqueous–organic interface. This crud can also adhere to various equipment surfaces. These solids can be largely grouped into three categories based on their composition and source origins: (1) precipitants of complexes with degradation products (e.g., dibutyl phosphate (DBP) and monobutyl phosphate (MBP)), (2) undissolved solids, and (3) silica-based gels and metal hydr(oxides) (Delegard and Casella 2016).

Chemical and radiolytic decomposition of the common organic extractant TBP occurs by acid hydrolysis in HNO_3 (Burger 1955; Lloyd and Fellows 1985) and by alpha and gamma radiation damage (Barney and Bouse 1977; Lloyd and Fellows 1985; Gao et al. 2015). The decomposition involves the loss of butoxy [-$O(CH_2)_3CH_3$] groups as butanol ($HO(CH_2)_3CH_3$) to make, in order, DBP, MBP, and phosphate, PO_4^{3-}. Complexes form from interaction of DBP with polyvalent metal ions including Zr^{4+} (or ZrO^{2+}), Pu^{4+}, Pu^{3+}, U^{4+}, UO_2^{2+}, Np^{4+}, and Th^{4+} (Hahn and Vander Wall 1964; Tsujino and Ishihara 1966; Zimmer and Borchardt 1986; Rubisov and Solovkin 1987; Sokhina et al. 1988; Uetake 1989; Lamouroux et al. 2000; May et al. 2000; Blazheva et al. 2006; McDonald et al. 2016). The complexes have been shown to form directly with DBP and displace TBP groups (Zimmer and Borchardt 1986; May et al. 2000). Other elements that have been found in crud include aluminum, iron, silicon, calcium, barium, manganese, magnesium, chromium, nickel, tin, ruthenium, and niobium (HAPO 1955). The solids appearing at the interface can also include undissolved solid plutonium dioxide, PuO_2, and other Pu-bearing solids based on flowsheet input streams (Delegard and Jones 2015; Delegard and Casella 2016).

Crud associated with TBP's degradation products can be minimized through proper solvent cleanup by alkaline washing, which removes water-miscible butanol and DBP, preventing further decomposition to MBP. This step is typically required due to the economic infeasibility and large waste volumes of only using the solvent once. The solvent is recycled to extend its longevity, which increases hydrolysis and degradation product buildup. Solvent washing can consist of an alkaline wash or an alkaline wash followed by an acid wash and is sometimes combined with treatment by surface-active solids (Long 1978). The removal of degradation products can be accomplished with a sodium carbonate scrub (Horwitz et al. 1980; Mailen and Tallent 1984; Sugai 1992). Other methods that have been researched include scrubs with hydrazine, hydroxylamine salts and various combinations with tartrate, and solid sorbents such as activated charcoal, silica gel, activated alumina, attapulgite clay, and macroreticular anion exchange resin (Mailen and Tallent 1984; Plumb 1984; Reif 1988). Additional advancements have incorporated a caustic soda scrub, an acidic wash to remove Zr and Tc, and vacuum distillation used in solvent cleanup (Fournier et al. 1990; Moyer 2010; Mishra et al. 2018).

Emulsions occur in the solvent extraction process when the two phases do not cleanly separate (or disengage) and one is finely dispersed in the other. Fine particles such as crud contribute to this phenomenon because they seek the organic–aqueous interface, stabilizing emulsions that slow down or prevent phase disengagement. Whether the emulsion is water-in-oil or oil-in-water depends on whether the crud-forming fines are wetted by the organic or the aqueous phase. Oil-in-water emulsions occur when partially hydrophobic fine solids are adsorbed on or coat droplets of the oil (organic) phase. Conversely, fine solids that are partially hydrophilic coat droplets of the aqueous phase. In either occurrence, the coating of the droplets inhibits coalescence.

In solvent extraction stages that are prone to crud and/or emulsion occurrences, such as during the first extraction stage, equipment is typically chosen that minimizes (via type of mixing), accommodates, and/or remediates the phenomena. If the emulsion-stabilizing crud continues to build, it can result in flooding of the stages as well as foaming that further contributes to improper phase disengagement. Pulsed columns provide an avenue for incorporating interface jet-out facilities to flush the crud/emulsions. This allows for continued operation without shutdown of the system. Solids can then be separated from the removed interface, via centrifugation or dissolution as well as washing of degradation products, and either be recycled back into the process or separated as waste (HAPO 1955; Delegard and Casella 2016). Research has also been performed exploring the use of microporous polymers for demulsification of surface-active species (Akay et al. 2012). Mixer-settlers and centrifugal contactors are less optimal in flowsheet streams prone to crud and emulsions due to the complexity of their moving parts; they are more susceptible to clogging and require downtime for maintenance (Drain, Vinoche, and Duhamet 2003; Nash and Lumetta 2011).

4.5.2 Third Phase Formation/Phase Splitting and Red Oil

Third phase formation is when the organic phase splits into two distinct phases, most often because the solubility of the solvate is exceeded. The solvate-rich organic phase is typically denser than the first organic phase and may also be denser than the aqueous phase. This phenomenon is of critical concern not only because it prevents proper continuous separation of phases, but because the denser organic phase contains the metal-rich extract species (with Pu, U, and Thorium [Th]) and an excess of nitrate that has implications for both criticality and downstream evaporation and recovery processes. This phenomenon has been studied historically to determine its effects on extraction as well as to prevent its formation (Rao and Kolarik 1996). Renewed interest has focused on elucidation of the species that comprise both the light and heavy organic phases. These studies have used nuclear magnetic resonance spectroscopy (NMR), Fourier transform infrared spectroscopy (FTIR), spectroscopic analysis, X-ray absorption spectroscopy, extended X-ray absorption fine structure measurements, small-angle neutron scattering, small-angle X-ray scattering, and two-phase calorimetry along with chemical analyses to determine the difference in speciation between the phases (Lefrancois et al. 2001; Borkowski et al. 2002; Jensen et al. 2002; Chiarizia et al. 2003; Nilsson 2015). Additional studies have worked toward developing mathematical predictive models of the species formed (Borkowski et al. 2002; S. Kumar and Koganti 2003).

The most notable third phase formation, red oil, has been high on the list of concerns for fuel fabrication and reprocessing facilities due to multiple occurrences of damage to equipment and facilities. While the red oil does separate from the other phases, it is not a classically defined third phase. The red color of the oil is commonly assumed to come from the nitrated organic species and the red fumes associated with nitrogen dioxide production. The nitrated organic species that form red oil are thought to be formed primarily by nitration of the degradation products of TBP. Red oil was first described in published literature in 1951 when it was found in aqueous-phase tanks used for uranium recovery studies requested by a memo dated March 24, 1950. While studies mentioned the presence of red oil as well as analyzed it, there were no known associated safety incidents. The studies concluded that the red oil was composed mostly of uranium complexes associated with TBP, DBP, and MBP along with hydrocarbon diluents, nitric acid, and other nitration products (Curtis 1951). The central focus of these studies was how to address the unstrippable uranium in the complexes and the emulsions they created in the line (Curtis 1951).

Incidents of safety concern involving red oil have occurred at the TNX facility at SRS in South Carolina, USA (1953); building 321 in Hanford, Washington, USA (1953); Oak Ridge National Laboratory, Tennessee, USA (1959); SRS, South Carolina, USA (1975); Ontario, Canada (1980); and at the Tomsk7 Site in Seversk, Russia (1993) (Davis, Baldwin, and Meservey 1960; James and Sheppard 1991; Hyder 1994b; IAEA 1998; Lumetta et al. 2001; Robinson et al. 2003; Usachev and Markov 2003; Swanson and Reed 2006).

Due to the aforementioned incidents, a number of reviews and investigations were performed in attempt to recreate red oil, characterize the reaction and energy release, and define safe operating parameters for the facilities where there is potential for the phenomena to occur (Colven, Nichols, and Siddall 1953; Hyder 1994a; Paddleford and Fauske 1995; Nichols 1960; Harmon et al. 1976). More recent reviews have been published summarizing these and other works (Usachev and Markov 2003; Srinivasan and Vasudeva Rao 2014). These studies evaluated scenarios involving hydrolysis, nitrolysis, pyrolysis, radiolysis, and oxidation of the TBP-nitrate system by itself, along with those involving metal-nitrate interactions with the TBP. The main products of concern from these processes, without the metal-nitrate interactions, are DBP, MBP, phosphoric acid, butanol, butyl nitrate, butyl nitrite, butene, various carboxylic acids, and the subsequent production of offgasses such as N_2, N_2O, NO, NO_2, N_2O_2, CO_2, CO, and hydrocarbons (Wagner 1951; Hyder 1994a, 1994b; Wright and Paviet-Hartmann 2010). The exothermic reactions, along with production of gasses (i.e., pressurization), become the prime concern for safety, especially with the presence of combustible (butyric acid and butyl nitrate) and flammable (butanol, butene, and butyl nitrite) species. Additional studies (J.S.V. Kumar et al. 2018) examined this system under irradiated conditions to elucidate the effects of radiolysis degradation products (Tahraoui and Morris 1995; Wright and Paviet-Hartmann 2010). The degradation products compete with TBP to complex metals in solution (primarily U, Pu, and Zr) and are somewhat soluble in both the aqueous and organic phases. In addition to unwanted TBP-nitrate complexes existing in the aqueous phase, metal-nitrates can oxidize TBP and produce CO, CO_2, N_2, and nitrogen oxides.

While there are inconsistencies in the kinetic parameters, main contributors to the reaction, and concentration of resultant species produced (Wright and Paviet-Hartmann 2010), the general conclusion is that temperatures must exceed 120°C–130°C before conditions exist for uncontrolled (i.e., runaway) gas and heat generation (Colven, Nichols, and Siddall 1953; Nichols 1960; Harmon et al. 1976; Hyder 1994a, 1994b; Robinson et al. 2003). The decomposition of the metal complexes can occur at these temperatures, specifically $(UO_2)(NO_3)_2(TBP)_2$, which is also exothermic and produces butene, CO, CO_2, steam, N_2, and nitrogen oxides (Harmon et al. 1976; Hyder 1994a, 1994b). This gas production is thought to be linked to the large over pressurization in closed vessels. Metal complex decomposition has also been linked to foaming and is thought to contribute to clogging of vent lines prohibiting proper depressurization (Hyder 1994a, 1994b; Usachev and Markov 2003).

Due to the severe safety concerns related to uncontrolled exothermic reactions, the Defense Nuclear Facilities Safety Board performed an assessment to minimize the hazard of red oil in 2003 and determined that the primary pieces of equipment in nuclear fuel reprocessing where red oil would be a concern are evaporators, acid concentrators, and denitrators (Robinson et al. 2003). In most cases, proper temperature, pressure, and mass flow control can be effectively utilized to prevent hazardous work conditions (Robinson et al. 2003). A later assessment by the Institute for Radiological Protection and Nuclear Safety (IRSN) in 2008 described additional measures that can be taken in reprocessing facilities to prevent red oil incidents in evaporators as the following: solvent flushing devices, systematic monitoring, not completely emptying aqueous transfer tanks (to prevent floating TBP in solutions from being transferred), solvent treatments with carbonate and soda washes, and limiting temperatures to well below 135°C (IRSN 2008).

4.6 CONCLUSIONS

Engineering surrounding solvent extraction requires a vast knowledge base on several technical areas. It is necessary for the engineer to understand computer modeling, criticality safety, mechanical equipment, and the underlying chemistry of a multitude of chemical processes/flowsheets. Perhaps paramount above all of these challenges is maintaining environmental, public, and worker safety in an environment rife with hazards such as the aforementioned red oil. Despite the challenges and due to the safety culture in the solvent extraction community, the worldwide track record for safely operating solvent extraction facilities is rather remarkable, particularly for the past 50 years.

As the reprocessing climate continues to change, and final disposition of used nuclear fuel is discussed, the solvent extraction community will need to play a major role in providing viable solutions for recycling used nuclear fuel. Quite simply, public acceptance is key to a nuclear renaissance, and without viable reprocessing strategies, widespread public acceptance for nuclear energy will never be achieved. To gain acceptance among the general population, it remains a critical need to streamline processes to target specific radionuclides while minimizing feed adjustments, reagent adjustments or additions, and the number of scrubbing and stripping steps. It must also be noted, however, that while separations processes and technologies are constantly changing, reactor technology is also in a state of constant change. Technologies developed for nuclear waste must be mindful of the constantly changing environment in order to remain relevant.

REFERENCES

Akay, G., T. Pekdemir, A. M. Shakorfow, and J. Vickers. 2012. "Intensified Demulsification and Separation of Thermal Oxide Reprocessing Interfacial Crud (THORP-IFC) Simulants." *Green Processing and Synthesis* 1 (1): 109–127. https://doi.org/10.1515/greenps-2011-0510.

Aneheim, E. H. K. 2012. "Development of a Solvent Extraction Process for Group Actinide Recovery from Used Nuclear Fuel." PhD diss., Department of Chemical and Biochemical Engineering, Chalmers University of Technology.

Ansari, S. A., P. Pathak, P. K. Mohapatra, and V. K. Manchanda. 2012. "Chemistry of Diglycolamides: Promising Extractants for Actinide Partitioning." *Chemical Reviews* 112 (3): 1751–1772. https://doi.org/10.1021/cr200002f.

Arm, S., and C. Phillips. 2011. "3–Chemical Engineering for Advanced Aqueous Radioactive Materials Separations." In *Advanced Separation Techniques for Nuclear Fuel Reprocessing and Radioactive Waste Treatment*, edited by Kenneth L. Nash and Gregg J. Lumetta, 58–94. Oxford, UK: Woodhead Publishing.

Baes, C. F., Jr. 2001. "Modeling Solvent Extraction Systems with Sxfit." *Solvent Extraction and Ion Exchange* 19 (2): 193–213. https://doi.org/10.1081/sei-100102691.

Baes, C. F., Jr. 1998. *Sxlsqi: A Program for Modelling Solvent Extraction Systems* (Oak Ridge, TN: Oak Ridge National Laboratory). ORNL/TM-13604.

Baes, C. F., Jr, W. J. McDowell, and S. A. Bryan. 1987. "The Interpretation of Equilibrium Data from Synergistic Solvent Extraction Systems." *Solvent Extraction and Ion Exchange* 5: 1–27.

Baes, C. F., Jr, B. A. Moyer, G. N. Case, and F. I. Case. 1990. "Sxlsqa, a Computer Program for Including Both Complex Formation and Activity Effects in the Interpretation of Solvent Extraction Data." *Separation Science and Technology* 25: 1675–1688.

Barney, G. S., and D. G. Bouse. 1977. *Alpha Radiolysis of Tributyl Phosphate—Effects of Diluents* (Richland, WA: Atlantic Richfield Hanford Company). http://www.iaea.org/inis/collection/NCLCollectionStore/_Public/09/357/9357134.pdf.

Bisson, J., B. Dinh, P. Huron, and C. Huel. 2016. "PAREX, A Numerical Code in the Service of La Hague Plant Operations," *Procedia Chemistry* 21: 117–124. https://doi.org/10.1016/j.proche.2016.10.017

Blazheva, I. V., Y. S. Fedorov, B. Y. Zilberman, and N. V. Ryabkova. 2006. "The Influence of Dibutyl Phosphoric Acid on Zirconium Extraction with Diluted Tributyl Phosphate and the Role of Uranyl Nitrate." *Czechoslovak Journal of Physics* 56 (4): D501–D508. https://doi.org/10.1007/s10582-006-0542-3.

Bonnesen, P. V., L. H. Delmau, T. J. Haverlock, and B. A. Moyer. 1998. *Alkaline-Side Extraction of Cesium from Savannah River Tank Waste Using a Calixarene-Crown Ether Extractant* (Oak Ridge National Lab., TN). ORNL/TM-13704; TRN: AH200112%%458 United States 10.2172/3296 TRN: AH200112%%458 OSTI as DE00003296 ORNL English. https://www.osti.gov/servlets/purl/3296.

Borkowski, M., J. R. Ferraro, R. Chiarizia, and D. R. McAlister. 2002. "FT-IR Study of Third Phase Formation in the U(VI) or Th(IV)/HNO_3, TBP/Alkane Systems." *Solvent Extraction and Ion Exchange* 20 (3): 313–330. https://doi.org/10.1081/Sei-120004807.

Bradley, K. 2013. *NEAMS: The Nuclear Energy Advanced Modeling and Simulation Program* (Lemont, IL: Argonne National Laboratory). ANL/NE-13/5.

Braley, J. C., J. C. Carter, S. I. Sinkov, K. L. Nash, and G. J. Lumetta. 2012. "The Role of Carboxylic Acids in TALSQuEAK Separations." *Journal of Coordination Chemistry* 65 (16): 2862–2876. https://doi.org.10.1080/00958972.2012.704551.

Braley, J. C., T. S. Grimes, and K. L. Nash. 2012. "Alternatives to HDEHP and DTPA for Simplified TALSPEAK Separations." *Industrial & Engineering Chemistry Research* 51: 629–638. https://doi.org/10.1021/ie200285r.

Brown, J., F. McLachlan, M. Sarsfield, R. Taylor, G. Modolo, and A. Wilden. 2012. "Plutonium Loading of Prospective Grouped Actinide Extraction (GANEX) Solvent Systems Based on Diglycolamide Extractants." *Solvent Extraction and Ion Exchange* 30 (2): 127–141. https://doi.org/10.1080/07366299.2011.609378.

Burger, L. L. 1955. *The Chemistry of Tributyl Phosphate–A Review* (Richland, WA: Hanford Atomic Products Operation). HW-40910. http://www.osti.gov/scitech/biblio/4334996.

Chiarizia, R., M. P. Jensen, M. Borkowski, J. R. Ferraro, P. Thiyagarajan, and K. C. Littrell. 2003. "Third Phase Formation Revisited: The U(VI), HNO3-TBP, n-Dodecane System." *Solvent Extr. Ion Exch.* 21 (1): 1–27. https://doi.org/10.1081/SEI-120017545.

Collins, E. D., G. D. Del Cul, and B. A. Moyer. 2011. "8—Advanced Reprocessing for Fission Product Separation and Extraction." In *Advanced Separation Techniques for Nuclear Fuel Reprocessing and Radioactive Waste Treatment*, edited by Kenneth L. Nash and Gregg J. Lumetta, 201–228. Oxford, UK: Woodhead Publishing.

Colven, T. J., Jr., G. M. Nichols, and T. H. Siddall. 1953. *TNX Evaporator Incident January 12, 1953. Interim Technical Report.* https://www.osti.gov/servlets/purl/4347079.

Curtis, M. H. 1951. *TBP Process Investigation-Formation of "Red Oil" in Extraction Column Studies* (Richland, WA: General Electric, Hanford Atomic Products Operation). HW-19997.

Davis, W. J., W. Baldwin, and A. B. Meservey. 1960. *Chemistry of the Intercycle Evaporator Incident of November 20, 1959* (Oak Ridge, TN: Oak Ridge National Laboratory). ORNL-2979.

Delegard, C., and A. J. Casella. 2016. *Literature Review: Crud Formation at the Liquid/Liquid Interface During TBP-Based Solvent Extraction Processes* (Pacific Northwest National Laboratory).

Delegard, C., and S. Jones. 2015. *Chemical Disposition of Plutonium in Hanford Site Tank Wastes* (Richland, WA: Pacific Northwest National Laboratory). PNNL-25888.

Delmau, L. H. 1999. *Improved Performance of the Alkaline-Side CSEX Process for Cesium Extraction from Alkaline High-Level Waste Obtained by Characterization of the Effect of Surfactant Impurities* (Oak Ridge, TN: ORNL Oak Ridge National Laboratory). ORNL/TM-1999/209; TRN: US200317%%246 United States 10.2172/814457 TRN: US200317%%246 ORNL English. https://www.osti.gov/servlets/purl/814457.

Delmau, L. H., J. F. Birdwell, J. McFarlane, and B. A. Moyer. 2010. "Robustness of the CSSX Process to Feed Variation: Efficient Cesium Removal from the High Potassium Wastes at Hanford." *Solvent Extraction and Ion Exchange* 28 (1): 19–48. https://doi.org/10.1080/07366290903408433.

Denniss, I. S., and C. Phillips. 1990. "The Development of a Three-Cycle Chemical Flowsheet to Reprocess Oxide Nuclear Fuel." *Proceeding International Solvent Extraction Conference*, Kyoto, Japan, July 16–21.

Dileep, C. S., P. Jagasia, P. S. Dhami, P. V. Achuthan, A. Dakshinamoorthy, B. S. Tomar, S. K. Munshi, and P. K. Dey. 2008. "Distribution of Technetium in PUREX Process Streams." *Desalination* 232 (1): 157–165, https://doi.org/10.1016/j.desal.2007.11.053.

Dozol, J. F., M. Dozol, and R. M. Macias. 2000. "Extraction of Strontium and Cesium by Dicarbollides, Crown Ethers and Functionalized Calixarenes." *Journal of Inclusion Phenomena and Macrocyclic Chemistry* 38 (1): 1–22. https://doi.org/10.1023/a:1008145814521.

Drain, F., J. L. Emin, R. Vinoche, and P. Baron. 2008. "COEX™ Process: Cross-Breeding between Innovation and Industrial Experience." *Waste Management*, Phoenix, AZ.

Drain, F., R. Vinoche, and J. Duhamet. 2003. "40 Years of Experience of Liquid-Liquid Extraction Equipment in the Nuclear Industry." *WM '03, Conference*, Tucson, AZ February 23-27 2003.

Fournier, W., D. Hugelmann, G. Dalverny, C. Bernard, P. Miquel, and A. Leudet. 1990. "PUREX Process Improvements for the UP3 Spent Fuel Reprocessing Plant at La Hague, France." *Proceedings of the International Solvent Extraction Conference.*

Frey, K., J. F. Krebs, and C. Pereira. 2012. "Time-Dependent Implementation of Argonne's Model for Universal Solvent Extraction." *Industrial & Engineering Chemistry Research* 51 (40): 13219–13226. https://doi.org/10.1021/ie301421d.

Gao, Y., W.-F. Zheng, X.-M. Cao, and S.-L. Chen. 2015. "Gamma and Alpha Radiolysis of TBP Solvent." *Nuclear Science and Techniques* 26 (1): 10303–S10303. https://doi.org/10.13538/j.1001-8042/nst.26.S10303.

Gelis, A. V., and G. J. Lumetta. 2014. "Actinide Lanthanide Separation Process—ALSEP." *Industrial & Engineering Chemistry Research* 53 (4): 1624–1631. https://doi.org/10.1021/ie403569e.

Gerber, M. S. W. H. 1996. *Plutonium Production Story at the Hanford Site: Processes and Facilities History*. https://www.osti.gov/servlets/purl/664389.

Gray, L. W., K. S. Holliday, A. Murray, M. Thompson, D. T. Thorp, S. Yarbro, and T. J. Venetz. 2015. *Separation of Plutonium from Irradiated Fuels and Targets*. https://www.osti.gov/biblio/1240057.

Grimes, T. S., G. Tian, L. Rao, and K. L. Nash. 2012. "Optical Spectroscopy Study of Organic-Phase Lanthanide Complexes in the TALSPEAK Separations Process." *Inorganic Chemistry* 51: 6299–6307. https://doi.org/10.1021/ic300503p.

Hahn, H. T., and E. M. Vander Wall. 1964. "Complex Formation in the Dilute Uranyl Nitrate-Nitric Acid-Dibutyl Phosphoric Acid-Tributyl Phosphate-Amsco System." *Journal of Inorganic and Nuclear Chemistry* 26 (1): 191–202. https://doi.org/10.1016/0022-1902(64)80245-1.

HAPO. 1955. *PUREX Technical Manual* (Richland, WA: General Electric Hanford Atomic Product Operation). HW-31000. https://www.osti.gov/biblio/4314635-purex-technical-manual.

Harmon, H. D., M. L. Hyder, B. Tiffany, L. W. Gray, and P. A. Soltys. 1976. *Behavior of Tributyl Phosphate in a-Line Processes* (Aiken, South Carolina: E.I. du Pont de Nemours and Company). DP-1418. https://inis.iaea.org/search/search.aspx?orig_q=RN:8289496.

Herbst, R. S., P. Baron, and M. Nilsson. 2011. "6–Standard and Advanced Separation: PUREX Processes for Nuclear Fuel Reprocessing." In *Advanced Separation Techniques for Nuclear Fuel Reprocessing and Radioactive Waste Treatment*, edited by Kenneth L. Nash and Gregg J. Lumetta, 141–175. Oxford, UK: Woodhead Publishing.

Herbst, R. S., J. D. Law, T. A. Todd, V. N. Romanovskiy, V. A. Babain, V. M. Esimantovskiy, I. V. Smirnov, and B. N. Zaitsev. 2002. "Universal Solvent Extraction (UNEX) Flowsheet Testing for the Removal of Cesium, Strontium, and Actinide Elements from Radioactive, Acidic Dissolved Calcine Waste." *Solvent Extraction and Ion Exchange* 20 (4–5): 429–445. https://doi.org/10.1081/SEI-120014366.

Heres, X., P. Baron, C. Hill, E. Ameil, I. Martinez, and P. Rivalier. 2008. "The Separation of Extractants Implemented in the DIAMEX-SANEX Process." *Atalante 2008: Nuclear Fuel Cycle for a Sustainable Future*, Montpellier, France.

Hill, C. 2011. "11—Development of Highly Selective Compounds for Solvent Extraction Processes: Partitioning and Transmutation of Long-Lived Radionuclides from Spent Nuclear Fuels." In *Advanced Separation Techniques for Nuclear Fuel Reprocessing and Radioactive Waste Treatment*, edited by Kenneth L. Nash and Gregg J. Lumetta, 311–362. Oxford, UK: Woodhead Publishing.

Hill, O. F., and V. R. Cooper. 1958. "Scale-up Problems in the Plutonium Separations Program." *Industrial and Engineering Chemistry* 50 (4): 599–602, https://pubs.acs.org/doi/pdfplus/10.1021/ie50580a026.

Holfeltz, V. E., E. L. Campbell, D. R. Peterman, R. F. Standaert, A. Paulenova, G. J. Lumetta, and T. G. Levitskaia. 2018. "Effect of HEH[EHP] Impurities on the ALSEP Solvent Extraction Process." *Solvent Extraction and Ion Exchange* 36 (1): 22–40. https://doi.org/10.1080/07366299.2017.1412111.

Holfeltz, V. E., G. B. Hall, E. L. Campbell, A. Paulenova, T. G. Levitskaia, and G. J. Lumetta. "Synergism in the ALSEP System." In Preparation.

Horwitz, E. P., M. L. Dietz, and D. E. Fisher. 1991. "SREX: A New process for the Extraction and Recovery of Strontium from Acidic Nuclear Waste Streams." *Solvent Extraction and Ion Exchange* 9 (1): 1–25. https://doi.org/10.1080/07366299108918039.

Horwitz, E. P., G. W. Mason, C. A. A. Bloomquist, R. A. Leonard, and G. J. Bernstein. 1980. "The Extraction of DBP and MBP from Actinides: Application to the Recovery of Actinides from TBP-Sodium Carbonate Scrub Solution." In *Actinide Separations*, In ACS Symposium Series, 475–496. Washington, DC: American Chemical Society.

Horwitz, E. P., and W. W. Schulz. 1990. *The Transuranium Extraction (TRUEX) Process: A Vital Tool for Disposal of Us Defense Nuclear Waste* (United States. http://inis.iaea.org/search/search.aspx?orig_q=RN:22067029.

Hurst, D. G. 1997. *Canada Enters the Nuclear Age, a Technical History of Atomic Energy of Canada Limited as Seen from Its Research Laboratories*. McGill-Queen's University Press.

Hyder, M. L. 1994a. *Safe Conditions for Contacting Nitric Acid or Nitrates with Tri-n-Butyl Phosphate (TBP)* (Aiken, South Carolina: Westinghouse Savannah River Company). WSRC-TR-94-059. https://inis.iaea.org/collection/NCLCollectionStore/_Public/26/023/26023675.pdf.

Hyder, M. L. 1994b. *Safe Handling of TBP and Nitrates in the Nuclear Process Industry* (Aiken, South Carolina: Westinghouse Savannah River Company). WSRC-TR-94-0372. https://inis.iaea.org/search/search.aspx?orig_q=RN:26032441.

IAEA. 1998. *The Radiological Accident in the Reprocessing Plant at Tomsk*. Edited by International Atomic Energy Agency. edited by G. Andreev, Y. Antipin, B. Bennett, V. Chizhikov, A.J. Gonzalez, A. Gordeev, L. Ilyin, G. Khandorin, O. Kochotkov, V. Kruchkov, A. Masljuk, I. Pollard, V. Popov, M. Savkin, A. Titov, A. Tsovyanov and G.A.M. Webb. Vienna. https://www-pub.iaea.org/MTCD/publications/PDF/P060_scr.pdf

IRSN. 2008. "Technical Note: Risks of Explosion Associated with "Red Oils" in Reprocessing Plants." Accessed December 7, 2018. https://www.irsn.fr/EN/newsroom/News/Pages/20080610_risks_of_explosion_associated_red_oils_in_reprocessing_plants.aspx.

James, N. J., and G. T. Sheppard. 1991. "Red Oil Hazards in Nuclear-Fuel Reprocessing." *Nuclear Engineering and Design* 130 (1): 59–69. https://doi.org/10.1016/0029-5493(91)90193-L.

Jensen, M. P., R. Chiarizia, J. R. Ferraro, M. Borkowski, K. L. Nash, P. Thlyagarajan, and K. C. Littrell. 2002. "New Insights in Third Phase Formation in the U(VI)-HNO_3, TBP-Alkane System." *International Solvent Extraction Conference 2002*, Cape Town, South Africa, March 18–21, 2002.

Kumar, J. S. V., V. S. Smitha, N. E. Sivanesh, M. Surianarayanan, H. Seshadri, and V. Lakshman. 2018. "Reactive Thermal Hazards of Irradiated Tributyl Phosphate with Nitric Acid." *Thermochimica Acta* 666: 18–26, https://doi.org/10.1016/j.tca.2018.05.022.

Kumar, S., and S. B. Koganti. 2003. "Speciation Studies in Third Phase Formation: U(IV), Pu(IV), and Th(IV) Third Phases in TBP Systems." *Solvent Extraction and Ion Exchange* 21 (4): 547–558. https://doi.org/10.1081/SEI-120022520.

Laidler, J. J., L. Burris, E. D. Collins, J. Duguid, R. N. Henry, J. Hill, E. J. Karell, S. M. McDeavitt, M. Thompson, M. A. Williamson, and J. L. Willit. 2001. "Chemical Partitioning Technologies for an ATW System." *Progress in Nuclear Energy* 38 (1–2): 65–79. https://doi.org/10.1016/S0149-1970(00)00096-2.

Lamouroux, C., C. Moulin, J. C. Tabet, and C. K. Jankowski. 2000. "Characterization of Zirconium Complexes of Interest in Spent Nuclear Fuel Reprocessing by Electrospray Ionization Mass Spectrometry." *Rapid Communications in Mass Spectrometry* 14 (19): 1869–1877. https://doi.org/10.1002/1097-0231(20001015)14:19<1869::Aid-Rcm108>3.3.Co;2-G.

Lanham, W. B., and T. C. Runion. 1949. *PUREX Process for Plutonium and Uranium Recovery*. Oak Ridge, TN: Oak Ridge National Laboratory.

Law, J. D., R. S. Herbst, D. H. Meikrantz, D. R. Peterman, C. L. Riddle, R. D. Tillotson, and T. A. Todd. 2005. "Development of Technologies for the Simultaneous Separation of Cesium and Strontium from Spent Nuclear Fuel as Part of an Advanced Fuel Cycle." United States, 2005-04-01.

Law, J. D., R. S. Herbst, and T. A. Todd. 2002. "Integrated AMP–PAN, TRUEX, and SREX Testing. II. Flowsheet Testing for Separation of Radionuclides from Actual Acidic Radioactive Waste." *Separation Science and Technology* 37 (6): 1353–1373. https://doi.org/10.1081/SS-120002615.

Law, J. D., V. Rutledge, C. Pereira, K. Frey, J. F. Krebs, L. E. Maggos, K. Nichols, et al. 2011. *Requirements for a Dynamic Solvent Extraction Module to Support Development of Advanced Technologies for the Recycle of Used Nuclear Fuel* (Idaho Falls, ID: Idaho National Laboratory). INL/EXT-11-22473. https://inldigitallibrary.inl.gov/sites/sti/sti/5094582.pdf.

Law, J. D., and T. A. Todd. 2008. *Liquid-Liquid Extraction Equipment* (Idaho National Laboratory). INL/CON-08-15151. https://inldigitallibrary.inl.gov/sites/sti/sti/4156452.pdf.

Lefrancois, L., J.-J. Delpuech, M. Hebrant, J. Chrisment, and C. Tondre. 2001. "Aggregation and Protonation Phenomena in Third Phase Formation: An NMR Study of the Quaternary Malonamide/Dodecane/Nitric Acid/Water System." *Journal of Physical Chemistry B* 105 (13): 2551–2564. https://doi.org/10.1021/jp002465h.

Leonard, R. A. 2010. Initial Flowsheet for the Next Generation CSSX Process in the MCU. Report for FY 2010 (Argonne National Laboratory). https://publications.anl.gov/anlpubs/2012/11/72864.pdf.

Leonard, R. A., G. J. Bernstein, A. A. Ziegler, and R. H. Pelto. 1980. "Annular Centrifugal Contactors for Solvent Extraction." *Separation Science and Technology* 15 (4): 925–43. https://doi.org/10.1080/01496398008076278.

Leonard, R. A., C. Conner, M. W. Liberatore, P. V. Bonnesen, B. A. Moyer, D. J. Presley, and G. J. Lumetta. 1997. "Developing and Testing an Alkaline-Side Solvent Extraction Process for Technitium Separation from Tank Waste." *Tenth Symposium on Separation Science and Technology for Energy Applications*, Gatlinburg, TN, October 20–24, 2997.

Lloyd, M. H., and R. L. Fellows. 1985. *Alpha Radiolysis and Other Factors Affecting Hydrolysis of Tributyl Phosphate* (Oak Ridge, TN: Oak Ridge National Laboratory). ORNL/TM-9565. https://www.osti.gov/servlets/purl/5534740.

Long, J. T. *Engineering for Nuclear Fuel Reprocessing. Nuclear Science and Technology Series*. La Grange Park, IL: American Nuclear Society, 1978.

Lumetta, G. J., J. R. Allred, S. A. Bryan, G. B. Hall, T. G. Levitskaia, A. M. Lines, and S. I. Sinkov. 2019. "Simulant Testing of a Co-Decontamination (CoDCon) Flowsheet for a Product with a Controlled Uranium-to-Plutonium Ratio." *Separation Science and Technology* 54 (12): 1977–1984. https://doi.org/10.1080/01496395.2019.1591899.

Lumetta, G. J., S. A. Bryan, G. B. Hall, F. D. Heller, T. G. Levitskaia, A. M. Lines, S. I. Sinkov, and J. R. Allred. 2018. *CoDCon Project: FY 2018 Status Report* (Richland, WA: Pacific Northwest National Laboratory). PNNL-27850.

Lumetta, G. J., K. J. Carson, L. P. Darnell, L. R. Greenwood, F. V. Hoopes, R. L. Sell, S. I. Sinkov, C. Z. Soderquist, M. W. Urie, and J. J. Wagner. 2001. *Caustic Leaching of Hanford Tank S-110 Sludge* (Richland, WA: Pacific Northwest National Laboratory). PNNL-13702.

Lumetta, G. J., A. J. Casella, B. M. Rapko, T. G. Levitskaia, N. K. Pence, J. C. Carter, C. M. Niver, and M. R. Smoot. 2015. "An Advanced TALSPEAK Concept Using 2-Ethylhexylphosphonic Acid Mono-2-Ethylhexyl Ester as the Extractant." *Solvent Extraction and Ion Exchange* 33: 211–233.

Lumetta, G. J., T. G. Levitskaia, A. Wilden, A. J. Casella, G. B. Hall, L. G. Lin, S. I. Sinkov, J. D. Law, and G. Modolo. 2017. "An Advanced TALSPEAK Concept for Separating Minor Actinides. Part 1. Process Optimization and Flowsheet Development." *Solvent Extraction and Ion Exchange* 35 (6): 377–395. https://doi.org/10.1080/07366299.2017.1368901.

Lumetta, G. J., and K. L. Nash. 2011. "Preface." In *Advanced Separation Techniques for Nuclear Fuel Reprocessing and Radioactive Waste Treatment*, edited by Kenneth L. Nash and Gregg J. Lumetta, 141–175. Oxford, UK: Woodhead Publishing.

Luther, T. A., R. S. Herbst, D. R. Peterman, R. D. Tillotson, T. G. Garn, V. A. Babain, I. V. Smirnov, E. S. Stoyanov, and N. G. Antonov. 2006. "Some Aspects of Fundamental Chemistry of the Universal Extraction (UNEX) Process for the Simultaneous Separation of Major Radionuclides (Cesium, Strontium, Actinides, and Lanthanides) from Radioactive Wastes." *Journal of Radioanalytical and Nuclear Chemistry* 267 (3): 603–613. https://doi.org/10.1007/s10967-006-0114-9.

Madic, C., M. Lecomte, P. Baron, and B. Boullis. 2002. "Separation of Long-Lived Radionuclides from High Active Nuclear Waste." *Comptes Rendus Physique* 3 (7): 797–811, https://doi.org/10.1016/S1631-0705(02)01370-1.

Magnusson, D., B. Christiansen, M. R. S. Foreman, A. Geist, J. P. Glatz, R. Malmbeck, G. Modolo, D. Serrano-Purroy, and C. Sorel. 2009. "Demonstration of a SANEX Process in Centrifugal Contactors Using the Cyme4-Btbp Molecule on a Genuine Fuel Solution." *Solvent Extraction and Ion Exchange* 27 (2): 97–106. https://doi.org/10.1080/07366290802672204.

Magnusson, D., A. Geist, and R. Malmbeck. 2013. "SX Process-a Code for Solvent Extraction Processes in Centrifugal Contactors Simulation." *Chemical Engineering Science* 99: 292–297, https://doi.org/10.1016/j.ces.2013.06.008.

Magnusson, D., A. Geist, R. Malmbeck, and U. Muellich. 2013. "Study of the Mass Transfer Behavior in a Centrifugal Contactor and Verification of the Solvent Extraction Model for the SX Process Program." *Solvent Extraction and Ion Exchange* 31 (6): 578–589. https://doi.org/10.1080/07366299.2013.806748.

Magnusson, D., and R. Malmbeck. 2012. "Development of a Solvent Extraction Model for Process Tests in Short Residence Time Centrifugal Contactors." *Solvent Extraction and Ion Exchange* 30 (2): 115–126. https://doi.org/10.1080/07366299.2011.581085.

Mailen, J. C., and O. K. Tallent. 1984. "Solvent Cleanup and Degradation: A Survey and Recent ORNL Studies." *International Topical meeting on Fuel Reprocessing and Waste Management*, Jackson Hole, WY, August 26–29.

Martell, A. E., R. S. Smith, and R. J. Motekaitis. 2004. *NIST Critically Selected Stability Constants of Metal Complexes*. Gaithersburg, MD: NIST.

May, I., R. J. Taylor, A. L. Wallwork, J. J. Hastings, Y. S. Fedorov, B. Y. Zilberman, E. N. Mishin et al., 2000. "The Influence of Dibutylphosphate on Actinide Extraction by 30% Tributylphosphate." *Radiochimica Acta* 88 (5): 283–290. https://doi.org/10.1524/ract.2000.88.5.283.

Mayo, J. 2018. "Mixer-Settlers: A Top Choice for Liquid/Liquid Extraction Processes." Accessed April 11. https://www.ddpsinc.com/blog-0/mixer-settlers-a-top-choice-for-liquid-liquid-extraction-processes.

McDonald, L. W., J. A. Campbell, T. Vercouter, and S. B. Clark. 2016. "Characterization of Actinides Complexed to Nuclear Fuel Constituents Using ESI-MS." *Analytical Chemistry* 88 (5): 2614–2621. https://doi.org/10.1021/acs.analchem.5b03352.

McKibben, J. M., D. F. Chostner, and E. G. Orebaugh. 1983. *Plutonium-Uranium Separation in the PUREX Process Using Mixtures of Hydroxylamine Nitrate and Ferrous Sulfamate* (United States). http://inis.iaea.org/search/search.aspx?orig_q=RN:17067644.

Miguirditchian, M., L. Chareyre, C. Sorel, I. Bisel, P. Baron, and M. Masson. 2008. "Development of the GANEX Process for the Reprocessing of Gen IV Spent Nuclear Fuels." *Atalante 2008: Nuclear Fuel Cycle for a Sustainable Future*, Montpellier, France.

Mishra, S., A. K. Soda, M. Sridhar, C. Mallika, N. K. Pandey, and U. K. Mudali. 2018. "Identification of Diluent Degradation Products in Radiolyzed PUREX Solvent." *Solvent Extraction and Ion Exchange* 36 (1): 54–65. https://doi.org/10.1080/07366299.2017.1417776.

Moyer, B. A. 2010. *Ion Exchange and Solvent Extraction: A Series of Advances, Volume 19*. Edited by B. A. Moyer. Boca Raton, FL: CRC Press, Taylor & Francis Group.

Moyer, B. A., C. F. Baes, F. I. Case, and J. L. Driver. 2001. "Liquid–Liquid Equilibrium Analysis in Perspective Ii. Complete Model of Water, Nitric Acid, and Uranyl Nitrate Extraction by Di-2-Ethylhexyl Sulfoxide in Dodecane." *Solvent Extraction and Ion Exchange* 19 (5): 757–790. https://doi.org/10.1081/sei-100107024.

Nash, K. L. 2015. "The Chemistry of TALSPEAK: A Review of the Science." *Solvent Extraction and Ion Exchange* 33 (1): 1–55. https://doi.org/10.1080/07366299.2014.985912.

Nash, K. L., and G. J. Lumetta. 2011. *Advanced Separation Techniques for Nuclear Fuel Reprocessing and Radioactive Waste Treatment.* Oxford, UK: Woodhead Publishing. http://app.knovel.com/hotlink/toc/id:kpASTNFRR7/advanced-separation-techniques/advanced-separation-techniques.

Nichols, G. S. 1960. *Decomposition of the Tributyl Phosphate-Nitrate Complexes* (Aiken, South Carolina: E.I. du Pont de Nemours and Company). DP-526.

Nilsson, M. 2015. *Microemulsions and Aggregation Formation in Extraction Processes for Used Nuclear Fuel Thermodynamic and Structural Studies.*

Nilsson, M., and K. L. Nash. 2007. "Review Article: A Review of the Development and Operational Characteristics of the TALSPEAK Process Review Article: A Review of the Development and Operational Characteristics of the TALSPEAK Process." *Solvent Extraction and Ion Exchange* 25: 665–701. https://doi.org/10.1080/07366290701634636.

Paddleford, D. F., and H. K. Fauske. 1995. "Safe Venting of "Red Oil" Runaway Reactions." *30th ASME/AICHE/ANS/AIAA National Heat Transfer Conference*, Portland, OR, August 5–9, 1995.

Paulenova, A. 2011. "2—Physical and Chemical Properties of Actinides in Nuclear Fuel Reprocessing." In *Advanced Separation Techniques for Nuclear Fuel Reprocessing and Radioactive Waste Treatment*, edited by Kenneth L. Nash and Gregg J. Lumetta, 23–57. Oxford, UK: Woodhead Publishing.

Pereira, C., G. F. Vandegrift, M. C. Regalbuto, A. Bakel, D. Bowers, A. V. Gelia, A. S. Hebden, L. E. Maggos, D. Stepinski, Y. Tsai, J. J. Laidler, and C. Engineering. Lab-Scale Demonstration of the UREX+1a Process Using Spent Nuclear Fuel. *Conference: Waste Management Symposium 2007*; February 25, 2007–March 1, 2007; Tucson, AZ: Argonne National Lab. (ANL), Argonne, IL, 2007.

Pereira, C., G. F. Vandegrift, and J. L. Swanson. 2002. *Preliminary Evaluation of Solvent-Extraction and/or Ion-Exchange Process for Meeting Aaa Program Multi Tier Systems Recovery and Purification Goals.* (Argonne National Laboratory, Lemont, IL). ANL-02/29. https://www.osti.gov/servlets/purl/805261-qJYYEJ/native/.

Peters, R., P. Seshadri, G. Aubert, T. Barracco, and L. Billes-Garabedian. 2006. *Economic Assessment of Used Nuclear Fuel Management in the United States.*

Plumb, G. R. 1984. "Review of Developments in U.K. Commercial Reprocessing." *Progress in Nuclear Energy* 13 (1): 3–18. https://doi.org/10.1016/0149-1970(84)90003-9.

Poinssot, C., B. Boullis, and S. Bourg. 2015. "2 - Role of Recycling in Advanced Nuclear Fuel Cycles." In *Reprocessing and Recycling of Spent Nuclear Fuel*, edited by Robin Taylor, 27–48. Oxford, UK: Woodhead Publishing.

Rao, P. R. V., and Z. Kolarik. 1996. "A Review of Third Phase Formation in Extraction of Actinides by Neutral Organophosphorus Extractants." *Solvent Extraction and Ion Exchange* 14 (6): 955–993. https://doi.org/10.1080/07366299608918378.

Regalbuto, M. C. 2011. "7 - Alternative Separation and Extraction: UREX+ Processes for Actinide and Targeted Fission Product Recovery." In *Advanced Separation Techniques for Nuclear Fuel Reprocessing and Radioactive Waste Treatment*, edited by Kenneth L. Nash and Gregg J. Lumetta, 176–200. Oxford, UK: Woodhead Publishing.

Reif, D. J. 1988. "Restoring Solvent for Nuclear Separation Processes." *Separation Science and Technology* 23 (12–13): 1285–1295. https://doi.org/10.1080/01496398808075630.

Rice, N. M., H. M. N. H. Irving, and M. A. Leonard. 1993. "Nomenclature for Liquid-Liquid Distribution (Solvent Extraction) (Iupac Recommendations 1993)." *Pure & Applied Chemistry* 65 (11): 2373–2396. http://dx.doi.org/10.1351/pac199365112373.

Robinson, R. N., D. M. Gutowski, W. Yeniscavich, J. S. Contardi, R. S. Daniels, and T. L. Hunt. 2003. *Control of Red Oil Explosions in Defense Nuclear Facilities.* (Washington D.C., Defense Nuclear Facilities Safety Board) DNFSB/TECH-33. https://www.dnfsb.gov/sites/default/files/document/9261/TECH-33.pdf.

Romanovskiy, V. N. 2002. "Radium Institute Research on Actinide Separation from High-Level Waste (Review)." *Journal of Nuclear Science and Technology* 39 (sup3): 8–13. https://doi.org/10.1080/00223131.2002.10875398.

Romanovskiy, V. N., I. V. Smirnov, V. A. Babain, T. A. Todd, R. S. Herbst, J. D. Law, and K. N. Brewer. 2001. "The Universal Solvent Extraction (UNEX) Process. I. Development of the UNEX Process Solvent for the Separation of Cesium, Strontium, and the Actinides from Acidic Radioactive Waste." *Solvent Extraction and Ion Exchange* 19 (1): 1–21. https://doi.org/10.1081/SEI-100001370.

Romanovskiy, V. N., I. V. Smirnov, V. A. Babain, and A. Yu Shadrin. 2011. "9 - Combined Processes for High Level Radioactive Waste Separations: UNEX and Other Extraction Processes." In *Advanced Separation Techniques for Nuclear Fuel Reprocessing and Radioactive Waste Treatment,* edited by Kenneth L. Nash and Gregg J. Lumetta, 229–265. Oxford, UK: Woodhead Publishing.

Rubisov, V. N., and A. S. Solovkin. 1987. "Simulating the PUREX Process and the Effects of Di-Normal-Butylphosphoric Acid on the Equilibrium in Weak-Acid Actinoid Reextraction." *Soviet Atomic Energy* 63 (1): 520–524. https://doi.org/10.1007/Bf01125150.

Schroeder, N. C., M. Attrep, and T. Marrero. 2001. *Technetium and Iodine Separations in the UREX Process.* (Los Alamos, NM: Los Alamos National Laboratory). LA-UR--01-6607. http://lib-www.lanl.gov/cgi-bin/getfile?00791847.pdf.

Shook, H. E. J. 1988. *Solvent Degradation Products in Nuclear Fuel Processing Solvents.* (Du Pont de Nemours (E.I.) and Co., Aiken, SC. Savannah River Lab.). DP-1759; Other: ON: DE89001169 United States 10.2172/7057142 Other: ON: DE89001169 NTIS, PC A03/MF A01; 1. SRS English. https://www.osti.gov/servlets/purl/7057142.

Sokhina, L. P., S. I. Rovnyi, L. V. Goncharuk, and F. A. Bogdanov. 1988. "Solubility of the Di-Normal-Butylphosphate of Tetravalent Plutonium." *Soviet Radiochemistry* 30 (3): 394–396, https://inis.iaea.org/search/search.aspx?orig_q=RN:20083354.

Srinivasan, T. G., and P. R. Vasudeva Rao. 2014. "Red Oil Excursions: A Review." *Separation Science and Technology* 49 (15): 2315–2329. https://doi.org/10.1080/01496395.2014.928321.

Sugai, H. 1992. "Crud in Solvent Washing Process for Nuclear Fuel Reprocessing." *Journal of Nuclear Science and Technology* 29 (5): 445–453. https://doi.org/10.1080/18811248.1992.9731550.

Svantesson, I., I. Hangstroem, G. Persson, and J. O. Liljenzin. 1979. "Separation of Americium and Neodymium by Selective Stripping and Subsequent Extraction with HDEHP Using DTPA-Lactic Acid Solution in a Closed Loop." *Radiochemical and Radioanalytical Letters* 37 (4–5): 215–22.

Swanson, M. T., and M. B. Reed. 2006. *CMX and TNX Savannah River's Pilot Plants* (Stone Mountain, GA: New South Associates). New South Associates Technical Report 1329.

Tachimori, S., T. Nakashima, Z. Kolarik, and R. Schuler. 1983. *Distribution of U(IV), U(VI), Pu(III) and Nitric Acid Between 30 Vol.% Tributyl Phosphate in Dodecane and Aqueous Nitrate Solutions.* Karsruhe, Germany.

Tahraoui, A., and J. H. Morris. 1995. "Decomposition of Solvent Extraction Media During Nuclear Reprocessing: Literature Review." *Separation Science and Technology* 30 (13): 2603–2630. https://doi.org/10.1080/01496399508013706.

Taylor, R. 2015. *Reprocessing and Recycling of Spent Nuclear Fuel*. Oxford, UK: Woodhead Publishing. https://doi.org/10.1016/B978-1-78242-212-9.09987-1.

Taylor, R., M. Carrott, H. Galan, A. Geist, X. Hères, C. Maher, C. Mason, R. Malmbeck, M. Miguirditchian, G. Modolo, C. Rhodes, M. Sarsfield, and A. Wilden. 2016. "The Euro-GANEX Process: Current Status of Flowsheet Development and Process Safety Studies." *Procedia Chemistry* 21: 524–529, https://doi.org/10.1016/j.proche.2016.10.073.

Taylor, R. J., I. S. Denniss, and A. L. Wallwork. 1997. "Neptunium Control in an Advanced PUREX Process." *Nuclear Energy* 36 (1): 39–46, http://inis.iaea.org/search/search.aspx?orig_q=RN:28035891.

Taylor, R. J., C. R. Gregson, M. J. Carrott, C. Mason, and M. J. Sarsfield. 2013. "Progress Towards the Full Recovery of Neptunium in an Advanced PUREX Process." *Solvent Extraction and Ion Exchange* 31: 442–462. https://doi.org/10.1080/07366299.2013.800438.

Todd, T. A., J. D. Law, and R. S. Herbst. 2001. "Separation of Radionuclides from Aqueous Mixed Waste Solutions Using Liquid-Liquid Extraction Processes." In *Hazardous and Radioactive Waste Treatment Technologies Handbook*, edited by Chang Ho Oh, 3.3–1. Boca Raton, FL: CRC Press.

Tsujino, T., and T. Ishihara. 1966. "Radiation Damage to TBP/Kerosene Solvent, (II)." *Journal of Nuclear Science and Technology* 3 (4): 144–149. https://doi.org/10.1080/18811248.1966.9732295.

Uetake, N. 1989. "Precipitation Formation of Zirconium-Dibutyl Phosphate Complex in PUREX Process." *Journal of Nuclear Science and Technology* 26 (3): 329–338. https://doi.org/10.1080/18811248.1989.9734314.

Ursu, I. 1985. "Nuclear Material Recovery from Irradiated Fuel and Recycling." In *Physics and Technology of Nuclear Materials*, Chapter 10, edited by Ioan Ursu, 318–366. Oxford, UK: Pergamon.

Usachev, V. N., and G. S. Markov. 2003. "Incidents Caused by Red Oil Phenomena at Semi-Scale and Industrial Radiochemical Units." *Radiochemistry* 45 (1): 1–8. https://doi.org/10.1023/A:1022353014980.

Vandegrift, G. F., M. C. Regalbuto, S. Aase, A. Bakel, T. J. Battisti, D. Bowers, J. P. Byrnes, et al. 2004. "Designing and Demonstration of the UREX+ Process Using Spent Nuclear Fuel" *Advances for Future Nuclear Fuel Cycles International Conference*, Nimes, France, June 21–25.

Wagner, R. M. 1951. *Hydrolysis Products of Tributylphosphate and Their Effect on the Tributylphosphate Process for Uranium Recovery*. (Richland, WA: Hanford Works). HW-19959. https://www.osti.gov/biblio/4163567-hydrolysis-products-tributyl-phosphate-effect-tributyl-phosphate-process-uranium-recovery.

Weaver, B., and F. A. Kappelmann. 1964. *TALSPEAK: A New Method of Separating American and Curium from the Lanthanides by Extraction from AN Aqueous Solution of AN Aminopolyacetic Acid Complex with a Monoacidic Organophosphate or Phosphonate* (Oak Ridge, TN: Oak Ridge National Laboratory). ORNL-3559.

Weaver, B., and F. A. Kappelmann. 1968. "Preferential Extraction of Lanthanides over Trivalent Actinides by Monoacidic Organophosphate from Carboxylic Acids and from Mixtures of Carboxylic and Aminopolyacetic Acids." *Journal of Inorganic and Nuclear Chemistry* 30: 263–272.

Wellock, T. "CRUD: Another Acronym Bites the Dust." *U.S. NRC Blog* (blog), United States Nuclear Regulatory Commission. March 31, 2015.

Woodfield, F. W. 1951. *Redox Technical Manual* (Richland, WA. HW-18700-DEL). HW-18700-DEL. https://www.osti.gov/opennet/detail?osti-id=16429093.

Wright, A., and P. Paviet-Hartmann. 2010. "Review of Physical and Chemical Properties of Tributyl Phosphate/Diluent/Nitric Acid Systems." *Separation Science and Technology* 45 (12–13): 1753–1762. https://doi.org/10.1080/01496395.2010.494087.

Zalupski, P. 2011. *Non-Ideal Behavior in Solvent Extraction* (Idaho Falls, ID: Idaho National Laboratory). INL/EXT-11-23462.

Zimmer, E., and J. Borchardt. 1986. "Crud Formation in the PUREX and THOREX Processes." *Nuclear Technology* 75 (3): 332–337. https://doi.org/10.13182/Nt86-A33845.

5 Filtration

Philip P. Schonewill

CONTENTS

5.1 INTRODUCTION TO FILTRATION

Filtration is a multiphase separation process that excludes solid particles suspended in a liquid or gas phase using a porous media or membrane. This specific subfield of filtration is also commonly called "membrane" filtration to distinguish it from other types of industrial filtration. The material passed through the porous media is a fluid stream (the filtrate or permeate) that is nominally free of solids; in some cases, it may contain solids that are size reduced compared to the size distribution of particles in the bulk fluid. Solid particles of sufficient size to be captured by the porous media are deposited on the filter media surface or entrapped in the porous media itself. The separation that is performed is based on physical principles; the filter is selected so that the solid particles in the fluid are larger than the filter media pore size(s) and are consequently size excluded from passing through. In traditional chemical engineering applications, filtration is used for a wide variety of separation processes; the primary objective is the isolation of a desired material by quarantining it from the undesired material in contact with it. The material of interest may be the solids, which then need to be recovered from the porous media, or it may be the solids-free liquid or gas stream. The versatility of the technique lends itself to several process functions: feed preparation or pretreatment, solid concentration, solid separation, polishing, contaminant control, and downstream equipment protection.

Filtration in nuclear processing applications is an important separation technology used to remove or contain certain constituents to produce a stream that has more desirable than undesirable properties or attributes. However, filtration in nuclear processing has to consider conditions, i.e., radiological hazards, applicable nuclear regulations, and waste volume minimization, that do not apply to filtration in other industrial settings. Even a mundane auxiliary function such as filtering exhaust from air spaces in nuclear facilities (which typically use High Efficiency Particulate Air (HEPA) filters to capture airborne radioactive solid particles) can be complicated by the requirements to protect the worker and environment from the impacts of radioactive particles. In general, the combination of traditional filtration principles and the restrictions inherent in operating filtration systems in a nuclear facility create unique engineering design and operation challenges.

In this chapter, filtration engineering principles applicable to nuclear processing will be discussed, with a focus on solid–liquid separations using filtration. Most nuclear fuel or waste processing is performed in the liquid phase. The majority of gas–solid separations occur in secondary or utility systems, i.e., air handling systems or off-gas management systems (emission control). There are several comprehensive descriptions and discussions of gas–solid filtration as it pertains to nuclear facilities that are useful to reference (Goossen, Eichholz, and Tedder 1991; International Atomic Energy Agency 1991; *Nuclear Air Cleaning Handbook* 2003; Vendel et al. 2009). After a discussion of general filtration principles, the phenomena that impact two general types of filtration (dead-end and crossflow) performance are discussed in detail. With that background in hand, engineering considerations regarding the use of filtration in nuclear applications are discussed. The last part of this chapter discusses some specific examples of filtration used in existing and planned nuclear processing facilities; these case studies may offer some insight into how the technologies have been or are in use.

5.2 GENERAL ASPECTS OF FILTRATION

Solid–liquid filtration is typically performed in one of the two basic geometries pictured in Figure 5.1. Although there is considerable variation in filtration equipment, the suspension is either carried by a bulk flow that dead-ends against a filter media (e.g., a dead-end filter, depicted in the top panel of Figure 5.1) or forced through the filter media despite a bulk flow that is tangential or has an alternative outlet (a crossflow filter, shown in the bottom panel of Figure 5.1). The dead-end filter, if properly sized, will retain all the solid particles in the suspension while permitting the liquid to permeate through. As the processing time increases, the thickness of the solids layer increases. Increased thickness of the solids layer (or a change in porosity or solids layer structure) makes the deposited layer more resistant to flow. This resistance builds over time, and obtaining the same permeate rate in the filter requires more driving force. Alternatively, the crossflow filter takes advantage of the tangential bulk flow to limit the amount of solids that

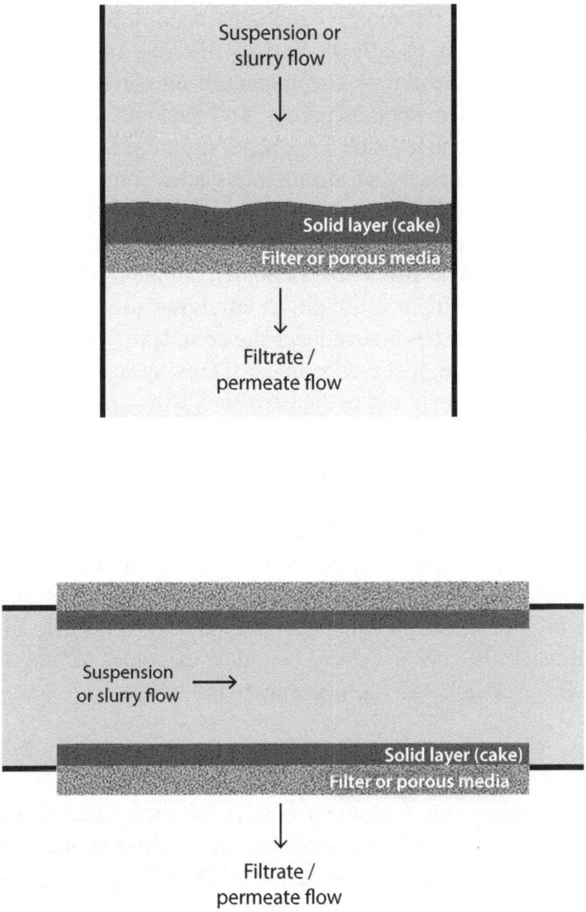

FIGURE 5.1 Representation of two basic filtration geometries: dead-end (top panel) and crossflow (bottom panel).

deposit on the filter media surface. This requires striking a balance between the force that drives the liquid through the filter media and the convective–diffusive phenomena that keep solid particles from depositing on the surface. In general, solids layer growth in a crossflow filter over time reaches a constant (or near-constant) value and, consequently, the permeate flow rate becomes constant for a fixed driving force.

The driving force for filtration is typically an applied pressure. The applied pressure can be generated by:

- Hydrostatic head (gravity);
- Pressure increased upstream of the filter, i.e., pumping;
- Pressure reduced downstream of the filter, i.e., drawing a vacuum;
- Centrifugal force across the filter.

Filter flux requirements in the more challenging solid–liquid separations encountered in nuclear processing usually necessitate the use of increased pressure to generate sufficient permeate flows. The increased pressure is generated by some combination of motive force, such as pumps, and backpressure, such as throttle or backpressure valves. In some industrial applications, the filter permeate stream may be discharged to an open vessel, but filtration in nuclear processing is performed in closed systems where the filtrate may still be at pressures higher than atmospheric in the system. Closed systems are used because the suspension being filtered may have a radiological hazard or dispersible radiological material present. For a similar reason, vacuum and centrifugal filtration are often not used in nuclear processing environments because their absence reduces the complexity of equipment needed to reliably operate in a radiological environment. Drawing a vacuum also introduces the possibility of volatilizing liquid in the permeate, which might increase the ventilation demands on a vessel.

The driving force is expressed as a pressure drop (also known as the transmembrane pressure [TMP]), which represents the difference between the solid–liquid (suspension or slurry) and liquid-only (permeate) streams. From a macroscopic perspective, the pressure drop is approximately the same everywhere across the media and is simply the difference between slurry-side and permeate-side pressure. However, there are variations in filter geometry that introduce local variations in TMP that can impact filter performance. For some examples of this, see Figure 5.2. In Figure 5.2a, the average TMP (ΔP) is expected to be

$$\Delta P = P_1 - P_2 \tag{5.1}$$

where P_2 is the permeate-side pressure and P_1 is the slurry-side pressure. Filtration only occurs if $P_1 > P_2$. As solids are deposited on the filter media surface (presumably homogeneously or nearly so), the actual TMP through the filter media alone will evolve with time to depend on the value of $P_{1,fc}(t)$, which is the value of P_1 at the filter–cake interface; the pressure will decrease as the fluid travels through the layer of solids in the surface cake. Figure 5.2a is a simplified dead-end filtration geometry; other representative dead-end geometries are shown in Figure 5.2b and c. These represent cases where the dead-end filter is a cartridge or finger geometry, which is a common configuration. In these configurations, there is still a differential transmembrane pressure as defined by Equation 5.1, but the cake layer may not be uniform across the filter media as shown. Consequently, the TMP may vary locally with position as well as with time due to the cake morphology. The diagrams shown in Figure 5.2b and c could have locations where the local driving pressure at the filter media surface $P_{1,fc}(t)$ would be different than in other locations depending on the cake characteristics.

Figure 5.2d represents a common crossflow filtration geometry. In this case, for a channel or tube of length L, the pressure drop across the filter tangential to the filter surface may be significant ($P_{1,\text{out}} - P_{1,\text{in}}$). Thus, an average TMP can be defined as

$$\Delta P = \frac{P_{1,\text{in}} + P_{1,\text{out}}}{2} - P_2, \tag{5.2}$$

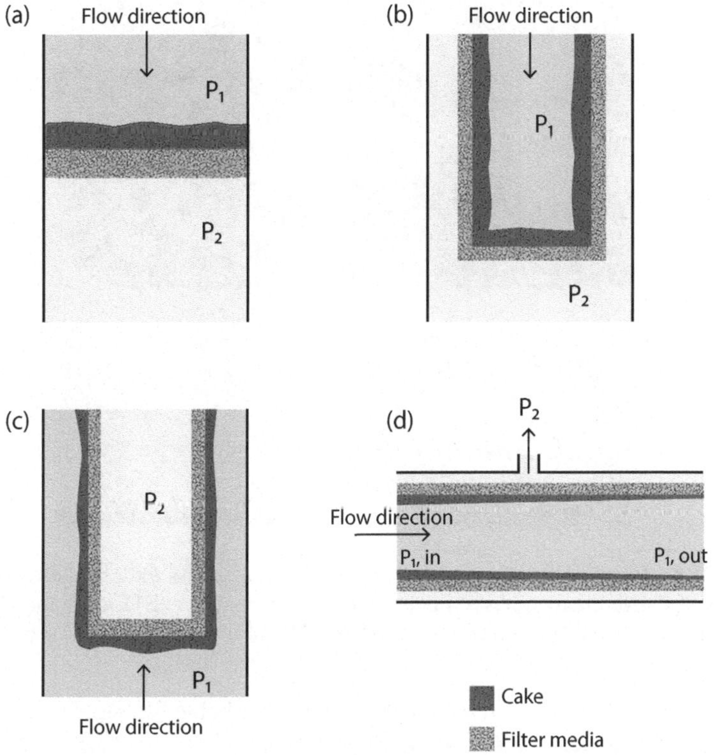

FIGURE 5.2 Local variations in applied pressure drop (TMP) during filtration: (a) plate (dead-end) geometry, (b) outflow finger (dead-end) geometry, (c) inflow finger (dead-end) geometry, and (d) horizontal crossflow geometry.

and assuming that the slurry flow is in the x direction and a constant pressure drop is present, the TMP ΔP_x varies with axial location as follows:

$$\Delta P_x = \left[\left(1 - \frac{x}{L}\right) P_{1,\text{in}} + \frac{x}{L} P_{1,\text{out}}\right] - P_2. \tag{5.3}$$

In addition, a layer of solids may accumulate on the surface and may vary with axial position (x) depending on the amount of elapsed time (or volume filtered). However, unlike the dead-end case, the layer thickness in crossflow filtration should eventually become constant with both time and position. While the layer is developing, a region of lower pressure at the filter surface exists where the layer is thicker (near the left-hand side of the filter as it is shown in Figure 5.2d).

The efficacy of the applied pressure in generating permeate through the filter media depends on the native resistance of the filter, R_m. The resistance is governed by the microscopic structure of the filter, including the size and distribution of the pores, as well as the thickness of the media. In some cases, the filter may have apparent increases in resistance due to poor wettability, chemical, or physical changes to the

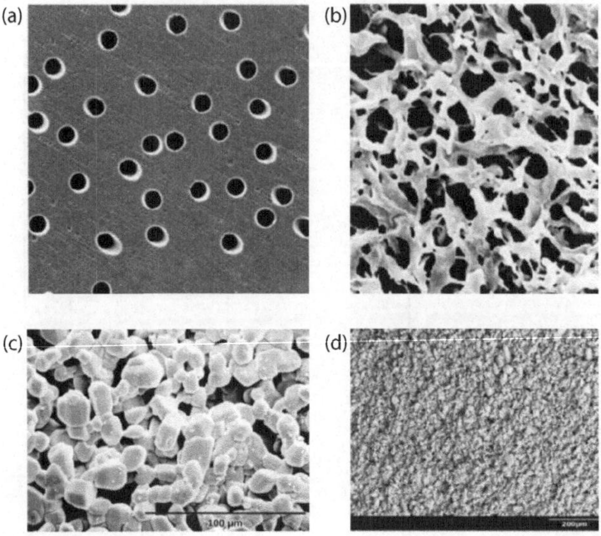

FIGURE 5.3 Examples of the variation in pore structure across filter media types, shown with representative micrographs: (a) polyethylene terephthalate (PETE), (b) polyvinylidene fluoride (PVDF), (c) silicon carbide (Source: DieselNet.com), and (d) sintered stainless steel.

media over time or unique interactions with the slurry being filtered. Some examples of filter microstructure are shown in Figure 5.3; there is a wide variety of potential filter media, each with different pore structures and permeabilities. The microstructure of the media has a significant role in the quantity and quality of solids captured. For nuclear processing applications, it is also important to verify that the filter material is robust with respect to being irradiated; there are several reference works containing in-depth material on this topic (Barbu and Martin 1992; Clough and Shalaby 1991; Kinoshita, Fukumoto, and Nakai 1991; Levy 1991; Zinkle 1999). The question of robustness has often led to the use of metal and metal alloy filters instead of polymers, ceramics, or other similar materials because filtration systems are often designed to be in use for long design lives (>20 years) in nuclear facilities.

The media selected for the filtration operation can also impact the nature or mechanism of the separation process. Filtration mechanisms are divided into two broad categories: (1) depth, deep bed, or clarifying filtration; and (2) surface or cake filtration. For simplicity, these will be referred to as depth and cake filtration throughout this chapter; these terms are descriptive of the unique filtration processes that distinguish them. Depth filtration separates solid particulate from the liquid stream by trapping the particles inside the filter media's pores at depths below the filter surface. In this case, the majority of the particles would be smaller than the filter pores and, critically, the pore openings on the filter surface. The penetration depth of the particles will be controlled by the morphology and configuration (tortuosity, isotropy/anisotropy) of the pore network in the filter, as well as the pressure gradient that induces the flow. Cake filtration separates solid particulate by retaining them on the filter surface in a cake layer of solid particles. The cake forms because there is

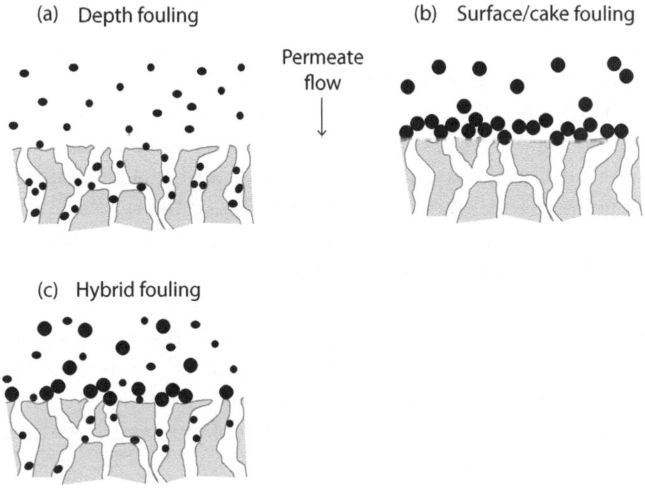

FIGURE 5.4 Representation of filtration fouling mechanisms: (a) depth filtration, (b) cake filtration, and (c) hybrid of depth and cake filtration.

a sufficient inventory of particles in the slurry that are larger in size than the filter pores on the media surface; these large particles are prevented from penetrating the media and instead pile up on top of each other. Figure 5.4 illustrates the physical difference between these two mechanisms and also shows a representation of a hybrid mechanism where both depth and cake filtration occurred. A hybrid mechanism could arise in a few plausible filtration scenarios:

- *Hybrid due to filtration initialization* occurs if some particles are retained at depth during initiation of the filtration operation before a cake can be established. This could occur if the driving pressure is sufficient to drive some particles into the filter media, or because there is a wide distribution in particle size in the slurry, including suspensions where the particles strongly agglomerate into (effectively) larger particles. At short time scales, the dominant mechanism is depth filtration; and at longer time scales, the controlling mechanism is cake filtration.
- *Hybrid due to high cake permeability* occurs if the particle distribution is broad enough for both mechanisms to be possible simultaneously. A cake is formed of primarily larger particles with enough open area to permit small particles to flow through the cake and deposit at depth in the filter media. In practice, the rate of depth filtration would become less pronounced over time: as the cake thickness increases, the probability that a smaller particle will pass through the cake (a filter in itself) decreases.

Filtration behavior that appears to have a hybrid mechanism could also arise from the filtration of soft, amorphous particles or gels. These more malleable materials may be stopped from penetrating into the filter but may also block pore openings at the surface or near-surface. This mechanism is sometimes described as "pore blocking" or "blocking." For well-controlled systems, e.g., a suspension of homogenous hard spheres of known size filtered by media with cylindrical, straight pores, a priori prediction of the type of filtration mechanism can be made by inspection. Most real systems have additional complexities that make a priori prediction challenging, and it is common to test filtration systems to determine performance empirically.

Filtration operations are often classified in terms of the type and size of species retained by the filter media. Three broad classifications are reverse osmosis (RO), ultrafiltration (UF), and microfiltration (MF):

- RO membranes target low-molecular-weight salt species for retention.
- UF membranes typically target the retention of high-molecular-weight salts, macromolecules, and colloidal materials.
- MF membranes are employed to separate colloids and larger particulate materials.

A summary of the membrane pore size and target applications of each filtration operation type is discussed in more detail elsewhere (Fane, Tang, and Wang 2011; Ripperger and Altmann 2002; Van Der Bruggen et al. 2004). The approximate ranges of application in terms of ΔP and particle size (d_p, the particle diameter) are summarized in Figure 5.5 (based on information in Schulz and Ripperger (1989) and Belfort, Davis, and Zydney (1994)). Generally, since the nuclear waste slurries or reprocessing solutions being filtered contain particles 10 μm in diameter or greater, most nuclear applications would be classified as microfiltration. This does not preclude the presence of sub-micron colloidal particles, which are quite common in nuclear filtration applications and may be present in a suspension with larger

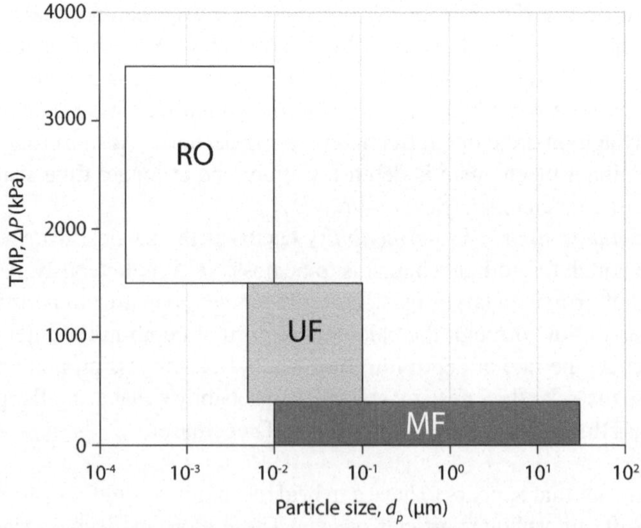

FIGURE 5.5 Range of particle size (d_p) and TMP (ΔP) for common filtration classifications.

particles. Colloidal particles can also be the major component of the solid phase in a low-solids concentration stream; these are generated by decanting liquid from a settled suspension or from precipitation occurring in super-saturated suspensions.

Filtration is sensitive to the nature of the solid particles being separated. Thus, in many industrial applications, the liquid and solid phases are well known and well controlled in several ways:

- Solids concentration fed to the filter is approximately constant;
- Speciation of the solids fed to the filter is well defined and unlikely to vary;
- The distribution of solid particle size fed to the filter is reasonably precise and does not change much during processing; and
- The suspending phase is fixed with known properties (density, viscosity, composition).

Variations in the aspects listed above can have significant impacts on performance. Nuclear applications, especially when treating nuclear waste, are notoriously uncertain regarding the definition of the liquid and solid phases and their variability over a processing campaign. The uncertainty arises from combining nuclear wastes of different compositions together in storage or lag vessels, extended contact of the wastes with radiation and thermal cycling that alters suspension chemistry, and adding liquids to manipulate pH, mitigate gas generation, or dilute concentrated mixtures. This uncertainty is further compounded because scale-up testing of a filtration process is often prohibitively expensive if it is tested with actual nuclear waste or post-reactor material. The extra expense is due to the cost of handling suspensions with radiological hazards, additional equipment and staff needed to conduct safe operations, and disposal of post-test contaminated material. Scale-up testing is usually conducted with surrogate suspensions that are commonly called waste (or radiological) simulants because, if designed well, these surrogate suspensions "simulate" the key characteristics of the actual material. Thus, nuclear filtration systems have to be designed for this additional uncertainty and must possess the ability to accommodate a wider range of feed suspensions while still delivering the necessary throughput. Some considerations for filter design and operation due to this limitation are discussed in more detail in Section 5.4.

Filtration is performed in either an intermittent or continuous mode depending on the separation needs of the nuclear process. In the case of nuclear waste treatment, the filtration operation is usually coordinated with a discrete volume of waste to be treated (say from a waste storage tank) and so may have periods where it is not operating while the next waste volume is prepared for feeding into the filtration system. While treating the discrete waste volume, the filtration system may be processing waste in a series of smaller batches (because the operating facility needs to perform other treatment steps with smaller volumes) or nearly continuously. This assumes that the nuclear waste being treated was generated by historical production campaigns and represents a legacy to be cleaned up rather than being treated as it is generated. As of 2019, all nuclear waste in the United States falls under this legacy category; it was generated during the weapons production mission conducted at several sites across the country. The United States does not have currently operating, used-fuel reprocessing plants, but reprocessing as practiced internationally typically relies on solvent extraction and precipitation processes (see Chapters 3 and 4) to perform key separations. In other nations—primarily France, Great Britain, Russia, Japan, and India—used nuclear fuel from reactors is reprocessed; any use of filtration (if used at all) for this application would be performed periodically as used fuel becomes available for processing. There are, however, some examples of filtration's use in nuclear power plants to treat wastewater or recycled chemical solutions (Asanuma et al. 2001; Chadwick and Swain 2014; Dang, Li, and Choo 2016; Efremenkov 1989; Nuclear Regulatory Commission 2012). Employing filtration in those processing applications is on the rise (Rana et al. 2013). Nevertheless, as already indicated, the majority of filtration systems in radiological service are employed in support of nuclear waste treatment objectives; there are several summary-level reviews on this topic (Ambashta and Sillanpaa 2012; Chmielewski, Harasimowicz, and Zakrzewska-Trznadel 1999; Dlouhy 1982; Rahman, Ibrahium, and Hung 2011; Zakrzewska-Trznadel 2006, 2013; Zakrzewska-Trznadel, Harasimowicz, and Chmielewski 2001). Regardless of the application, it is important to carefully select the filter media and operating parameters needed for operation over long periods of time in either a batch or continuous configuration because maintenance or cleaning of the filter system may not be possible without significant effort.

5.3 THEORY OF FILTRATION

Filtration theory is formulated based on a low Reynolds number treatment that describes the transport and flow in porous media. The Reynolds number in this instance is usually defined as:

$$Re = \frac{\rho U_F d}{\mu} = \frac{Jd}{v} \tag{5.4}$$

where ρ, μ, and v are the density, dynamic viscosity, and kinematic viscosity of the fluid, respectively. The linear velocity in the filter media U_F is usually approximated using the filter flux J. The flux is the flow rate of permeate (Q) generated through the cross-sectional filter area (A_F). The length-scale d is commonly selected to be the grain (or particle) diameter of the filter media to represent the media's nominal pore size. In the low Reynolds number regime, Darcy's law (Darcy 1856) governs the flow behavior of a fluid. Darcy's law is only valid if the flow is laminar (low Re), and this holds when $Re < 10$. There is a transition zone up to a Re of ~100 where the flow is still laminar but the assumptions in Darcy's law no longer hold; however, the application of Darcy's law when $10 < Re < 100$ should still provide a reasonable estimate of performance (Bear 1972). Darcy's law describes the relationship between pressure and flow rate in a porous media; the original construction was based on describing this relationship as:

$$Q = KA_F \frac{\Delta h}{\Delta_F}, \tag{5.5}$$

where K is a constant and $\Delta h/\Delta_F$ is a change in fluid head over a filter thickness. The definition of filter flux arises directly from Equation 5.5. The original expression is often rearranged in various ways, but one common form is:

$$J = \frac{\Delta P}{\mu R}, \tag{5.6}$$

where R is the filter media's resistance. Here, the constant K is recast as $\sim\Delta_F/\mu R$. Note that J, ΔP, R, and μ can all be functions of t. It is also common to write Equation 5.6 in differential terms, with the following substitution:

$$J = \frac{1}{A_F}\frac{dV}{dt}, \tag{5.7}$$

where V is the volume filtered. Equation 5.7 describes an instantaneous rate; most empirical data contributing to filter design and operation will be collected in a manner that approximates the rate at discrete times, e.g.,

$$J\left(\frac{t_{n+1}+t_n}{2}\right) = \frac{1}{A_F}\left(\frac{V_{n+1}-V_n}{t_{n+1}-t_n}\right), \tag{5.8}$$

where the subscripts n and n+1 refer to discrete points in time with t_{n+1} occurring at some increment after t_n. The flux J would then be reported at the mid-point of t_{n+1} and t_n and represent the "average" flux over that entire period.

In writing Equation 5.6 algebraically, the assumptions are (a) that the pressure gradient is uniform and linear, (b) that the viscosity is not a strong function of location or time, and (c) that a macroscopic resistance term represents the flow behavior through the entire filter media, i.e., the media is isotropic and of uniform thickness. If these assumptions break down for a given filtration operation, the expression(s) governing the pressure–flow relationship quickly become complex and difficult to handle analytically. Most filtration models or descriptions of performance start with Equation 5.6 as a basis and modify that expression to handle departures from the assumption set. Darcy's law does not distinguish between variations in the physics of the suspension being fed to the filter, i.e., the extent of mixing, the flow regime, and the chemical and physical nature of the suspension itself (aside from the viscosity). Many of these aspects are more important for estimating and predicting crossflow filtration performance than dead-end filtration performance because they govern what state the suspension is in as it interacts with the filter media. The impact of the physics of the suspension on crossflow filtration will be discussed in more detail in Section 5.3.2.

Equation 5.6 also mathematically illustrates the two primary modes of typical filter operation: constant pressure and constant flow (flux). If the transmembrane pressure is fixed, then the filter flux of a suspension with constant viscosity will change over time in response to changes in the filter resistance. Likewise, fixing the flux requires that the TMP changes over time in response to changes in the filter resistance. The mode of operation that is used is a design choice dependent on the needs of the process. For production facilities, filtration is more likely to be operated in constant flux mode in order to meet the throughput targets of the facility.

Filtration performance data is commonly presented as flux (when operating in constant pressure mode) or pressure (when operating in constant flux mode) versus time to exhibit the temporal behavior; the volume filtered can be used instead of time. The curvature in the performance data provides information about the nature of the fouling that occurs during the filter's operation. Knowledge of the dominant fouling mechanism helps determine what operating strategies should be used to recover performance when the increase in required pressure or decrease in filter flux falls outside the required range. The performance data collected depends on the filter type, and impacts the operating strategies that are relevant for performance recovery. The two primary types of filters used in nuclear separations are discussed in detail in Sections 5.3.1 and 5.3.2. For additional information, many review articles covering a range of topics on the theory and practice of both filtration types have been published; the reader is referred to several of these reviews (Bacchin, Aimar, and Field 2006; Belfort, Davis, and Zydney 1994; Bowen and Jenner 1995; Davis 1992; Kim and DiGiano 2009; Matthiasson and Sivik 1980; Patel et al. 2010), among others.

5.3.1 Dead-End Filtration

As described previously, dead-end filtration is one of the two general filtration types used to separate solid particles from suspensions. The filter media is usually selected in order to obtain complete rejection of solids above a certain size to produce a liquid filtrate or permeate. In many cases, the objective is to produce a solids-free filtrate to simplify the handling of the liquid stream in subsequent processes downstream of the filtration step. In dead-end filtration, all of the suspension sent to the filter interacts with the filter media, whether as filtrate that is pushed through the filter and moves downstream or as solid particulate that remains behind and accumulates in the filter cake (or embeds in the filter media itself). Filter selection is driven by several factors, but particularly important parameters include the particle size distribution of the solids, which governs the pore size and rating of the filter media, and the permeability of the filter media, which controls the baseline (i.e., clean filter) pressure drop and the maximum achievable flux.

Dead-end filtration is more suitable for processing scenarios in which accumulation of solids on the filter will be relatively slow, thus maintaining a low resistance to flow. Since all solid particles are captured during the filtration step, the solid layer on the filter grows with time and increases the overall resistance of the system. Examples of scenarios where this would be common are:

1. Filtering nuclear waste suspensions with low solids concentrations; what defines a suspension of "low" solids concentration will vary depending on the particle size and speciation of the particulate that is present but will generally be less than a few mass percent.
2. Filtering to protect other processing equipment; because the primary function is to protect or guard downstream equipment, filters selected for this

purpose are likely to target total rejection of any residual or unexpected solids fed to it and will not necessarily be optimized for throughput.

3. Filtering small-volume batches of nuclear waste suspensions with sufficient time between campaigns to perform cleaning or maintenance; the time required to remove the solids that have accumulated on the surface will be on the order of hours to days depending on the nature of the solids being filtered.

The expectation in dead-end filtration is that solid particles will develop a cake on the filter surface that grows with the processing time (or volume of suspension filtered). The theory relating the growth in cake thickness to flux when filtering at constant pressure was originally proposed by Ruth and others (Ruth 1935a, 1935b; Ruth, Montillon, and Montonna 1933a, 1933b). The development of Ruth's theory starts from the Darcy's law formulation of Equation 5.6 and expands the resistance to include a membrane resistance (R_m) and a resistance due to the formation of the cake (R_c):

$$J = \frac{\Delta P}{\mu\left(R_c + R_m\right)} = \frac{\Delta P}{\mu\left(\dfrac{\alpha V}{A_F} + R_m\right)}. \tag{5.9}$$

R_m is assumed to remain constant and is a characteristic of the chosen filter media, whereas α is the specific cake resistance and is defined as:

$$\alpha = \frac{k\left(1-\varepsilon\right)S_o^2}{c_s \rho_p \varepsilon^3}, \tag{5.10}$$

where k is a constant, ε is the porosity of the cake, S_o is the specific surface area of the particles, c_s is the mass of solids per unit volume of filtrate, and ρ_p is the particle density. Equation 5.10 is derived from the Carman-Kozeny relation (Carman 1937; Kozeny 1927), which is valid for packed beds of particles with laminar flow. The constant $k = 4.17$ for random particles of definite size and shape may be different for other particle types. The specific cake resistance is expected to be a function of pressure because the cake porosity may be impacted by ΔP as the cake compresses. One way this is empirically formulated is:

$$\alpha = \alpha_o \left(\Delta P\right)^s, \tag{5.11}$$

with both α_o and s being constants determined from a series of filtration experiments at different constant pressures. From these experiments the pressure functionality of the cake resistance can be obtained; a plot of $\ln(\alpha)$ vs. $\ln(\Delta P)$ will yield a slope of s and intercept of $\ln(\alpha_o)$. The cake is incompressible (and not impacted by pressure) when $s = 0$. Typical values for s when the cake is compressible are between 0.1 and 0.8.

If the specific cake resistance is assumed to be constant, i.e., for an incompressible cake or a case of constant ΔP, then Equation 5.9 can be inverted and rearranged in the following form:

$$\frac{1}{J} = \frac{dt}{dV} = \left(\frac{\mu a}{\Delta P A_F^2}\right) V + \frac{R_m}{\Delta P A_F} \tag{5.12}$$

for simplicity, the two terms can be written $\mu\alpha(\Delta P A_F{}^2)^{-1} = \beta_1$ and $R_m(\Delta P A_F)^{-1} = \beta_2$. Figure 5.6 shows how data can be used to estimate β_1 and β_2 for a dead-end filter, i.e., cake filtration, that follows Ruth's law behavior. The intercept β_2 is governed by the resistance of the membrane (R_m), and the slope β_1 is proportional to α.

In some instances, dead-end filtration does not follow behavior that aligns with Ruth's law, meaning that the data has some non-linearity or curvature when plotted as in Figure 5.6. Classically, non-linear behavior is described by a series of blocking laws originally proposed by Hermia (Hermia 1982; Hlavacek and Bouchet 1993; Iritani et al. 2015), one of which is a cake filtration law analogous to Ruth's law or Equation 5.12. The four blocking laws describing fouling in dead-end filtration systems are: (1) complete blocking, (2) standard blocking, (3) intermediate blocking, and (4) cake filtration. These are usually described in the context of a membrane comprised of straight, cylindrical pores with uniform diameter and length. A graphical illustration of each blocking law is shown in Figure 5.7.

Mathematically, these four blocking laws are useful because they are reduced down to a single differential equation with an exponent n that specifies the law. The equation can be written either as:

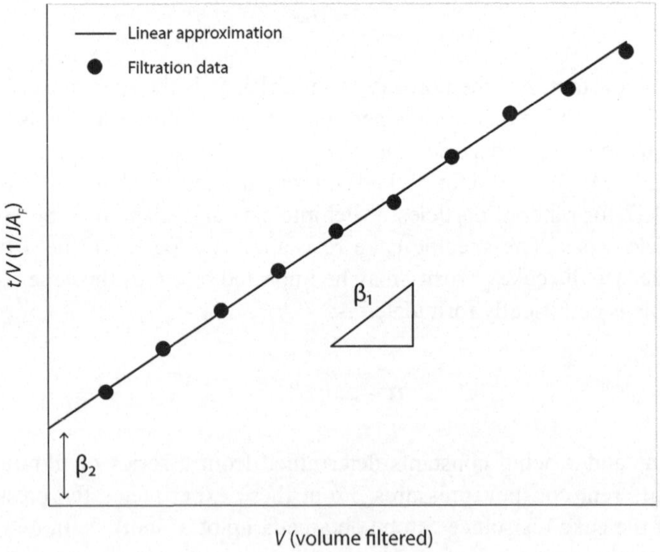

FIGURE 5.6 Example of dead-end filter data exhibiting Ruth's law behavior. The slope β_1 and the intercept β_2 are obtained by a linear fit (or approximation) of empirical data.

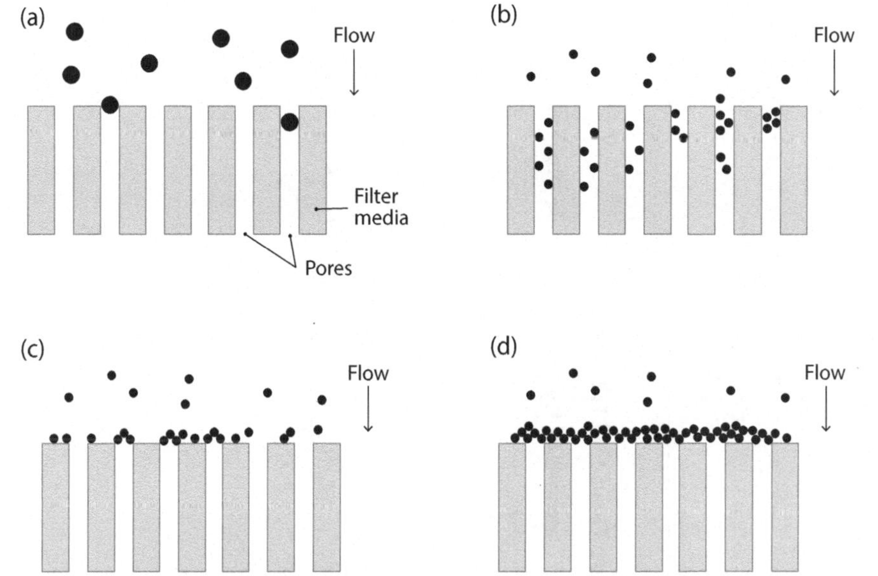

FIGURE 5.7 Schematic representation of the four classical blocking laws proposed by Hermia (1982).: (a) Complete blocking, (b) standard blocking, (c) intermediate blocking, and (d) cake blocking. (From Hermia, J., *Transactions Institution Chem. Eng.*, 60, 183–187, 1982.)

$$\frac{d^2t}{dv^2} = K\left(\frac{dt}{dv}\right)^n, \tag{5.13}$$

where $v = V/A_F$ and K is the blocking constant, or

$$\frac{dJ}{dt} = -KJ^{3\ n}, \tag{5.14}$$

which uses the flux construction ($J = dv/dt$) first devised by Sun et al. (Sun et al. 2003). Table 5.1 shows the exponent n for each blocking law and the analytical solution for Equations (6.13) and (6.14) for each form. The constant K is given a law-specific subscript because it has a unique value that depends on which blocking law is assumed. There have been many first-principles efforts to estimate the blocking constants a priori, but they are overwhelmingly assigned values through empiricism. For example, take the intermediate blocking law. The $J(t)$ solution in Table 5.1 can be rearranged and written as:

$$\frac{1}{J} = K_{ib}t + \frac{1}{J_o}, \tag{5.15}$$

where J_o is the initial flux. Using Equation 5.15, a filter that exhibits intermediate blocking behavior at constant pressure would generate data such that a plot of $1/J$

TABLE 5.1
Blocking Law Exponents and Analytical Solutions for Constant Pressure Dead-End Filtration

Blocking Law	Complete Blocking	Standard Blocking	Intermediate Blocking	Cake Filtration
Value of exponent n	2.0	1.5	1.0	0.0
Constant	K_{cb}	K_{sb}	K_{ib}	K_{cf}
$v(t)$ solution	$v=\left(\frac{J_o}{K_{cb}}\right)\left\{1-\exp\left(-K_{cb}t\right)\right\}$	$v=\frac{1}{\frac{K_{sb}}{2}t+\frac{1}{J_o}}$	$v=\frac{1}{K_{ib}}\ln\left(1+K_{ib}J_o t\right)$	$v=\frac{t}{\frac{K_{cf}}{2}v+\frac{1}{J_o}}$
$J(t)$ solution	$J=J_o\exp(-K_{cb}t)$	$J=\frac{J_o}{\left(\frac{K_{sb}J_o}{2}t+1\right)^2}$	$J=\frac{1}{K_{ib}t+\frac{1}{J_o}}$	$J=\frac{J_o}{\left(1+2K_{cf}J_o^2t\right)^{1/2}}$

Source: Adapted from Iritani, E. and Katagiri, N., *Kona Powder and Part. J.*, 33, 179–202, 2016.

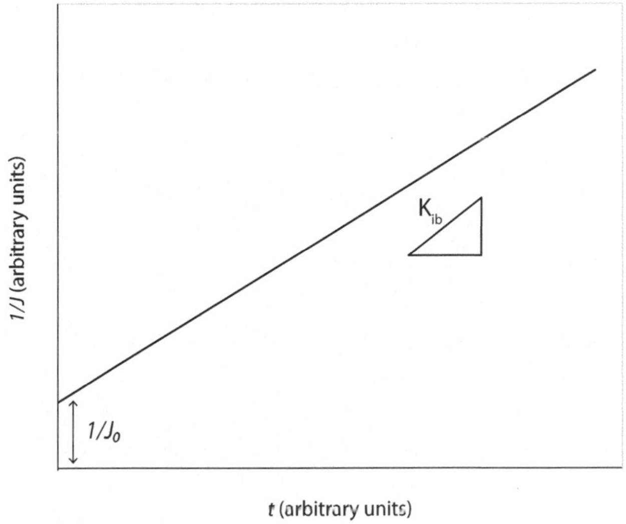

FIGURE 5.8 Example of an empirical determination of a blocking law constant at constant pressure.

versus t would have a slope of K_{ib} and an intercept of $1/J_o$ as shown in Figure 5.8. For a useful summary of these types of diagnostic plots for filter data, refer to Iritani and Katagiri (2016) or Grenier et al. (2008).

As the governing blocking law equations indicate, for the same values of ΔP, J_o, and t, the change in flux with time increases as n gets larger. A dead-end filter with complete blocking behavior fouls much faster and has a much lower flux than a dead-end filter with cake filtration behavior. For comparison, see Figure 5.9, which was generated using the $J(t)$ solutions in Table 5.1 where J_o is held constant ($J_o = 1$) for each blocking law and $K_{cb} = K_{sb} = K_{ib} = K_{cf} = 0.5$. This makes intuitive sense because a mechanism that cuts off flow to entire pores (complete blocking) reduces filtrate flow faster than mechanisms that do not. Standard and intermediate blocking are also more reductive than cake filtration because they impact the permeability of the filter media and increase its resistance, whereas the cake effectively adds a second filter media (with its own lesser permeability in most cases) constructed of solid particles. This is also clear from the mathematical expression in Equation 5.9: if R_m remains constant with time, then only the increasing cake resistance is forcing the filter flux to decrease at constant pressure.

Many real dead-end filter systems do not exhibit the flux-time performance of only a single blocking mechanism. Effectively, the filtration has a hybrid performance that combines multiple mechanisms at the same time, or it transitions from one mechanism to the other during operation. A common hybrid mechanism occurs during start-up of the filtration operation when the suspension being separated contains at least some particles that are smaller than at least some of the filter media's pores. The dead-end filter surface is initially clear of any solids, and when the suspension to be separated is presented to the surface, the exposed pore structure permits

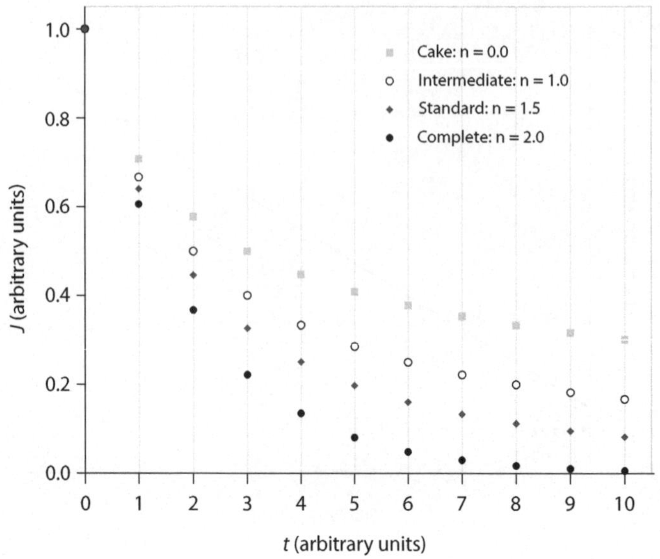

FIGURE 5.9 Comparison of the impact of dead-end filtration blocking laws on flux with operating time. The comparison assumes the same J_o and blocking constant for each law.

some particles to penetrate into the media until they become lodged somewhere in the pore network. Over time, the pore openings available to the solids decrease as a cake begins to form, and eventually the penetration of particles into the pore network becomes negligible. A filter experiencing this would not initially have linear t/V versus V behavior, such as that shown in Figure 5.6, until the cake began to build up appreciably enough to be the dominant resistance in the system. If the initial particle penetration occurred over a relatively short time frame, it may just be realized in a higher implied R_m due to a larger β_2.

The constant pressure operating mode is typical of lab-scale and research systems due to the ease of implementing a fixed-pressure delivery to the filter. However, constant flux operation is far more common in industrial systems and facilities such as nuclear waste plants. In this case, changes in filter resistance are interpreted via the changes in ΔP that are required to generate the flux. The same blocking laws that were used to build constant pressure models can also be applied to the constant flux case, starting from the differential equation:

$$\frac{dP}{dv} = K'P^n, \tag{5.16}$$

where P represents the applied filtration pressure (=ΔP) and K' is a constant. The prime indicates that the constants from constant pressure and constant flux data sets are not absolutely interchangeable, though they may be quantitatively similar. Thus, to provide the best information for design and scale-up, it is preferable to gather empirical data operating in the same mode as the intended nuclear application. Note that the exponent n is unchanged, owing to the fact that the value of n is the same in both cases for each blocking law. Table 5.2 summarizes the analytical solutions for each blocking law in constant flux operation. In each case, the solution requires knowledge of P_o, which is the initial driving pressure needed as filtration begins. In two of the four blocking laws, J_o also needs to be known in order to fully describe the pressure behavior. Since the filtered volume is fixed, writing the laws in terms of t or v is equivalent.

As for the case of constant-pressure blocking laws, a diagnostic plot approach will also reveal if a dead-end filter's performance is predicted by one of the mechanistic analytical solutions. Using the linearized analytical solution, each slope will be related to the blocking law constant and the intercept to a function of P_o. Taking the standard blocking solution as an example, a plot of $(1/P^{1/2})$ versus v will have a slope of $-K_{sb}'/2$ and an intercept of $(1/P_o^{1/2})$.

Whether the dead-end filter is operated in constant pressure or constant flux mode, in nuclear applications there is a strong preference for selecting the filter such that it is predominantly operating in the cake filtration regime. Unlike traditional industrial applications, recovery of filter performance is complicated by the safety and health concerns of maintenance and corrective action on equipment in a nuclear facility. Physical maintenance by plant workers is often not feasible, and any cleaning solutions coming into contact with a filter will be deemed radiologically contaminated and have to be handled accordingly. Approaches to performance recovery in nuclear facilities are discussed in more detail in Section 5.4, but cake filtration dead-end filters can operate longer prior to intervention or shutdown and are less likely to need aggressive cleaning protocols. Therefore, selecting a dead-end filter that is likely to operate in the cake filtration regime (when possible considering other design criteria or constraints) is an important objective in dead-end filter design in a nuclear waste application.

5.3.2 Crossflow Filtration

The second filtration type often used in nuclear processing is crossflow filtration, where a suspension is passed over a filter media and the filtrate is removed in the direction that is transverse to the bulk flow. The suspension can be passed over the filter media in a variety of geometries: flat plate, disk, channel, tube, or a bundle of tubes/channels. Flow is usually driven by either pumping, gravity, or the rotation of filter elements. The presence of the transverse flow makes crossflow filtration more suitable than dead-end filtration for suspensions with higher solids concentrations and higher throughput (larger waste volume) processing. The higher complexity of crossflow systems makes them less suitable for use as a method to protect process equipment or support small batch processes, as well as making them more equipment-intensive and costly than a dead-end system.

Fundamentally, the behavior of the solids in crossflow filtration is different than in dead-end filtration because the particles do not necessarily interact with the filter media. The degree to which solid particles come into contact with the filter surface at all is dependent on several factors:

TABLE 5.2
Blocking Law Exponents and Analytical Solutions for Constant Flux Dead-End Filtration

Blocking Law	Complete Blocking	Standard Blocking	Intermediate Blocking	Cake Filtration
Value of exponent n	2.0	1.5	1.0	0.0
Constant	K_{cb}'	K_{sb}'	K_{ib}'	K_{cf}'
Linearized analytical solution	$\frac{P_o}{P}=1-\frac{K_{cb}'}{J_o}v$	$\left(\frac{P_o}{P}\right)^{1/2}=1-\frac{K_{sb}'}{2}v$	$\ln\left(\frac{P}{P_o}\right)=K_{ib}'v$	$\frac{P}{P_o}=1+K_{cf}'J_ov$
Rearranged for $P(v)$	$P=\frac{P_o}{\left(1-\frac{K_{cb}'}{J_o}v\right)}$	$P=\frac{P_o}{\left(1-\frac{K_{sb}'}{2}v\right)^2}$	$P=P_o\exp\left(K_{ib}'v\right)$	$P=P_o\left(1+K_{cf}'J_ov\right)$

Source: Adapted from Iritani, E. and Katagiri, N., *Kona Powder and Part. J.*, 33, 179–202, 2016.

- Fluid properties: density, rheology, solid concentration
- Solid properties: particle size, morphology, speciation, and density
- Operating parameters: pressure, fluid velocity, temperature
- Filter media: pore size distribution, porosity

All of these factors play a role in the multiphase flow physics occurring at and near the filter surface. The physics are important because there are two physical phenomena happening simultaneously: the transmembrane pressure that drives the filtrate through the filter media (which pushes or drags solid particles toward the surface); and the shear gradient that is produced by the transverse flow, which imparts a lift or repulsion force on the particles and directs them away from the surface.

In Figure 5.10, the relevant crossflow filtration physics are illustrated at an arbitrary moment in time looking downstream from a bulk flow Q_B imparted in the positive x direction. The crossflow filter is separating a bulk suspension with solids concentration (volume fraction) of ϕ_b at an average velocity of U. Initially, the filter media is free of solids. Once the bulk transverse flow begins, solid particles present in the bulk suspension will concentrate closer to the media surface as permeate is removed; the particles are driven toward the surface by the applied pressure. If the particles are proximate enough to interact with the surface prior to flowing past it, they begin to accumulate on the surface in a cake. The cake layer grows along the length as this process continues to a maximum concentration of ϕ_w or ϕ_c (the terms are often used interchangeably, with subscript "w" standing for wall and "c" for cake), adding resistance to flow and reducing the local flux rate through the filter media. This in turn impacts local hydrodynamics and mass transfer, as indicated by the representative velocity $u(x, z)$ and concentration $\phi(x, z)$ profiles that are shown. These profiles evolve with operating time and have a corresponding effect on the filtration rate J. The expectation during crossflow filter operations is that the evolution continues until a balance is struck between the mass transfer to the filter surface and the hydrodynamic forces keeping solids

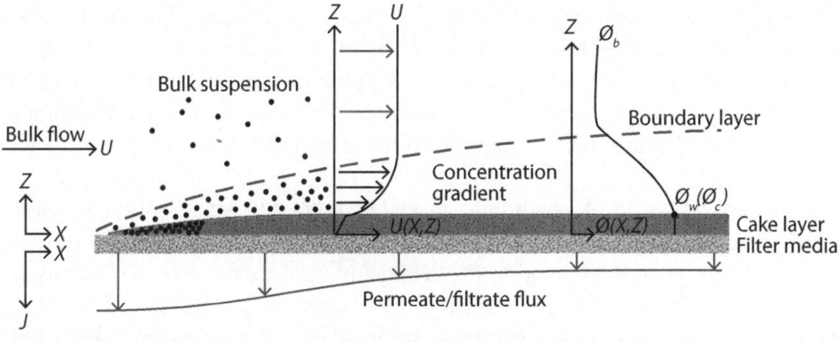

FIGURE 5.10 Important physics occurring at an arbitrary time during crossflow filtration along the filter length (in the direction of flow) as a function of position.

aloft (or resuspending them) and carrying them downstream. At this point, the cake layer achieves a steady-state thickness (often called δ_c) and the flux has a corresponding steady-state value J_{ss}, the momentum and concentration boundary layers are more or less fixed, and filtration will continue as long as the operating parameters are held constant (U, ΔP, ϕ_b). Similarly, if the crossflow filter was in constant flux mode, the expectation would be that a steady-state transmembrane pressure P_{ss} would be achieved once the cake layer achieves a steady-state thickness. For simplicity (and because the bulk of the models in literature are developed assuming constant pressure), the discussion of crossflow filtration in this section will be restricted to constant pressure operation. It is worth remarking that the pressure behavior in constant flux mode would, qualitatively, have a response that is the inverse of the flux: if the flux decreases due to fouling, pressure would increase due to similar fouling, etc.

Assuming the applied transmembrane pressure is constant, the achievement of a steady state is governed by the shear rate at the cake layer surface that assists in sweeping particles downstream. The steady-state permeate flux will depend on the mechanism by which back-transport, i.e., the particles moving from the cake layer "back" into a region of the transverse sweeping flow, occurs in the system. In some cases, predictions of steady-state flux can be built from the back-transport mechanism that is assumed to be dominant in the system of interest. Several mechanisms have been proposed, but three commonly referenced cases are discussed in the following paragraphs.

The first mechanism is Brownian diffusion, whereby solid particles being filtered diffuse back from the cake surface and are carried away by the flow. If diffusion is the transport mechanism, then the mass transfer of solid particles can be described by a concentration polarization model. The key physical parameter is the Brownian diffusion coefficient, which is defined as:

$$D_o = \frac{k_B T}{6\pi\mu a}, \tag{5.17}$$

where T is the absolute temperature, k_B is the Boltzmann constant (1.38×10^{-23} m^{2}kg s^{-2} K^{-1}), and a is the radius of the solid particle. For most micron-sized particles in fluids similar to water, D_o is on the order of 10^{-13} m^{2}s^{-1} or less. Compared to macromolecules that might be separated in an ultrafiltration or reverse osmosis process, this is orders of magnitude smaller. Thus, Brownian diffusion is a relatively minor driving force, except when the solid particles or fluid viscosity are very small.

In the case where the formed cake layer has a height that is only a fraction of the half-height of the channel, tube, or other geometry in which the filtration is taking place, an analytical solution for the steady-state flux can be derived assuming Brownian diffusion is the controlling back-transport mechanism. The solution is broken into two asymptotes: (a) the solution when the suspension being filtered is concentrated ($\phi_w - \phi_b \ll \phi_w$), and (b) a solution when the suspension is dilute ($\phi_b \ll \phi_w$). They are written as:

$$(a)\ J_{ss} = 0.81\left(\frac{\dot{\gamma}_m D_o^2}{L}\right)^{1/3} ln\left(\frac{\phi_w}{\phi_b}\right), \text{ and}$$

$$(b)\ J_{ss} = 1.31\left(\frac{\dot{\gamma}_m D_o^2 \phi_w}{L\phi_b}\right)^{1/3}, \tag{5.18}$$

where L is the length of the filter surface, and $\dot{\gamma}_m$ is the shear rate at the filter surface. The shear rate can be written algebraically for laminar tube flow in a tube of radius R as $4Q_B/\pi R^3$, but it is challenging to describe analytically for more complex geometries or turbulent flow. The solid concentration at the wall (ϕ_w) is often taken to be the maximum packing density for the particles: this ranges from 0.6 to 0.9 depending on the nature of the particles (~0.6 for rigid spheres of similar size and up to ~0.9 for compressible particles of polydisperse size).

The second mechanism is shear-induced diffusion, which was developed primarily in response to the use of the models given in Equation 5.18. Upon comparison with most experimental data, it was found that these expressions underpredicted the flux. This was likely due to the magnitude of D_o not being representative of the back-transport physics. The proposal, originally by Zydney and Colton (Zydney and Colton 1986), was that the concentration polarization formulation could be used with a shear-induced diffusivity in place of D_o. The shear-induced diffusion coefficient D_s accounts for the random displacements that particles undergo as they interact with the local shear field and other particles in the suspension and is usually written as:

$$D_s = \lambda_s \dot{\gamma}_m a^2, \tag{5.19}$$

where λ_s is a constant with a value of approximately 0.1 (in Zydney and Colton, they use $\lambda_s = 0.03$ following Eckstein, Bailey, and Shapiro (1977)). The shear-induced diffusivity has a different functionality then D_o, with a dependence on the shear rate at the membrane surface and a stronger dependence on particle size. Since D_s is a direct substitution for D_o, flux expressions corresponding to Equation 5.18 can be written for the same asymptotes:

$$(a)\ J_{ss} = \lambda_{s,a}\dot{\gamma}_m\left(\frac{a^4}{L}\right)^{1/3} ln\left(\frac{\phi_w}{\phi_b}\right), \text{ and}$$

$$(b)\ J_{ss} = \lambda_{s,b}\dot{\gamma}_m\left(\frac{\phi_w a^4}{\phi_b L}\right)^{1/3}. \tag{5.20}$$

The constants $\lambda_{s,a}$ and $\lambda_{s,b}$ are also both approximately O(0.1).

The third mechanism is inertial lift, which proposes that the back diffusion of particles is due to the inertial migration of particles. Inertial migration supplies a migration velocity toward the centerline of the flow that is sufficient to balance and/or overcome the velocity toward the wall due to permeate removal through the filter media. In laminar flow, the inertial lift velocity u_L has the form:

$$u_L = \lambda_L \frac{\rho a^3 \dot{\gamma}_m^2}{16\mu}, \tag{5.21}$$

where λ_L is an O(1) constant that depends on the position of a particle relative to the wall. The value of λ_L is greatest near the wall and decreases as the particle's position moves toward the centerline of the flow. To generate an expression for the steady-state flux when inertial lift is the relevant back-transport mechanism, the permeate flux is set equal to the lift velocity described in Equation 5.21. Upon balancing the velocities, the steady-state flux prediction is written as:

$$J_{ss} = 0.036 \frac{\rho a^3 \dot{\gamma}_m^2}{\mu}. \tag{5.22}$$

The inertial lift prediction is not a function of filter length or solids concentration like the diffusion-based models and is a stronger function of the fluid properties (μ, ρ), shear rate, and particle size. Though it is not a function of solids concentration, Equation 5.22 was generated assuming that the suspension is dilute and that a more concentrated suspension would have particle–particle interactions that would likely modify the inertial lift velocity as well as the model for J_{ss}.

Each back-transport mechanism described in this section contains a different dependence on some of the important physical parameters of the crossflow filter and the suspension being filtered, especially particle size (a), shear rate $(\dot{\gamma}_m)$, and viscosity (μ). Estimates of the steady-state flux from the three back-transport mechanisms will vary by orders of magnitude in some cases. For example, see the comparison between the three mechanisms as a function of particle diameter d_p (=2a) in Figure 5.11. For a fixed set of fluid properties, the impact of particle radius a strongly influences the magnitude of J_{ss} over a range of sizes that encompass particles encountered in nuclear wastes. The region where the three mechanisms agree is relatively narrow and restricted to large particle size (>100 μm).

It is critical to note that the expressions to estimate J_{ss} are all derived assuming laminar flow, dilute concentration ($\phi_b \ll 0.1$), and a single characteristic particle size; thus, fluid and particle properties are essentially constant, and the flow profile can be described analytically. For many filtration applications in nuclear waste separations, the flow is fully turbulent, $\phi_b \sim$ O(0.1), and the particle size distribution is polydisperse; for these conditions, the validity of applying the back-transport models may come into question. However, they do supply an order-of-magnitude estimate of what the crossflow steady-state flux could be, assuming that a thin cake layer is dominating the resistance to the permeate flow and the solid particle size is reasonably described by a single diameter/radius.

The three back-transport mechanisms that were just discussed can be used to generate predictions of steady-state flux in crossflow filters when the cake layer is relatively thin. Counterintuitively, in cases where the cake is expected to be thicker (with respect to some dimension of the flow geometry, such as a tube radius or channel height), the membrane resistance may also be important in determining the flux that a filter ultimately achieves. Physically, this is because thicker cakes are usually

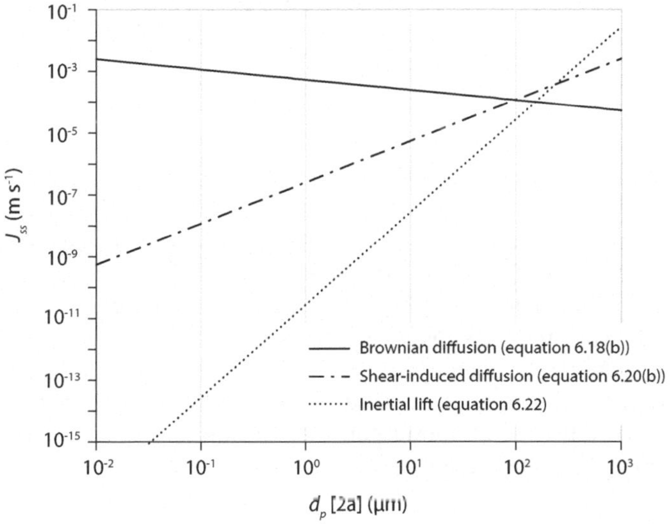

FIGURE 5.11 Comparison of steady-state flux predicted by three back-transport mechanisms for $Q_B = 10$ gpm, $\phi_w = 0.6$, $\phi_b = 0.05$, $\mu = 3$ cP, $T = 25°C$, $R = 0.127$ m (0.5 in), $L = 1$ m, and $\rho = 1200$ kg m^{-3} for a range of particle diameters.

made up of larger particles and, subsequently, are considerably more permeable such that the cake resistance is not the only significant contributor to the total system resistance. For the thicker cake case, crossflow filtration theory often "borrows" the expression from the dead-end filtration theory described in Equation 5.9, and many of the same concepts still apply. Equation 5.9 is applied to determine the steady-state flux expected from the system by solving for an equilibrium or critical cake thickness, which distinguishes it from dead-end filtration (which does not have an expected steady-state filtrate rate). The distinction is that R_c in dead-end filtration increases indefinitely with the amount of volume filtered, whereas R_c in crossflow filtration is expected to eventually reach a constant value (even for thicker cakes) because the cake can only grow to a finite thickness in the flow geometry. One method to express this is by scaling cake resistance per unit depth (R_c') using the cake thickness:

$$R_c = R_c'\delta_c = \delta_c\left(\frac{\alpha}{\rho_p(1-\varepsilon)}\right), \tag{5.23}$$

where δ_c is the cake thickness and α is defined in Equation 5.10. What remains is to find the critical value of cake thickness where its growth due to particle deposition is balanced by hydrodynamic forces. Note that Equation 5.23 can accommodate compressibility of the cake by using the relation in Equation 5.11, if appropriate.

Equation 5.23 was written assuming that the geometry is flat, i.e., parallel plate or channel flow, and does not account for the curvature found in tube flow. Curvature impacts the cake resistance expression because the cake area changes with radial position. Therefore, the cake resistance equation must be modified accordingly:

$$R_c = \left(\frac{\alpha}{\rho_p(1-\varepsilon)}\right)\left[R\ln\left(\frac{R}{R-\delta_c}\right)\right]. \quad (5.24)$$

Whether the crossflow filtration is occurring in a flat or tubular geometry, when the cake is thicker, predicting the flux requires establishing the value of δ_c in order to calculate J_{ss}. Equation 5.9 and Equations 5.23 or 5.24 are used together once δ_c is known. The cake thickness builds up over time until it reaches a critical value. The critical value can be determined indirectly by experiments (direct observation of cake thickness is a difficult measurement to perform for all but the simplest filtration system), but the back-transport mechanisms can also be used to predict it. The example of using inertial lift to solve for a critical cake thickness will be discussed here where it applies to filtration processes with larger particles and shear rates, which are more typical (but not exclusively so) of nuclear waste filtration scenarios.

Starting with Equation 5.9, it is convenient to recast the expression in the following form:

$$J = \frac{\Delta P}{\mu(R_m + R_c)} = \frac{J_m}{1+\theta\delta^*}, \quad (5.25)$$

where $J_m = \Delta P/\mu R_m$ (the flux in the absence of a cake layer or the initial flux when filtration begins), $\theta = \alpha_c/R_m$ (resistance ratio), and $\delta^* = \delta_c/R$ (the dimensionless cake thickness). The quantity α_c is a measure of the cake resistance for a layer filling the entire flow geometry. It is related to other parameters already discussed, i.e., it is equivalent to the right-hand side of Equation 5.23 with R substituted for δ_c. Equation 5.25 applies for channel flow, with R representing the half-height of the channel. As discussed before, the curvature of the tube geometry requires a modification to Equation 5.25 as follows:

$$J = \frac{J_m}{(1-\delta^*)\left[1+\theta\ln(1-\delta^*)^{-1}\right]}. \quad (5.26)$$

The flux expression depends on δ^* and θ, neither of which are known a priori in most filtration systems. Recall that the inertial lift velocity is described in Equation 5.21, but in this case, the shear rate varies depending on the cake thickness as the solid layers grow and reduce the size of the channel or tube cross-section. For a constant volumetric flow rate Q_B, the shear rate increases with δ_c by a factor of $R^p/(R-\delta_c)^p$, where $p = 2$ for a channel and $p = 3$ for a tube. The corresponding adjustment to the inertial lift velocity results in the expression

$$u_{L,\delta} = \frac{u_L}{\left(1-\delta^*\right)^{2p}}, \tag{5.27}$$

where u_L is given generally by Equations 5.21 or specifically by Equation 5.22 and is the lift velocity when the flow is not impacted by the presence of a finite cake layer. The solution of Equations 5.25 or 5.26 with Equation 5.27 is simultaneous. The opposing velocities J and $u_{L,\delta}$ are set equal to one another at steady state, permitting determination of J and δ_c for specific values of θ. It is interesting to consider the asymptotic behavior of filtration governed by inertial-lift back transport. Note that when $\theta >> 1$, which is implied by $J_m > u_L$, the cake layer is the more significant resistance in the system. The cake thickness is simply:

$$\delta_c = \frac{R_m}{\alpha_c}\left(\frac{J_m}{u_L} - 1\right), \tag{5.28}$$

which is inversely proportional to θ. However, when $\theta << 1$, the membrane is the major player in the system and the flux will be governed solely by R_m, e.g., $J = J_m$. The cake thickness becomes:

$$\delta_c = R\left(1 - \left(\frac{u_L}{J_m}\right)^{\frac{1}{p+2}}\right). \tag{5.29}$$

There is a range of solutions at intermediate θ that fall between these two extrema. For intermediate values, δ_c increases as J_m/u_L increases until J_m/u_L approaches large values and nearly the entire flow geometry is filled with solids. In this limit, the channel geometry still has a constant (but small) flux that is almost reduced to zero within the tube geometry as the curvature constricts the surface area, resulting in a negligible value. The inertial lift model was discussed since it can be described in relatively simple mathematical terms that assist in building a conceptual understanding of the relevant phenomena. Not all applications of crossflow filtration to nuclear waste processes adhere to or are well described by a simple model framework; other physical models increase in complexity rapidly (for instance, combining two or more back-transport mechanisms) and often warrant numerical solutions (or simulations) if performance predictions are needed.

Every crossflow filtration steady-state flux performance model discussed to this point has assumed that the membrane resistance R_m is constant or almost constant. This assumption was also applied in the dead-end filtration models, all of which usually start with Equation 5.9 or a related formulation as described in Section 5.3.1. For dead-end filtration, where a cake builds up a finite thickness at relatively short operating times and continues to grow with additional operating time, this is a reasonable assumption. There are situations in which a change in a dead-end filter R_m at longer operating time is plausible: e.g., smaller particles could migrate through the cake (if it is comprised of mostly larger particles) and foul the filter, or cake maturation processes such as compaction and attrition could generate finer particles that infiltrate a short way into the filter's surface pores. However, these scenarios are expected to be rare; the expectation is that the cake layer protects the dead-end filter from experiencing appreciable fouling.

A steady state flux will not be achieved, or the flux may only reach an apparent steady state if the membrane resistance changes over time. A crossflow filter does not necessarily have the same level of protection from its cake because the crossflow is intended to prevent a significant cake from forming. Therefore, the cake may be very thin, highly porous, unevenly distributed on the filter surface, or unstable. Any one of these possible configurations could permit easier penetration of solid particles into the filter pore structure than in a dead-end filter (recall from Section 5.2 the distinction between depth and cake fouling). This is a counterintuitive aspect of crossflow filtration: the transverse flow can limit one type of fouling (cake) only to promote another (depth). As would be expected, the prediction of membrane resistance changes with operating time is both uncertain and difficult and can really only be confirmed by empirical observation. A common observation that indicates membrane resistance has changed is the comparison of a clean flux before and after filtering a suspension; in this case, "clean" simply means that fluid, e.g., water, is nominally devoid of undissolved solids. Membranes can also be examined for evidence of solids intrusion after being used to process a suspension, but the methods to achieve this are usually destructive and involve dismantling the filter element. Models that have been developed to describe the transience of the membrane resistance largely agree that the resistance decays from a starting value to a final value in an exponential fashion, i.e.,

$$R_m = R_{m,o} + \left(R_{m,f} - R_{m,o}\right)\left(1 - \exp\left(-\frac{t}{\tau_m}\right)\right), \tag{5.30}$$

where $R_{m,o}$ is the initial membrane resistance, $R_{m,f}$ is the final membrane resistance, and τ_m is a time constant that governs the membrane fouling. However, though several studies have shown Equation 5.30 to be phenomenologically descriptive for these cases (Aimar, Baklouti, and Sanchez 1986; Ethier and Kamm 1989a, 1989b; Matsumoto, Nakao, and Kimura 1988; Petzny and Quinn 1969; Robertson and Zydney 1990), experiments are needed to elucidate the appropriate values for τ_m and $R_{m,f}$.

Rather than outright empiricism, a more common approach to address transient flux behavior in crossflow filtration is to lump the transience into a single term using a model developed from dead-end filtration assumptions. This is constructed based on a mass balance at the edge of the cake layer, which is assumed to be growing during the transient period. Combining Equations 5.9 and 5.23, the resulting differential equation becomes:

$$\left(J + \frac{d\delta_c}{dt}\right)\phi_b = \phi_c \frac{d\delta_c}{dt}, \tag{5.31}$$

yielding a separable differential equation that can be integrated for constant pressure operation. The integrated result is quadratic in δ_c and is solved to give

$$\delta_c = \frac{R_m}{R_c'}\left[\left(1+\frac{2R_c'\phi_b\Delta Pt}{(\phi_c-\phi_b)\mu R_m^2}\right)^{1/2}-1\right]; \tag{5.32}$$

recall that R_c' is defined in Equation 5.23 and can be used to solve for the flux by substitution:

$$J = J_m\left(1+\frac{2R_c'\phi_b\Delta Pt}{(\phi_c-\phi_b)\mu R_m^2}\right)^{-1/2} \tag{5.33}$$

The result in Equation 5.33 assumes that both the cake and membrane resistance are constants. Crossflow filters following this flux model will start from the membrane-only flux (J_m), build up cake at short times, and continue to decay as $t^{-1/2}$ at long times. The significance of Equation 5.33 is that it can be used to predict flux without experiments if all the parameters can be properly estimated.

Other mechanistic models have been built assuming that a classical dead-end blocking law or some combination thereof can be applied to the crossflow case. There are several variations on this theme available in the literature (Bolton, LaCasse, and Kuriyel 2006; Ho and Zydney 2000; Peng and Tremblay 2008; Yee, Wiley, and Bao 2009). Figure 5.12 shows one example from Peng and Tremblay that illustrates

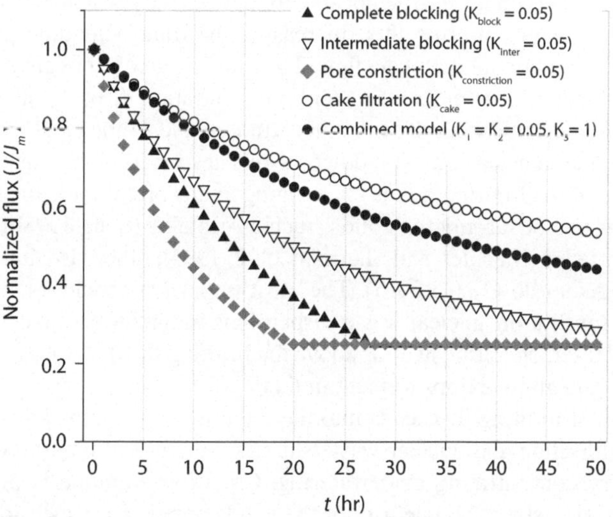

FIGURE 5.12 Representative transient models for crossflow filtration based on expressions in Peng and Tremblay (2008). All constants were fixed at a value of 0.05 h^{-1} with the exception of K_3, which is equal to unity. The expressions were modified with an assumed steady-state flux of $J/J_m = 0.25$.

the variation in transience for the different models when the constants are of the same order. Note the similarity to Figure 5.9 (dead-end filtration models). The use of dead-end blocking laws may represent the initial transience well but will not predict a steady state (or pseudo steady state). Figure 5.12 takes a common approach to adjust blocking law models for crossflow filtration, which is to marry a transience model with an appropriate asymptotic steady-state flux for long time periods.

Other models have incorporated more complex dynamics to formulate a transient description of the flux (Song 1998a; Zhang and Song 2000) or have used statistical approaches (Asaadi and White 1992; Vincent-Vela et al. 2012), but almost all the models suffer from the same dependence on empiricism to estimate at least one parameter. Because of this limitation, the development of relevant crossflow filtration models has remained an active area of investigation, including numerous forays into the use of numerically solved equation systems (Fradin and Field 1999; Kazemi, Soltanieh, and Yazdanshenas 2012; Lu, Hwang, and Ju 1993).

One other aspect of crossflow filtration performance that is used to characterize operation instead of a flux model is the concept of critical and limiting flux. The critical flux is typically described as the conditions for which the filter first experiences fouling (deviates from filtering as if it is a clean membrane). The critical flux nominally marks the transition between concentration polarization in the boundary layer and actual membrane fouling. Thus, operating at or near this point is considered "critical" for achieving the best performance. Critical flux is also a rich research area that has many nuances (see Bacchin, Aimar, and Field (2006) for helpful discussions), but the time scales involved are usually short with respect to operating times that are of interest to nuclear waste processing, which makes operating at "critical" conditions impractical. More relevant is the limiting flux, usually described as the maximum flux achievable for a given suspension and set of operating parameters; at the limiting flux, increasing the transmembrane pressure does not result in any additional permeate flow. The limiting flux is thought to occur when the cake has formed at its maximum solids concentration ϕ_c or ϕ_w (also sometimes interchangeably referred to as the gel concentration) and so the pressure drop across the cake becomes constant. As with other models described in this subsection, a priori estimation of the limiting flux is challenging and is often a suitable approach for dilute suspensions of sub-micron solids, such as proteins (Song 1998b) or spherical polymeric particles (Mackley and Sherman 1992, 1994); otherwise, it is approached empirically (Astudillo-Castro 2015). The most important conclusion to draw from limiting flux for filtering nuclear waste suspensions is that once a cake has formed in the crossflow filter, operating at a ΔP above the limiting flux threshold is expected to produce only a negligible improvement in flux.

The notion of limiting flux leads into the final important topic area with respect to crossflow filtration and nuclear waste slurries—the use of crossflow filtration to perform slurry concentrating or dewatering. Crossflow filtration is preferred over dead-end filtration since the majority of the solids remain in suspension during the filtration process; a concentration operation can be performed by recirculating back to a vessel after the solid–liquid separation occurs in the crossflow filter. In an application of this type, the bulk solids concentration fed to the filter often increases with time until it approaches a maximum undissolved solids concentration (the gel concentration ϕ_g). The gel concentration is approximately similar to the centrifuged solids concentration of the slurry. As ϕ_b approaches ϕ_g, the filter flux is often well-described by

$$J = K_{gp} \ln\left(\frac{\phi_g}{\phi_b}\right), \tag{5.34}$$

where K_{gp} is the gel-polarization mass transfer coefficient (usually empirically determined). Song (1998b) gives an expression based on a Leveque solution where K_{gp} is written in terms very similar to Equation 5.18(a); the equivalent expression for shear-induced diffusion would be Equation 5.20(a). Mathematical descriptions of this process based on gel polarization or steady-state models, though they have a simplistic view of the physics, are often found to be applicable over a reasonably wide range of solids concentrations.

A crossflow filtration dewatering curve is often depicted using an asymptotic approach. As the dewatering begins, $\phi_b << \phi_g$, and the flux is approximately controlled by the membrane resistance only. In this limit (R_c ~ 0) and via inspection of Equation 5.9, $J = \Delta P/\mu R_m = J_m$. At higher solids concentrations and as ϕ_b approaches ϕ_g, Equation 5.34 applies until ϕ_g is reached. The point at which the flux transitions between these two regions is called the "knee" of the curve. An example of a family of asymptotic curves is shown in Figure 5.13. The filter flux is normalized by J_m in each case, with the value of K_{gp} changing with respect to J_m to illustrate the impact of the gel mass transfer coefficient. The "typical data" is included to be representative of what real filtration data might look like in a system where K_{gp} ~ $0.5J_m$. Filters with

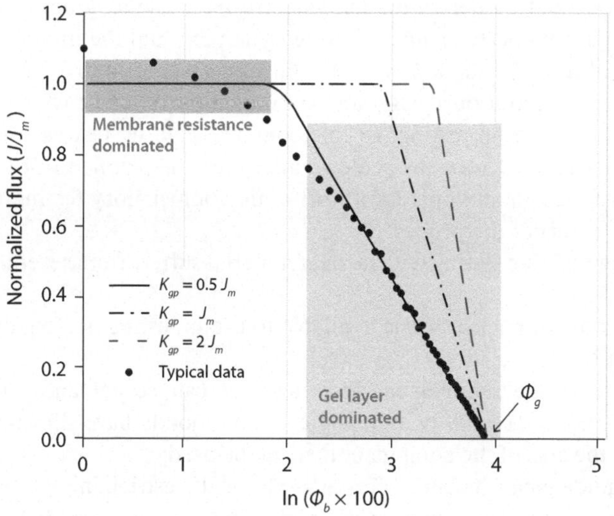

FIGURE 5.13 Example asymptotic dewatering curves for suspensions with $K_{gp} = 0.5J_m$, $K_{gp} = J_m$, and $K_{gp} = 2J_m$. The typical data is shown as a representation of what "real" data might look like for a suspension/filter system with a K_{gp} of ~0.5.

non-constant R_m or rapid cake fouling would exhibit a more significant slope in the membrane-resistance-dominated region; these factors can smear the "knee" region and complicate the estimation of K_{gp}. The filter flux can theoretically be driven to zero when the concentration reaches ϕ_g, as shown in Figure 5.13, but in practice, filtration operations would cease prior to that point—either the target solids concentration is reached, or the flux becomes low enough to require intervention.

As a final note, the increase in ϕ_b during a dewatering operation conducted at constant pressure would manifest as a decrease in J with time. Decreasing flux with time is also observed when feeding at constant ϕ_b (or approximately so). Thus, use of various flux performance models for crossflow filtration need to consider the rate of change ($d\phi_b/dt$) in solids concentration fed to the filter, which impacts the use of a predictive model. The same would also be true of constant flux operations. Either way, crossflow filtration models are most useful for making qualitative judgments about performance. As of this writing, a priori quantitative performance prediction has proven to be difficult because even the simpler models with the fewest assumptions are mathematically complex and often solved numerically.

5.4 FILTRATION PERFORMANCE IN NUCLEAR WASTE APPLICATIONS

Given the discussion of filtration theory and performance in the preceding sections, what can be said regarding filtration performance when nuclear waste suspensions are being processed? Recall from Section 5.2 that knowledge of the filtered suspension's properties and composition can be imprecise and/or wide ranging. Without a well-defined suspension, the use of the models discussed in Sections 5.3.1 and 5.3.2 (which already have limitations) becomes even more challenging. Theory and its derivative models still have use as guides to help make judgments regarding performance, fouling dynamics, and the impact of changing operating conditions. In the absence of robust and representative predictive models, nuclear waste filtration design and optimization relies heavily on empirical correlations and scale-up testing. In this context, scale-up testing implies taking filter performance data from any scale smaller than the planned full-scale deployment to inform design and operation; full-scale confirmatory testing may also be performed as a part of the process.

Performing scale-up testing with nuclear material suffers from several limitations:

- The amount of nuclear waste available to use in testing is often only a few liters or less;
- The size of the radiological facilities that can accept and process the waste—most hot cells or radiological fume hoods have dimensions that restrict the size of the equipment that can be used;
- The radiological hazards present—physical restrictions on access to equipment and, depending on the radiological hazard, traditional process equipment may not be suitable (not radiation hardened);
- The cost to conduct the testing is high—due to the preceding limitations, extensive testing with radiological material is expensive at larger scales.

These limitations have an important result that was discussed previously—the majority of the scale-up testing is performed using surrogate suspensions, i.e., waste simulants. These surrogate suspensions should be designed to mimic aspects of the waste suspensions so that data collected with the simulants are useful for design and optimization efforts (for some American Society for Testing and Materials [ASTM] guidance on simulant development and property measurement, see ASTM standards C1750 and C1752 (ASTM 2017a, 2017b)). The development and verification of simulants is a rich subject area that could easily be the subject of its own chapter and presents significant challenges due to the complexity of the waste (Peterson et al. 2018; Stewart 1998; Wilmarth et al. 2011). For the purposes of discussing filtration performance, the key point is that it is important to link simulant performance to actual waste performance (if available) whenever possible, i.e., filter flux obtained with a simulant should be similar to the flux obtained with the waste it is designed to represent. Many nuclear wastes have been tested historically, and two starting points for locating relevant waste data were produced by Pacific Northwest National Laboratory (PNNL) (Daniel et al. 2010b) and Savannah River National Laboratory (SRNL) (Johnson and Duigan 2011). This is often more than a matter of simply matching the waste suspension chemistry or creating a suspension with the same bulk physical properties, introducing additional complexity into the process. Unsurprisingly, performance data collected with waste simulants has an inherent uncertainty, and it is recommended that any filtration system considers that uncertainty when design and operating decisions are made.

In general, filtration performance for a certain suspension and set of operating parameters scales up proportionally with filter area. That is, the relationship

$$J = \frac{Q_1}{A_{F,1}} = \frac{Q_2}{A_{F,2}} \tag{5.35}$$

holds, where the subscripts 1 and 2 represent two different scales of filter. The flux measured in the two filters would be nominally the same, and so the expected flow rate in scale 2 could be estimated as simply $J \times A_{F,2}$. Equation 5.35 should hold as long as the geometries are similar enough to render the local filtration conditions (*Re*, pressure differentials, back-transport mechanisms, etc.) similar between the scales, and the ratio of the solids loading (in a volumetric sense) to the surface area in the two suspensions is more or less the same, i.e.,

$$\frac{\phi_{b,1} Q_B}{A_{F,1}} \approx \frac{\phi_{b,2} Q_B}{A_{F,2}}. \tag{5.36}$$

There are examples in which Equation 5.35 may not be accurate even for cases where Equation 5.36 is satisfied. For instance, a dead-end filter processing a suspension at a fixed ΔP in a configuration similar to that shown in Figure 5.2a with a filter area of $A_{F,1}$ and a larger dead-end filter with area $A_{F,2} = 10A_{F,1}$ but in a configuration similar to Figure 5.2c may not scale with one another due to differences in how the solids interact with the filter geometry. Likewise, a crossflow filter with a single tube of diameter d_1 and length L_1 operated at a fixed ΔP would not scale up with area if the second filter also had diameter d_1 but $L_2 = 10L_1$. The pressure drop across the longer length would alter the local dynamic interactions because there are regions in the filter with length L_2 where the local ΔP is lower and higher than in the filter with length L_1. This suggests that, to simplify the scale-up process, it is useful to break the filtration operation into repeated modules that yield more confidence in using the relationship of Equation 5.35 because the geometry is preserved across the modules. This is fairly common, and modules could be a series of filter cartridges, a bundle of same-size tubes, or a series of parallel plates/discs.

5.4.1 Operating Conditions

If the filter media (including the associated configuration) and waste suspension is fixed, only a few parameters can be manipulated operationally to impact filtration performance: pressure (ΔP), axial velocity/bulk flow rate (U/Q_B), and temperature (T). The pressure has the largest impact of these three parameters and is the parameter of greatest interest. Equation 5.6 indicates that flux should increase proportionally with increased ΔP; this relationship may hold over small pressure changes in some filter systems, but there is typically a non-linear relationship between flux and ΔP. This is because the Darcy's law description of filtration depends on the resistances in the system remaining constant, the suspension fed to the filter staying unchanged, and solids being uniformly deposited onto the filter surface (or in its pores). What is observed from waste simulant test data and filter operations is that operating at a lower ΔP is usually preferable because it keeps filter operations sustainable over longer time periods. Application of a higher ΔP results in a larger instantaneous flux, but it also tends to depth-foul filters more quickly and/or compact solid cake layers, leading to faster flux decline over time. For an example of this behavior with a nominal 5 wt% suspension in a tubular crossflow filter, see Figure 5.14 (for more information, refer to Schonewill et al. (2015)). Note that even initially, the normalized flux is actually larger at the lowest pressure of 20 psid in this case, and after 100 h of operation, the normalized flux remains greater by approximately a factor of five. The instantaneously measured fluxes for the same three data sets are given in Table 5.3 at various time points to illustrate how each evolved with time. From Table 5.3, it is apparent that the higher pressures impart a much greater decay in performance with time than the lowest pressure, resulting in a similar instantaneous flux after 5 h of operation and a progressively lower instantaneous flux at longer times even though the pressure is a factor of two or three greater. This "real" fouling behavior is representative of nuclear waste suspension filtration and demonstrates how the performance models can only be used as a guide and are not a good substitute for empirical data.

Another interesting aspect of pressure in filtration systems is that process history can impact how filtration will respond to changes in pressure. For example, a filter could be operated in a mode where the initial pressure is lower (to avoid a higher rate of depth fouling initially) and then step-increased in an attempt to get a higher flux later in the operating period. Conversely, the initial pressure could be high (yielding an increased initial instantaneous flux) and then dropped to a lower pressure later to avoid the long-term fouling that has been observed at higher ΔP. Figure 5.15 shows

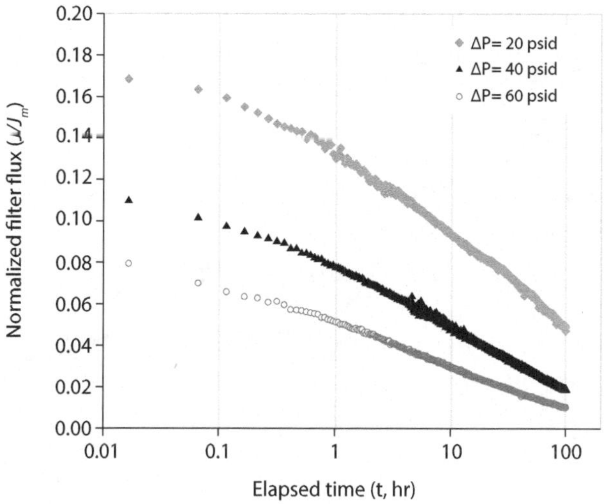

FIGURE 5.14 Normalized flux measured in a crossflow filter at three constant transmembrane pressures over a 100-h operating period. The filter was used to process a 5-wt% suspension with precipitated metal oxides in a simulated waste supernatant. (Adapted from Schonewill, P.P. et al., *Chem. Eng. Res. Des.*,102, 69–79, 2015.)

TABLE 5.3
Instantaneous Fluxes Measured at Three Transmembrane Pressures in a Tubular Crossflow Filter Corresponding to the Normalized Data in Figure 5.14

Time (hr)	J (m s^{-1}) at $\Delta P = 20$ psid	J (m s^{-1}) at $\Delta P = 40$ psid	J (m s^{-1}) at $\Delta P = 60$ psid
0	3.88×10^{-5}	5.35×10^{-5}	6.29×10^{-5}
1	3.01×10^{-5}	3.54×10^{-5}	3.59×10^{-5}
5	2.46×10^{-5}	2.51×10^{-5}	2.50×10^{-5}
10	2.13×10^{-5}	2.26×10^{-5}	2.00×10^{-5}
20	1.87×10^{-5}	1.79×10^{-5}	1.59×10^{-5}
50	1.47×10^{-5}	1.25×10^{-5}	1.06×10^{-5}
100	1.12×10^{-5}	8.68×10^{-6}	7.18×10^{-6}

data from both of those cases using the same 5-wt% suspension that was used to generate the data in Figure 5.14 and Table 5.3. What is observed is that beginning the filtration at a lower pressure initially and then increasing it by a factor of three does not prevent an acceleration in flux decline. This is likely brought about by a combination of depth fouling and cake compaction/maturation (depending on the thickness and characteristics of the initial cake deposited). In the opposite case, filtration at 60 psid initially fully matures the system such that when operation conditions are switched to a lower pressure, the flux remains steady and unchanged. The filter cake

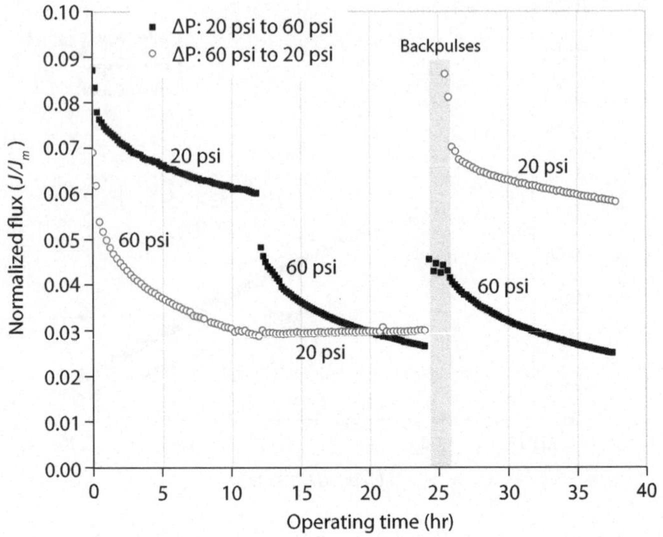

FIGURE 5.15 Normalized flux measured in a crossflow filter using reciprocal pressure histories. The filter was used to process a 5-wt% suspension with precipitated metal oxides in a simulated waste supernatant (same slurry as in Schonewill et al. (2015)). The flux was measured in three segments of ~12 h at the ΔP indicated. The third segment was proceeded by a series of backpulses to clear the solids layer before filtering.

in that operating scenario may be conditioned by the initially higher-pressure operation, enabling the achievement of a pseudo steady state. Notice that both scenarios end up at nearly the same normalized flux after 24 h; even so, the 60 → 20 psi case obtained a steady flux, while the 20 → 60 psi case still experienced flux decline. In this sense, the process history of the filter can matter with regard to long-time performance. As an aside, following a recovery operation after 24 h, as indicated by the "Backpulses" region in Figure 5.15, the fluxes returned to their expected values based on the initial 12 h of operation despite the difference in performance over time. The data collected in this example suggests that the extent of fouling in the filter media itself was likely similar between the cases, and the majority of the difference in performance was likely due to particle–solid interactions at the filter surface and in the cake layer.

The bulk flow rate used in filtration systems typically has a negligible impact when dead-end filtration is in use because all the solids are being brought into proximity to the filter regardless. However, the axial velocity (and more importantly, the shear rate) generated by the bulk flow in crossflow filtration systems is critical to helping limit cake growth and maturation; depending on the relevant back-transport mechanism that is dominant in the system (as described in Section 5.3.2), the effect of the shear rate on flux will vary. In the same system that was used to collect the data in Table 5.3, 100-h operating runs were also performed at three different axial velocities. Table 5.4 summarizes some of the instantaneous fluxes over time. For this filter system, increased velocity causes a small increase in the filter flux that remains

TABLE 5.4
Instantaneous Fluxes Measured at Three Axial Velocities in a Tubular Crossflow Filter at ΔP = 40 psid

Time (hr)	J (m s^{-1}) at U = 4.0 m s^{-1}	J (m s^{-1}) at U = 4.6 m s^{-1}	J (m s^{-1}) at U = 5.2 m s^{-1}
0	4.31×10^{-5}	5.35×10^{-5}	5.58×10^{-5}
1	3.11×10^{-5}	3.54×10^{-5}	3.77×10^{-5}
5	2.34×10^{-5}	2.51×10^{-5}	2.89×10^{-5}
10	1.91×10^{-5}	2.26×10^{-5}	2.51×10^{-5}
20	1.55×10^{-5}	1.79×10^{-5}	2.06×10^{-5}
50	1.06×10^{-5}	1.25×10^{-5}	1.54×10^{-5}
100	7.08×10^{-6}	8.68×10^{-6}	1.07×10^{-5}

approximately constant over the operating time; hence, the shear rate limits cake deposition and other fouling mechanisms in a similar way over the 100-h period. This aligns with the expectations from performance models and intuition for crossflow filtration. Higher shear rates should increase back transport and be more effective at limiting solids deposition on the filter or cake surface.

Another important consideration regarding the bulk flow rate in nuclear waste suspensions is keeping the solids suspended in the system and not permitting them to settle out in the pipeline or flow channel that delivers them. Bulk flow rates and their corresponding axial velocities should be selected to avoid settling so that the possibility of pipeline plugging is significantly reduced. Solid deposition in pipelines is particularly undesirable for nuclear applications as it can increase radiation dose, the generation of flammable gases, and the potential for criticality accidents in unexpected system locations. Pipeline deposition of solids is a rich and complex hydrodynamic topic area that has implications for operations in an array of industries; for an initial exploration of the topic as it relates to nuclear waste transport, the reader is referred to Poloski et al. (2009, 2010) and Welch (2001).

Operating temperature indirectly impacts the filtration performance in two ways. First, temperature impacts the liquid viscosity, which is indirectly proportional to flux as described by Darcy's law (Equation 5.6). The relationship between viscosity and temperature is non-linear; a common method used to approximate the relationship is an Andrade correlation of the form:

$$J_{ref} \sim J_T \exp\left(\lambda\left(\frac{1}{T} - \frac{1}{T_{ref}}\right)\right), \tag{5.37}$$

where T_{ref} is a reference temperature, J_{ref} is the flux at that reference temperature, J_T is the flux at temperature T, and λ is an empirical constant. This is described in more detail in works authored by Geeting and Daniel (Daniel et al. 2010a; Geeting and Hallen 2005; Geeting et al. 2003). Figure 5.16 shows how the use of Equation 5.37 accounts for the variation of filter flux when filtering a nuclear waste suspension

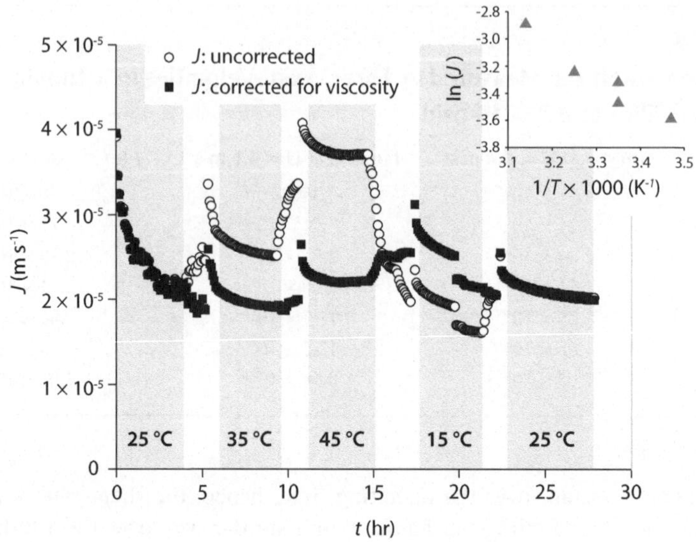

FIGURE 5.16 Filter flux measured at feed temperatures ranging from 15°C to 45°C. The flux is shown both uncorrected (as-measured) and corrected using a viscosity correlation of the form given in Equation 5.37. The inset figure shows the natural logarithm of the average flux for each operating temperature plotted against the inverse of the absolute temperature.

with reasonable accuracy. The uncorrected data is the as-measured flux at a series of different feed temperatures and used an Andrade correlation to normalize the flux to a reference temperature of 25°C. The corrected flux data is more or less constant (ignoring the initial transience at each different temperature) over the filtering period after the correction is applied, indicating that the correlation describes the viscosity functionality in the 15°C–45°C range well. Figure 5.16 (inset) shows that the linear relationship ln(J) vs. $1/T$ is obtained when an Andrade correlation is applicable. The use of a correlation is usually required because nuclear waste suspensions have complex chemistries and dissolved species that do not permit accurate estimation of the liquid viscosity from the suspension's individual components.

The second impact of operating temperature is that changes in the solution chemistry occur when the temperature changes, resulting in events such as the precipitation of solids or shifts in solution pH. This is also true of changes in ΔP but to a lesser extent. Flux behavior as described by Equation 5.37 or a similar expression may not hold if the temperature changes create suspensions that are different in character as well as liquid viscosity. Decreasing the solution temperature increases the chance of precipitating additional solids (thereby increasing ϕ_b) because many nuclear waste suspensions are saturated with multiple chemical components. Increasing the temperature would generally be expected to improve filtration performance because the viscosity is lowered and the risk of precipitation is lessened, but increased temperature can also alter chemical equilibria between species, adjust interaction potentials (both particle–particle and particle–filter), promote gas evolution in solution, and change solution pH. The impact on filter flux due to these temperature-driven changes is not easily predicted, and operating temperature changes should be carefully considered before they are implemented in a filtration system.

5.4.2 Suspension Properties

The properties of the suspension being filtered, as already discussed, play a critical role in the filtration performance that is achieved in a given system. In many nuclear waste treatment applications, there are severe restrictions on the manipulation of these properties during filtration operations. The sources of the limitations are two-pronged: (1) the requirement to keep processes as simple and robust as possible, and (2) the desire to keep the volumes of waste to be processed at the minimum. The implication of the first element is that process steps to alter suspension properties will be avoided unless they are absolutely necessary; complicated process steps requiring significant process instrumentation and control are not ideal in a radiological environment. The second element limits the addition of other chemical solutions, catalysts, or other agents because anything added to the system either increases waste volume or introduces a secondary waste stream that must be handled (and will become radiologically contaminated during processing).

The solids concentration in the suspension (ϕ_b), as it increases, will tend to reduce the filter flux achieved in dead-end and crossflow filters as described in Equations (6.9), (6.33), and (6.34). Gel polarization theory also predicts a solids concentration above which the flux is effectively zero, so operating below that threshold (ϕ_g) is important for maintaining throughput in a filtration system. The solids concentration in the feed can be decreased using dilution with water (note this also simultaneously decreases liquid viscosity), but this comes at the cost of processing increased waste volume or introducing a condensing step that would occur downstream of the filter system. Lowering the solids concentration too significantly (in some instances, depending on particle size distribution and morphology) increases the chance that a filter experiences more significant depth fouling because there is not an appreciable cake layer to help "protect" the filter media.

In addition to concentration, several other properties of the solids will affect filtration performance. These can include particle size distribution, density, morphology, speciation, interaction potentials, and hardness. Most models treat particles as hard spheres of a fixed size and density. Rarely are the solids in a nuclear waste suspension represented by a single component with a tight size distribution. In reality, the solid phase is a mixture of particles with various sizes, densities, shapes, and chemical composition. Particles will also undergo, to varying degrees, physicochemical changes such as aggregation, particle–particle interactions, chemisorption on the filter media, compaction, or attrition in the filter system. One crucial observation relevant to nuclear waste filtration is that distinct particles present in solution at a similar ϕ_b and possessing the same nominal size and density can have a different effect on the magnitude of the filter flux at the same operating conditions. The disparate impact on performance appears to occur solely because the particles are distinct chemical species, i.e., a Na_3PO_4 particle filters differently than an $Fe(OH)_3$ particle of the same physical size (for a recent example of this, see Daniel et al. 2018). This type of behavior is not predicted by conventional filtration models. Consequently, the potential parameter space is vast, which is why nuclear waste suspensions are often thoroughly tested empirically to established expected performance.

The properties of the suspending phase also have a role in filter performance; the role of viscosity (as it changes with temperature) has already been briefly discussed in Section 5.4.1. The viscosity of the suspending phase can range over a couple orders of magnitude (0.5–50 cP) for nuclear waste suspensions due to high dissolved solids content and liquid phases that tend to have high concentrations of sodium hydroxide, salts, or organics. At a given temperature and pressure, the higher viscosity suspensions will have lower filter fluxes. Nuclear waste separation applications also involve the processing of non-Newtonian suspensions, particularly as ϕ_b increases above ~0.05 (5% by volume). If, for example, the waste behaves as a Bingham plastic fluid with a yield stress, then once the fluid is yielded by the bulk flow, the consistency stands in for the suspension viscosity (the suspension is pseudo-Newtonian). There are other cases where non-Newtonian fluid behavior may add wrinkles to the filtration application, but in general, the challenges are centered on fluid handling, i.e., providing sufficient power to pump the fluid, applying the necessary pressure to generate the target flux, and reducing regions where the waste is stagnant (the waste is not yielded, or is shear-thickening). Note that the performance models discussed in this chapter assumed a Newtonian suspension was being filtered, thus they would need to be adapted for use with a non-Newtonian suspension.

The viscosity of the suspending phase is the physical property with the most influence on filtration performance. The suspending phase density is also of interest for crossflow filtration, but only with respect to the density difference between the suspending phase and the solids it carries. The density difference will have a minor role in determining the probability that a particle interacts with the filter media/cake layer or passes by with the bulk fluid. The effect of suspension-phase chemistry has been studied in some detail, resulting in a large body of work that is challenging to summarize here. For some examples of work in that area, see Palecek, Mochizuki, and Zydney (1993), Nazzal and Wiesner (1994), Singh and Song (2005), and Wang et al. (2014). A relevant recent Hanford waste study is described in Daniel et al. (2011).

5.4.3 Flux Recovery Techniques and Performance

The use of a filtration system in a nuclear waste treatment process is unique compared to most other industrial processes because once the filter media goes into service, access to the equipment becomes limited due to the radiological hazards. Standard maintenance and physical cleaning of the system is much more complicated; even replacing filter elements can be a significant undertaking. Thus, when a filter is not delivering the needed throughput, it is preferable to have robust mitigation approaches built into the system. The need for a flux recovery technique would be precipitated by exhausting the available parameter adjustments (ΔP, U, T, etc., as applicable) because implementing the techniques costs operating time.

Two simple flux recovery options that are not commonly used in nuclear facilities are mentioned here for completeness. One option is adding a filter aid to the suspension to increase the permeability of the cake layer (referred to as "body feed"). The filter aid is usually an inert material, such as diatomaceous earth. In nuclear waste operations, it is preferable to avoid adding additional materials that need to be incorporated into a downstream disposition process. Thus, even though the filter aids are typically inert materials, their use adds uncertainty to the waste treatment flowsheet that the filter system is a part of. In addition, the majority of inert filter aids are not compatible with the alkaline wastes that are characteristic of the majority of process streams. For this same reason, filter aids are not used as precoats (to establish a protective layer on the filter media) and other filter media pretreatments are not used in nuclear applications. The second option is using a clean solvent such as water to rinse the filter surface of solids. This type of approach is not used because water rinses end up creating more secondary waste volume that needs to be dealt with by using aggressive cleaning methods.

In some cases, scouring has been effective in nuclear waste filtration systems (Duignan and Nash 2012). Scouring is effective only for crossflow filters and relies on an amplified shear rate (higher bulk flow) to scour the filter surface of deposited particles while filtrate is not being removed. In the absence of transmembrane pressure, the force on the particles in the direction of the filter media (perpendicular to the flow) is reduced, and particles should be more susceptible to being swept away by the elevated hydrodynamic force. This is an attractive approach due to its ease of implementation, but it is not likely to mitigate performance issues where the majority of the particles have penetrated into the filter media itself (depth fouling). For certain suspensions, scouring may effectively clear out most of the cake layer responsible for flux decline only to expose the filter media to particles when filtration is reinitiated after the scouring operation.

The primary flux recovery techniques used in waste treatment systems are backflushing, or backpulsing, the filter and chemical cleaning. Backflushing utilizes solids-free liquid to flush the filter, typically by forcing the liquid through the filter's pores. To reduce the accumulation of secondary wastes, it is convenient to use recently generated liquid permeate. Backflushing requires a hold-up vessel to collect permeate for use in the backflush or a charge vessel if a separate backflush liquid will be used (e.g., water or other solvents). The purpose of the backflush is to break up and carry away the solids accumulated on and in the filter. Forcing fluid through the filter pores can require modest pressures, so pressurized gas is often used to generate the required force. It is fairly common to pair backflushing with dead-end filtration systems, and the backflushed solids can be routed back into a feed vessel or sent to another vessel for separate (later) treatment.

For example, consider a dead-end filter that is operating at a constant flux and allows the transmembrane pressure to vary to meet its throughput target. Perhaps the operational decision is made to backflush with permeate every 24 h. Figure 5.17 shows how the system might perform depending on the nature of the suspension begin filtered and the operating conditions. Cake filtration will have a linear pressure behavior, whereas depth filtration (a standard blocking model was used in this case) will have a non-linear pressure behavior. If the backflush recovers essentially 100% of the solids, pressure response will continue periodically every 24 h as exhibited in the first two curves in Figure 5.17. However, it is more likely that depth filtration would not be completely recovered after a backflush, as indicated by the third curve. In this instance, the starting pressure is 5% higher after each backflush. Over time,

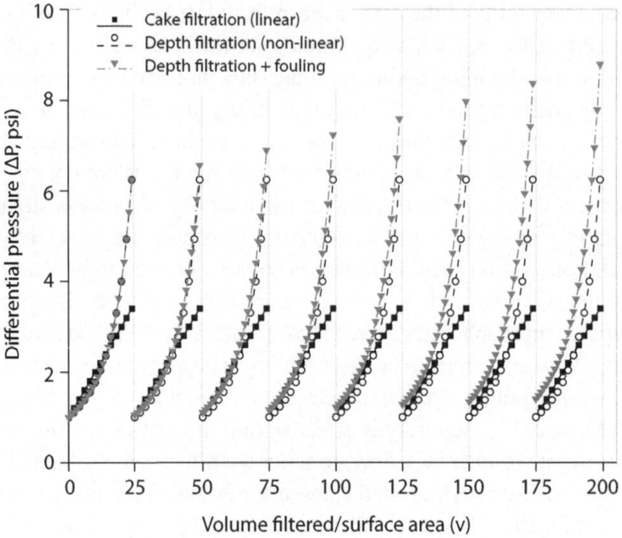

FIGURE 5.17 Differential pressure response in a dead-end filter with periodic flux recovery every 24 h. The pressure response is based on cake filtration behavior, standard blocking filtration behavior, and standard blocking behavior with some irrecoverable fouling occurring after each 24-h period.

that curve will continue to grow until it eventually overcomes the pressure constraints of the system. This also suggests that careful observation of the filter's performance with time can help guide some operating decisions. The cake filtration case in Figure 5.17 has a modest pressure increase over 24 h, and it could be prudent to increase the time elapsed between backflushes.

Another variation of this effect is if the trigger for a backflush reaches a target differential pressure. The effectiveness of the backflush will govern the length of operating periods subsequent to some reference (or starting) operating period. Figure 5.18 presents a simple example of this, where the operating time was solved for using the cake-standard model given in Bolton et al. (Bolton, LaCasse, and Kuriyel 2006):

$$\frac{P}{P_o} = \left[\left(1 - \frac{K_s J_o t}{2} \right)^{-2} + K_c J_o^2 t \right] \tag{5.38}$$

where K_s and K_c are constants for standard and cake filtration, respectively. To generate Figure 5.18, equation 5.38 was rearranged to solve for t when $P = 20$ psi and P_o has an initial value of 5 psi.

In the full recovery case, the period is always the same length because the initial pressure always returns to $P_o = 5$ psi. But if there is a small (0.2 psi) or modest (1.0 psi) increase in starting pressure P_o in each subsequent operating segment, the operating time begins to dramatically reduce. Permeate is not being produced during a backflush since it takes time to perform, representing a reduction in treatment facility throughput.

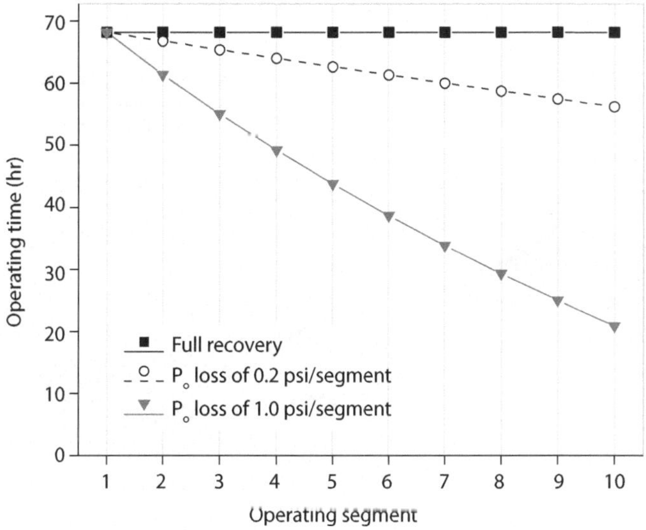

FIGURE 5.18 Impact of flux recovery on operating time in a dead-end filter assuming intervention (a flux recovery technique) occurs at $\Delta P = 20$ psi with an initial pressure $P_o = 5$ psi in the first operating segment.

It is important to point out that the recovery effectiveness for a backflush (or a backpulse, a very similar technique) will not necessarily be reduced ad infinitum. Intuitively, this is probably because there are regions of the filter that are irreversibly fouled and some that are reversibly fouled; for each filter, the relative volume of each of those regions will determine when the filter is saturated with irreversibly removed particles. This has been observed in tests with waste simulants (see Figure 5.19). Figure 5.19 shows the flux measured in a crossflow filter system normalized by the initial flux in the first operating segment. By the fourth operating segment, the flux no longer changes segment to segment after flux recovery (a backpulse) is performed at the end of the 4-h operating period. One might conclude from this data that the filter media is as fouled as it can be in that processing environment.

Backpulsing is, in many ways, still a backflush. The distinction is that a backpulse is relatively short and may be more likely to use higher pressures to drive the backflow. It still involves forcing liquid back through the filter pores from the permeate side to the bulk (feed) side. Backpulsing is more common in crossflow systems because the pressurized pulse can be applied while the transverse flow is still ongoing, thereby sweeping away the solids dislodged from the cake layer and the filter media. In a crossflow system, a backpulse is even more likely to use permeate than a backflush since the bulk suspension is still being recirculated as the backpulse occurs. In either a backflush or a backpulse, the use of a pressurized gas phase by itself is discouraged due to concerns with trapping gas in the filter pores, introducing gas into the processing system, and potentially impacting the wettability of the filter media.

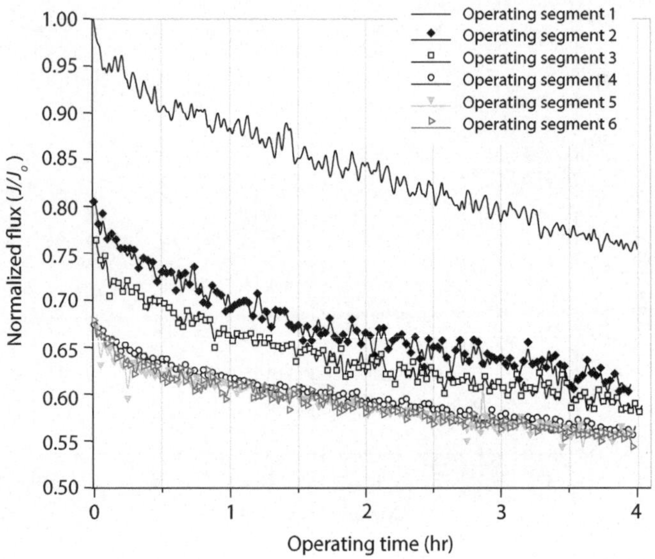

FIGURE 5.19 Example of the impact of repeated backpulsing every 4 h in a crossflow filter at fixed ΔP, T, and ϕ_b.

Backpulsing has some interesting impacts on crossflow filtration performance because how it is used can impact the filter performance over longer time periods. Refer to Figures 5.20 and 5.21 for examples. Figure 5.20 is similar in concept to Figure 5.17, except that it only considers a crossflow filter experiencing cake filtration. Cake filtration in this case is modeled simply as

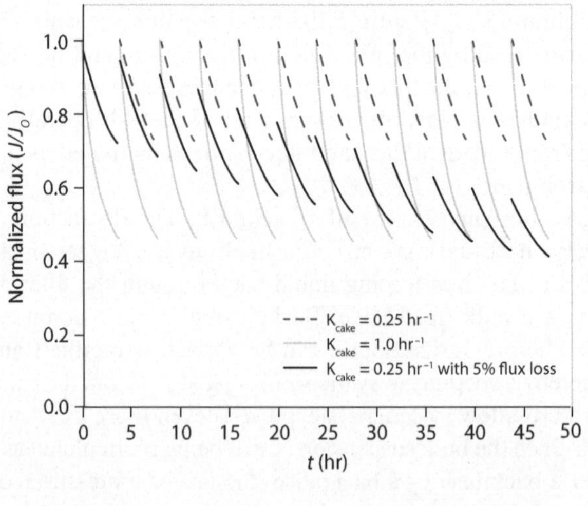

FIGURE 5.20 Comparison of cake filtration in a backpulsed crossflow filter with different cake deposition rates (as determined via the value of K_{cake}) and the impact of small repeated flux loss (5% per backpulse). The flux is normalized by the flux at $t = 0$, J_o.

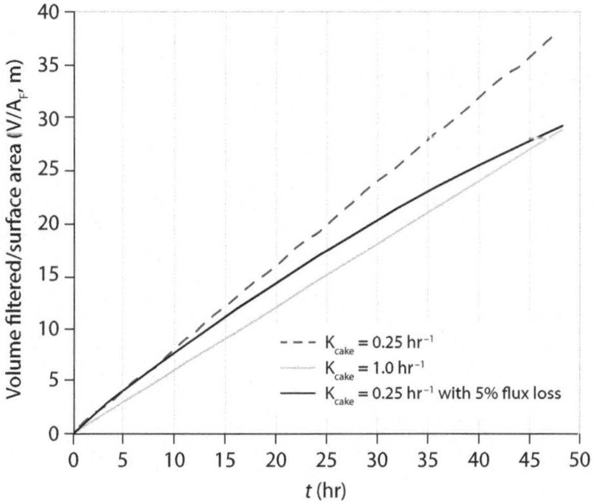

FIGURE 5.21 Cumulative effect of the various behaviors shown in Figure 5.20 on volume filtered/surface area over time.

$$\frac{J}{J_o} = \left(1 + K_{cake}t\right)^{-1/2}, \tag{5.39}$$

with K_{cake} being a cake time constant. There are three curves present in the figure, each with a cake constant (as in Equation 5.39) indicated in the legend. Assuming a backpulse occurs every 4 h and takes 10 minutes to conduct, the impact of a cake constant that is a factor of four larger is clear from examining the first two data series in Figure 5.20. However, the act of backpulsing periodically exposes the filter media to the suspension. If a few solids foul the filter irreversibly before a cake is formed on the surface, a small amount of area available for permeate production may be lost with each backpulse. If a 5% flux loss per backpulse is assumed, the normalized flux for the third data series is lower than that of the system with a much larger cake constant by the end of the twelfth operating period. Viewed another way in Figure 5.21, the volume filtered/surface area over the entire operating period becomes similar despite the higher inherent resistance of the filter system where $K_{cake} = 1.0\ h^{-1}$. Thus, the application of a backpulse strategy may reduce the long-term performance of a filter significantly. This example also shows the value of having a simple model (even if empirically based) to describe the expected filtration behavior and using it to understand what impact flux recovery techniques might have on the modeled performance over time.

Another aspect of long-term performance and backpulsing is the concept of optimum backpulse frequency. An optimum frequency arises because there is a balance between the flux that is recovered and the amount of time it takes to conduct the backpulse operations (during which a filter is not creating permeate). Backpulse frequencies that are too short have (in a relative sense) too much downtime, whereas

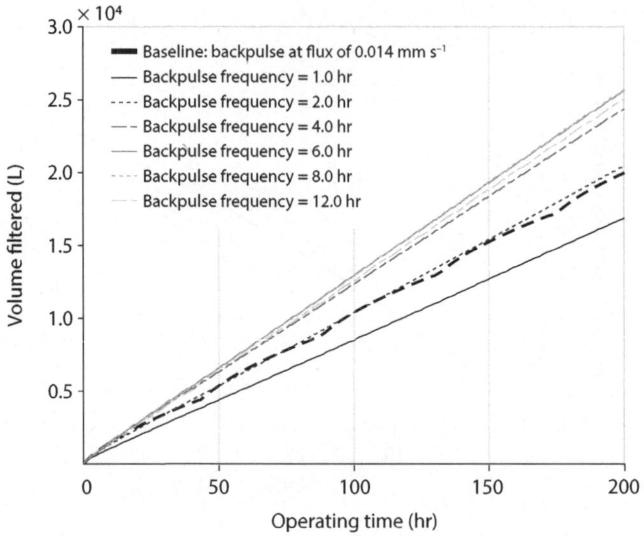

FIGURE 5.22 Example of a backpulse optimization calculation for a crossflow filtration scenario (U = 15 ft s^{-1}, ΔP = 40 psid). The baseline case backpulses when the flux reaches a value of 0.014 mm s^{-1}, whereas the other cases backpulse on a fixed frequency as indicted.

a backpulse frequency that is too long does not restore the flux often enough. Figure 5.22 illustrates a series of different backpulse frequencies and their effect on the total volume filtered, i.e., permeate generated, over time. These various frequencies are compared to a baseline case of backpulsing only when a target filter flux is reached, which represents the minimum flux acceptable for the processing scenario. In this example, the assumed backpulse downtime was 30 minutes and a flux recovery saturation effect (similar to the effect exhibited in Figure 5.19) was implemented in the simulation. The result is that a backpulse frequency in the neighborhood of 6–8 h appears to be the optimum for this filtration operating paradigm.

Figure 5.23 summarizes this result by comparing the "extra" volume filtered over the baseline case (backpulsing when a minimum acceptable flux is reached) for cases with different operating conditions. This is represented as the percent of additional volume filtered after 200 h. The U = 15 ft s^{-1}/ΔP = 40 psid data is derived from the information in Figure 5.22. Similar curve shapes are obtained for a lower velocity (U = 13 ft s^{-1}) and higher pressure (ΔP = 60 psid) case, though the higher pressure appears to have an optimum shifted closer to 6 h than the other two cases. Analyses of this type derived from scaled testing or pilot facilities can help guide operations in a nuclear waste facility.

If any of the preceding flux recovery techniques either do not work or have had serious reductions in effectiveness, cleaning the filter is the flux recovery technique of last resort. This is particularly true of filtration processes that experience appreciable depth fouling because backpulsing or backflushing lose their effectiveness over time in recovering flux, and cleaning becomes inevitable. Cleaning is used as sparingly as possible because it generates new waste streams and shuts down

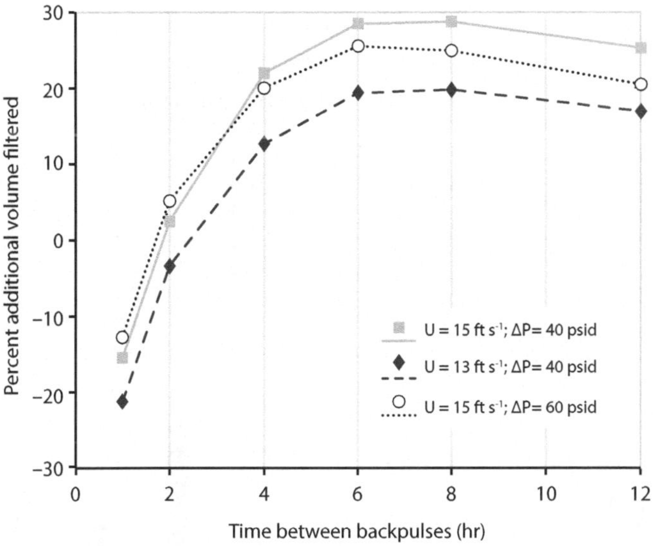

FIGURE 5.23 Optimization results for the data in Figure 5.22 ($U = 15$ ft s^{-1}, $\Delta P = 40$ psid) and two other similar crossflow cases. The percent additional volume filtered is referenced to the baseline case of backpulsing only when a minimum flux target is reached.

production for an appreciable period of time. Cleaning for nuclear waste filtration is almost always chemical (as opposed to mechanical or manual) due to the radiological hazards present. The chemicals can range from water to mild caustics to mild acids. The majority of nuclear wastes that could be filtered are alkaline due to the presence of hydroxide and dissolved salts. It is also fairly common to have numerous metal oxides present. Though caustic solutions may successfully dissolve some of these compounds, chemical cleaning of filters is primarily performed with acidic solutions such as nitric (HNO_3), oxalic ($H_2C_2O_4$), or sulfuric (H_2SO_4).

The methodology used to clean filters is wide-ranging and often established using trial and error. Some options include:

- Soaking the filter in cleaning solution for an extended period of time;
- Recirculating the cleaning solution across the filter (more common for crossflow);
- Actively filtering the cleaning solution;
- Backflushing, or backpulsing, the cleaning solution through the filter.

In most cases, the filter needs to be thoroughly rinsed to remove residual cleaning solution from the system. This avoids the possibility of significant neutralization chemistry occurring at or inside the filter media (caustic waste in contact with acidic cleaning solution, or vice versa).

In scaled test systems, filter cleanliness can be measured by comparing the flux of a standard, such as a suspension of known makeup or water. Figure 5.24 presents some test system data collected using water before a 100-h filter operation

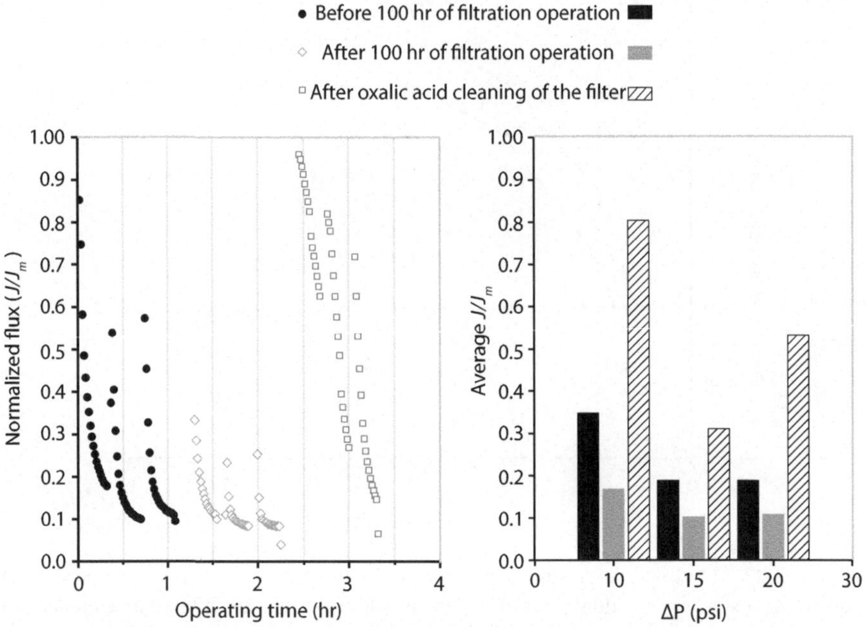

FIGURE 5.24 Water flux data obtained prior to and after a 100-h filtration operation, as well as after a chemical cleaning was performed. The left panel shows the instantaneous flux data collected at 10, 20, and 15 psi (in that order) with an arbitrary time scale for convenience, and the right panel shows the average normalized water flux for each pressure.

with a waste simulant after the operation was concluded (but before it was cleaned) and again after the filter was cleaned with 0.5 M oxalic acid. The left panel of Figure 5.24 shows the instantaneous flux data at 10, 20, and 15 psi for each condition. Despite the rapid decay in the water flux (likely related to contamination in the filter system itself), the recovery after acid cleaning is evident from this data. The right panel of Figure 5.24 averages the instantaneous flux over operation at each pressure and demonstrates that cleaning can return the flux to the performance observed prior to starting the operation (the higher average values are probably an artifact of the short operating periods during the water flux measurements). The efficacy of the cleaning regimen will vary and depends on the strength of the chemical(s) used, contact time, and methodology. It is important to select chemical cleaning approaches that are effective but create as little additional liquid waste as possible, since anything contacting a filter in nuclear service will also be handled as radioactive waste.

5.5 EXAMPLES OF FILTRATION IN NUCLEAR WASTE PROCESSING

Filtration has been deployed at several nuclear waste sites since the Cold War era. In this section, five examples of filtration's use for processing is described for four waste sites in the United States and one in the United Kingdom. These examples are not exhaustive, and research and exploration of filtration technology in nuclear applications continues around the world. As mentioned early in this chapter, filtration is usually not the preferred technology to perform separation steps in nuclear fuel reprocessing or spent fuel handling, so these examples are focused on the treatment of nuclear waste. The examples are from different levels of process maturity, from planned process flowsheets where the use of filtration is proposed to deployments of filtration in which the treatment mission at the site has been successfully completed. The approach for the provided examples is not to exhaustively describe the process but rather to provide illustrations of how filtration has been used in nuclear waste treatment contexts.

A final note on these examples: the majority of filtration applications process waste that is caustic/alkaline (or at minimum, neutral pH) and has high ionic strength. This occurs either due to heavy use of sodium hydroxide during waste-generating chemistry (weapons production, fuel reprocessing) or using hydroxide as a reagent/pH adjustment in process steps upstream. Filtration is not restricted to caustic service only; in fact, the waste site in Idaho near Idaho Falls (which has primarily acidic waste) has considered the use of crossflow filtration to support waste treatment operations at their site (Law et al. 2014; Mann and Todd 1998).

5.5.1 Melton Valley Storage Tank Wastewater Triad Project

Waste from the Melton Valley Storage Tanks in Oak Ridge, Tennessee was processed between 1997 and 2000; additional storage volume was required for continuing research and reactor operations at Oak Ridge National Laboratory. The waste was classified as low-level liquid waste containing several soluble salts, notably sodium nitrate, potassium nitrate, and sodium carbonate. There were 14 total campaigns during the operation of the integrated Wastewater Triad Project, during which 268,000 gallons of waste were treated. The process used crossflow filtration to perform a separation step before ion exchange to remove cesium, as illustrated in Figure 5.25. Note the use of multiple recirculation loops in conjunction with the filter

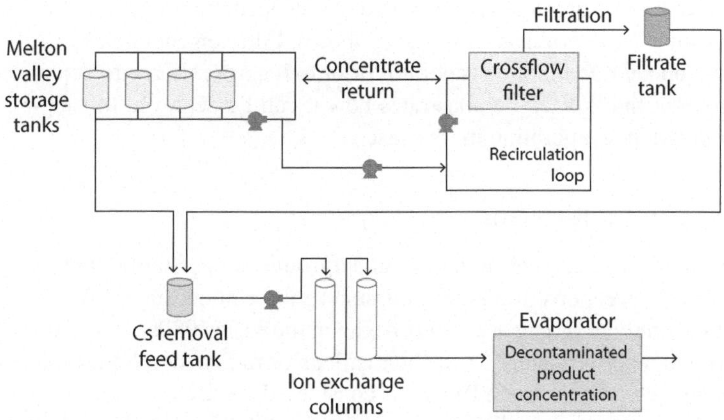

FIGURE 5.25 Simplified schematic of the Melton Valley Storage Tank Wastewater Triad Project showing the key unit operations.

in order to concentrate the solids (which were not sent to be processed downstream) and send a cesium-rich supernatant to an ion exchange decontamination step.

The crossflow filter system that was used contained a pair of sintered stainless steel tubular bundles rated to 0.5 microns. Each tube in the bundle had an outer diameter of 0.75 inches and a length of 5 feet. The total surface area for each bundle was approximately 50 ft^2. The system included a backpulse capability to periodically recover flux. Nominal operating conditions included an axial velocity U ~ 10 ft s^{-1} and ΔP ranging from 10 to 30 psi. The design flux range of 0.02 to 0.1 gpm ft^{-2} was generally met during operation (reported as 0.012 to 0.16 gpm ft^{-2}). Additional information on the Wastewater Triad Project and work supporting its implementation and design can be found in Fowler and Hewitt (1989), Walker and McCabe (1997), Walker et al. (1998), and Walker and Kent (2001).

5.5.2 West Valley Liquid Waste Treatment System

At West Valley, New York from 1966 to 1982, commercial nuclear spent fuel was received and reprocessed. Reprocessing operations generated both high- and low-level waste that ultimately totaled approximately 600,000 gallons. A low-level waste treatment facility was constructed in the early 1970s to treat lightly contaminated wastewater (<0.005 μCi/mL); this process included a filtration step prior to decontamination by ion exchange. A second low-level treatment facility was brought online later as the West Valley site shifted from a reprocessing mission to a treatment and remediation mission. The treatment started in earnest after 1982 and is known as the West Valley Demonstration Project (WVDP). The West Valley site has completed most of its waste treatment and moved into a decommissioning phase (Milner and Watters 2002).

The WVDP was an integrated radwaste treatment system that included treatment of both supernatant and solid-bearing waste, with the waste being dispositioned into either glass or cementitious waste forms (McIntosh et al. 1988). Crossflow filtration was deployed in the liquid waste treatment step as indicated in Figure 5.26. The deployed filters protected an evaporator from accumulating solids by working in concert with a demineralization vessel as shown. Filtrate generated by the crossflow filter was concentrated in the evaporator prior to incorporation into cement. The use of filtration in the WVDP demonstrates how it can be deployed to support another unit of operation (evaporation in this case).

5.5.3 Savannah River Site

The Savannah River site has numerous underground storage tanks containing legacy waste from the weapons production mission in the United States. Waste treatment, which is currently on-going at the site, began in the early 1980s. As of 2019, there are 43 tanks with approximately 35 million gallons of radioactive liquid waste remaining to be processed. Savannah River has considered the use of filtration technologies in a variety of potential applications over the past few decades, including crossflow filters to support an in-tank precipitation process through existing risers to treat liquid waste (as discussed in Wilmarth et al. (2011)) and deploying rotary microfilters,

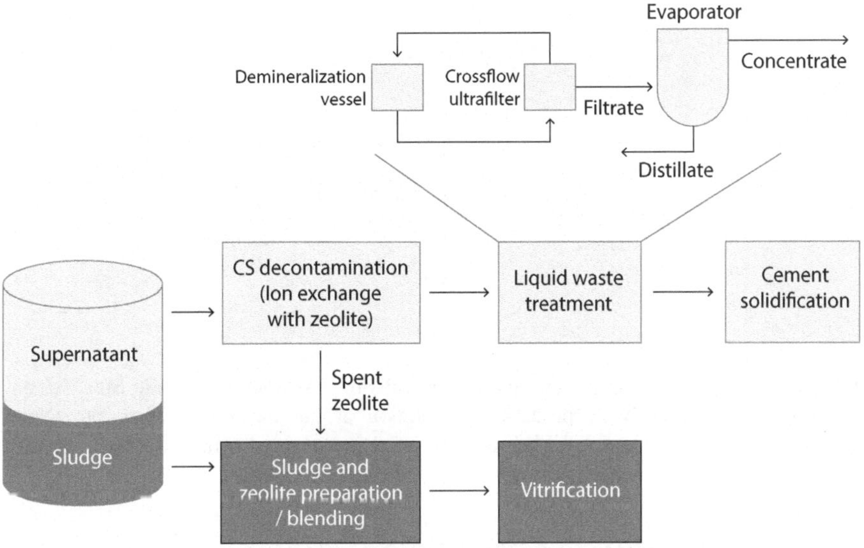

FIGURE 5.26 Schematic illustrating the process flow of the West Valley Demonstration Project integrated radwaste treatment system. Filtration was primarily used in the liquid waste treatment system as shown.

which are a more compact alternative to crossflow filtration and are also suitable for in-tank installation and operation (Herman et al. 2003, 2009; Herman, Poirier, and Fink 2006; Poirier et al. 2004).

The primary and on-going application of filtration at Savannah River is exemplified in the Actinide Removal Process (ARP)/Modular Caustic Side Solvent Extraction (MCU) flowsheet, the objective of which is to segregate actinides (primarily strontium) and separate the actinide-rich solid phase from the depleted liquid phase. The depleted liquid phase is sent to a solvent extraction process, and the concentrated solid phase is sent to the Defense Waste Processing Facility where it is incorporated into waste glass (see Figure 5.27). The actinides are captured using particles of monosodium titanate, which is the majority of the solid phase that is captured via crossflow filtration. A dead-end guard filter is also used to protect the liquid waste targeted for the MCU. The crossflow filter used in the system was originally a bundle of 144 sintered stainless steel tubes of 0.1-micron grade. The tubes had an inner diameter of 5/8 inch, a length of 10 feet, and a total surface area of 235 ft^2. Subsequent operations have used a filter with the same geometric configuration but a grade of 0.5 microns. The crossflow filter has typically produced fluxes ranging from 0.02 to 0.04 gpm ft^{-2}; however, recently, waste streams that do not contain monosodium titanate have been processed that exhibit higher fluxes. The ARP/MCU process has experienced some of the challenges described in previous sections, including flux decline, precipitation in the permeate phase, dead-end filter plugging, and difficulty with chemical cleaning approaches. Some additional information on filtration in this application can be found in Mann et al. (2004), Hobbs, Herman, and Poirier (2012), Martino et al. (2014), and Herman (2016).

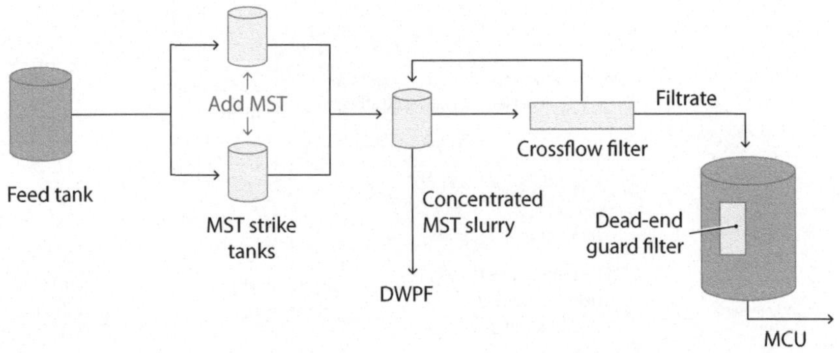

FIGURE 5.27 Diagram of the Actinide Removal Process/Modular Caustic Side Solvent Extraction Unit (ARP/MCU) process in operation at the Savannah River site. Note: MST = monosodium titanate, DWPF = defense waste processing facility.

5.5.4 Sellafield Enhanced Actinide Removal Plant

The Sellafield site has been at the core of the nuclear industry in the United Kingdom; it housed both the initial weapons production facility of that country and the first commercial nuclear reactor in the world (Calder Hall). Nearly all of the spent nuclear fuel generated by the country's nuclear reactors are reprocessed there, as well as waste from various decommissioned nuclear facilities. One of the facilities that handles low- and medium-activity effluents is the Enhanced Actinide Removal Plant (EARP), which began operation in 1994. The process flowsheet of the EARP is shown in Figure 5.28.

In the primary processing loop, the EARP uses the combination of a flocculation/precipitation process followed by filtration in order to concentrate the actinides into a solid phase that can be encapsulated. The primary loop contains nine ultrafilters with 800 tubes each that dewaters the suspension by a factor of 8–10. The secondary loop, with two ultrafilters of 500 tubes each, performs the final dewatering that concentrates the suspension still further. Permeate removed by the filtration systems is held in delay tanks until it can be discharged into the environment. The dewatering performed by the ultrafilters in the EARP proceeds until the solids content is in excess of 10 wt%, occasionally reaching 20 wt% (this represents a total volume reduction factor of approximately 10–60). The solids content varies based on the feed stream because the plant has an upper-limit viscosity of 700 cP (0.7 Pa·s). Above this limit, the suspension becomes very difficult to pump and filter. The ultrafilters are backflushable, but to avoid both excessive backflushing and/or chemical cleaning, quite a bit of effort has been poured into designing the flocculation process so that the precipitated solids have a large particle size, contain very few fines, and are resistant to attrition in the EARP (Hildred et al. 2000; Weatherill 2017). If necessary, the ultrafilter bundles were designed to be replaceable, and entire units can be swapped out.

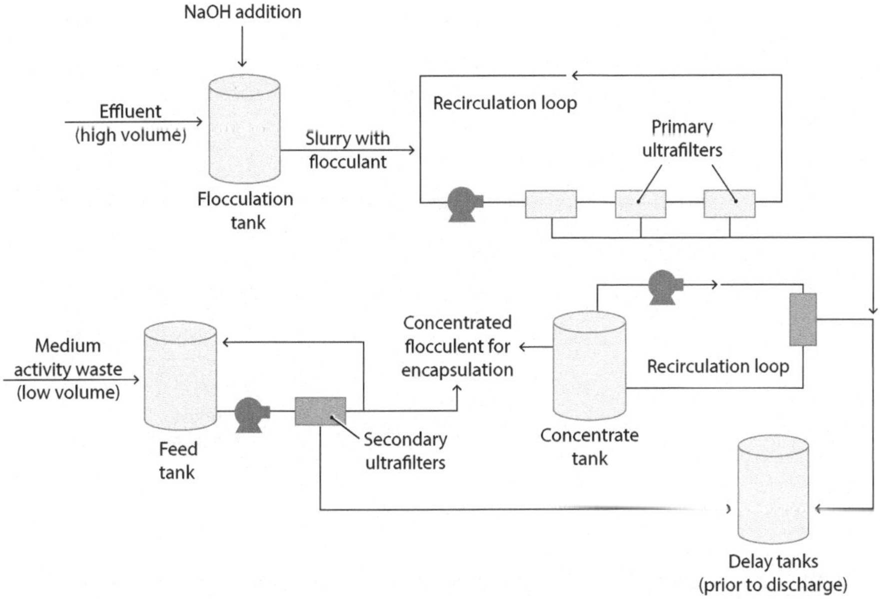

FIGURE 5.28 Schematic of the Enhanced Actinide Removal Plant (EARP) used at the Sellafield site in the United Kingdom.

5.5.5 Hanford Site

The Hanford site in the state of Washington is the location of the largest volume of legacy nuclear waste in the United States. The waste is distributed across 177 underground tanks and is estimated to encompass a volume of approximately 55 million gallons. Processing of the waste has yet to begin at a production scale on the Hanford site; the original proposal was to build one of the largest and most complex treatment facilities in the world—the Hanford Waste Treatment and Immobilization Plant (WTP). This facility is still in the construction stage, but when it is operable, the Pretreatment Facility in the WTP plans to depend heavily on crossflow filtration to separate the low-activity and high-level waste streams. The initial steps of that process are presented in Figure 5.29.

In the Pretreatment Facility, nuclear waste that is retrieved from the waste storage tanks will undergo a caustic leaching step to solubilize aluminum solids and then will be concentrated up to approximately 20 wt% using a series of crossflow filtration bundles. Permeate removed by the filters becomes the low-activity waste stream (to be vitrified after cesium ion exchange), whereas the concentrated suspension is destined for the high-level waste vitrification facility. The filters are 0.1-micron grade sintered stainless steel tubes in bundles of 8 and 10 feet in length. Permeate is collected in pulse pots that contain sufficient volume to conduct backpulses with pressurized gas supplying the motive force as needed.

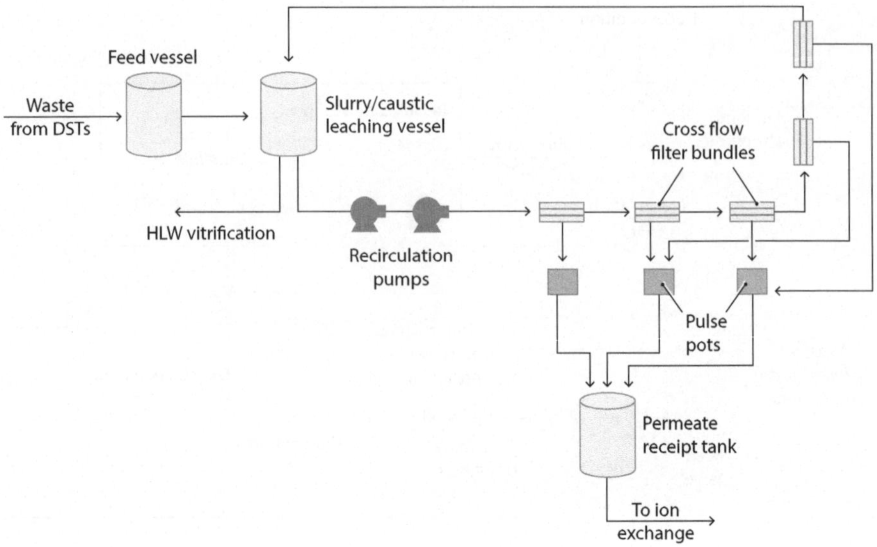

FIGURE 5.29 Proposed crossflow filtration subsystem to separate low-activity and high-level wastes in the Hanford Waste Treatment and Immobilization Plant. Note: HLW = high-level waste, DST = double shell tank.

The feed suspension is expected to be non-Newtonian and relatively high in solids (~5 wt%), making the prediction of its performance relatively complex. This filtration problem has been studied extensively, but refer to Foust, Holton, and Demick (2005), Foust and Ghosehajra (2010), Daniel et al. (2011), and Schonewill et al. (2015) for some recent examples.

A more recent treatment concept is to process only low-activity supernatant from the Hanford waste tanks by placing a modular system at or near the tanks themselves. A proposed flow sheet for this system, the Tank Side Cesium Removal (TSCR) system, is given in Figure 5.30. The TSCR process is very similar to another system also installed at-tank on the Savannah River site called the Tank Closure Cesium Removal (TCCR) system. The TCCR system had, at the time of this writing, begun operating to process liquid supernatant in the Savannah River tank farm (see Keefer et al. 2017 for a description). In TSCR, the filtration is accomplished using a pair of sintered stainless steel dead-end filters that are capable of being backflushed using compressed air. Only one filter is operated at a time with the other on standby. Dead-end filtration was chosen because the decanted liquid from the Hanford double shell tanks (DSTs) is expected to have very small solids loadings. The filters are also planned to be operated at low pressure differentials (<2 psi) in order to generate a permeate flow rate of 5 gpm. The primary function of the filters is to protect the ion exchange columns where the decontamination of the waste is occurring. The collected solids will be discharged to a different DST following backflushing to be processed at a later date.

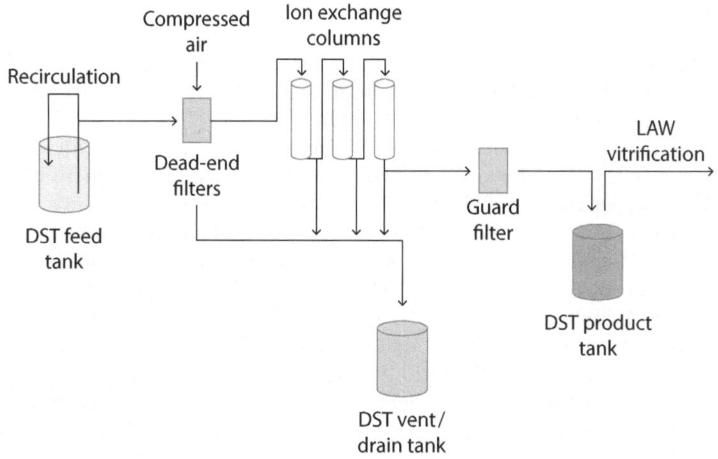

FIGURE 5.30 Process flow diagram for proposed Tank Side Cesium Removal (TSCR) system to process liquid waste on the Hanford site. Note: LAW = low-activity waste, DST = double-shell tank.

5.6 SUMMARY

In this chapter, filtration has been discussed as a separation technology that can be used in the processing of nuclear waste. Despite the availability of extensive literature on the subject, the performance of filters in nuclear environments is extraordinarily hard to predict a priori. The compiled filtration literature does serve as a useful guide for thinking through specific uses of filter technology, but in almost every case, empirical data is the best source of information to use when designing or implementing filtration in a nuclear waste facility. In contrast with typical uses of filtration in other industries, the feed suspension is chemically complex, not always well characterized or controlled, and has variable physical and chemical properties over the operating lifetime of a treatment facility. The presence of radiological hazards also encourages the selection of robust and reliable filter systems that can withstand irradiation over their design life, remain operable with minimal required maintenance, and generate as little secondary waste as possible.

Filtration has been successfully deployed in several historic and active nuclear treatment facilities and will continue to be considered for deployment in future facilities. Interest in identifying other nuclear applications for filtration is reflected in recent published research (Bushar, Tran, and Asay 2002; Dulama et al. 2009; Oliveira, Barbosa, and Afonso 2013; Poncelet et al. 2001; Rosell 2015). Filters are hardly the only well-studied solid–liquid separation unit operation; alternatives such as hydrocyclones, centrifuges, clarifiers, classifiers, and mixer/settlers may also be appropriate depending on the needs of the process. However, filtration is well-suited for nuclear waste operations due to its simple operating paradigm, flexibility, relatively small footprint, and low cost.

REFERENCES

Aimar, P., S. Baklouti, and V. Sanchez. 1986. "Membrane—Solute Interactions: Influence on Pure Solvent Transfer During Ultrafiltration." *Journal of Membrane Science* 29 (2): 207–224. doi:10.1016/S0376-7388(00)82470-9.

Ambashta, R. D., and M. E. T. Sillanpaa. 2012. "Membrane Purification in Radioactive Waste Management: A Short Review." *Journal of Environmental Radioactivity* 105: 76–84. doi:10.1016/j.jenvrad.2011.12.002.

Asaadi, M., and D. A. White. 1992. "A Model for Determining the Steady-State Flux of Inorganic Microfiltration Membranes." *Chemical Engineering Journal and the Biochemical Engineering Journal* 48 (1): 11–16. doi:10.1016/0300-9467(92)85002-Q.

Asanuma, N., M. Harada, Y. Ikeda, and H. Tomiyasu. 2001. "New Approach to the Nuclear Fuel Reprocessing in Non-Acidic Aqueous Solutions." *Journal of Nuclear Science and Technology* 38 (10): 866–871. doi:10.1080/18811248.2001.9715107.

ASTM. 2017a. *Standard Guide for Development, Verification, Validation, and Documentation of Simulated High-Level Tank Waste* (ASTM C1750-17. West Conshohocken, PA: ASTM International).

ASTM. 2017b. *Standard Guide for Measuring Physical and Rheological Properties of Radioactive Solutions, Slurries, and Sludges.* C1752-17. West Conshohocken, PA: ASTM International.

Astudillo-Castro, C. L. 2015. "Limiting Flux and Critical Transmembrane Pressure Determination Using an Exponential Model: The Effect of Concentration Factor, Temperature, and Cross-Flow Velocity During Casein Micelle Concentration by Microfiltration." *Industrial & Engineering Chemistry Research* 54 (1): 414–425. doi:10.1021/ie5033292.

Bacchin, P., P. Aimar, and R. W. Field. 2006. "Critical and Sustainable Fluxes: Theory, Experiments and Applications." *Journal of Membrane Science* 281 (1–2): 42–69. doi:10.1016/j.memsci.2006.04.014.

Barbu, A., and G. Martin. 1992. "Radiation Effects in Metals and Alloys." *Solid State Phenomena* 30–31: 179–228. doi:10.4028/www.scientific.net/SSP.30-31.179.

Bear, J. 1972. *Dynamics of Fluids in Porous Media* (New York: American Elsevier Publishing Company, Inc.).

Belfort, G., R. H. Davis, and A. L. Zydney. 1994. "The Behavior of Suspensions and Macromolecular Solutions in Crossflow Microfiltration." *Journal of Membrane Science* 96 (1): 1–58. doi:10.1016/0376-7388(94)00119-7.

Bolton, G., D. LaCasse, and R. Kuriyel. 2006. "Combined Models of Membrane Fouling: Development and Application to Microfiltration and Ultrafiltration of Biological Fluids." *Journal of Membrane Science* 277 (1–2): 75–84. doi:10.1016/j.memsci.2004.12.053.

Bowen, W. R., and F. Jenner. 1995. "Theoretical Descriptions of Membrane Filtration of Colloids and Fine Particles: An Assessment and Review." *Advances in Colloid and Interface Science* 56: 141–200. doi:10.1016/0001-8686(94)00232-2.

Bushar, S., P. Tran, and R. Asay. 2002. "Low-Level Liquid Waste Processing Pilot Studies Using a Vibratory Shear Enhanced Process (VSEP) for Filtration." *Waste Management 2002 Conference*, Tucson, AZ, February 24–28, 2002.

Carman, P. C. 1937. "Fluid Flow through Granular Beds." *Transactions, Institution of Chemical Engineers* 15: 150–166.

Chadwick, C., and A. Swain. 2014. "Nuclear Power: Filtration in Nuclear Decommissioning." *Filtration + Separation* 51 (1): 20–22. doi:10.1016/S0015-1882(14)70031-X.

Chmielewski, A. G., M. Harasimowicz, and G. Zakrzewska-Trznadel. 1999. "Membrane Technologies for Liquid Radioactive Waste Treatment." *Czechoslovak Journal of Physics* 49 (1 Supplement): 979–985. doi:10.1007/s10582-999-1027-y.

Clough, R. L., and S. W. Shalaby. 1991. *Radiation Effects on Polymers.* Washington, DC: American Chemical Society.

Dang, T. T. H., C. W. Li, and K. H. Choo. 2016. "Comparison of Low-Pressure Reverse Osmosis Filtration and Polyelectrolyte-Enhanced Ultrafiltration for the Removal of Co and Sr from Nuclear Plant Wastewater." *Separation and Purification Technology* 157: 209–214. doi:10.1016/j.seppur.2015.11.019.

Daniel, R. C., C. A. Burns, R. L. Russell, P. P. Schonewill, S. D. Hoyle, and E. J. Antonio. 2018. *Crossflow Filtration of LAWPS Simulants* (Richland, WA: Pacific Northwest National Laboratory). PNNL-27251.

Daniel, R. C., J. M. Billing, C. A. Burns, R. A. Peterson, R. L. Russell, P. P. Schonewill, and R. W. Shimskey. 2011. *Filtration Understanding: FY10 Testing Results and Filtration Model Update* (Richland, WA: Pacific Northwest National Laboratory). PNNL-20299.

Daniel, R. C., J. M. Billing, J. R. Bontha, C. F. Brown, P. W. Eslinger, B. D. Hanson, J. L. Huckaby, et al. 2010a. *EFRT M-12 Issue Resolution: Comparison of Filter Performance at PEP and CUF Scale* (Richland, WA: Pacific Northwest National Laboratory). PNNL-18498 Rev 1, WTP-RPT-185 Rev 1.

Daniel, R. C., P. P. Schonewill, R. W. Shimskey, and R. A. Peterson. 2010b. *A Brief Review of Filtration Studies for Waste Treatment at the Hanford Site* (Richland, WA: Pacific Northwest National Laboratory). PNNL-20023.

Darcy, H. 1856. *Les Fontaines Publiques De La Ville De Dijon* (Paris, France: Dalmont).

Davis, R. H. 1992. "Modeling of Fouling of Cross-Flow Microfiltration Membranes." *Separation and Purification Methods* 21 (2): 75–126. doi:10.1080/03602549208021420.

Dlouhy, Z. 1982. "Processing Liquid and Solid Radioactive Wastes." In *Studies in Environmental Science*, 65–97. New York: Elsevier Science.

Duignan, M. R., and C. A. Nash. 2012. "Experiments on Cake Development in Crossflow Filtration for High Level Waste." *Journal of Fluids Engineering-Transactions of the ASME* 134 (8). doi:10.1115/1.4007055.

Dulama, M., N. Deneanu, M. Pavelescu, and L. Pasare. 2009. "Combined Radioactive Liquid Waste Treatment Processes Involving Inorganic Sorbents and Micro/Ultrafiltration." *Romanian Journal of Physics* 54 (9–10): 851–859.

Eckstein, E. C., D. G. Bailey, and A. H. Shapiro. 1977. "Self-Diffusion of Particles in Shear Flow of a Suspension." *Journal of Fluid Mechanics* 79 (1): 191–208. doi:10.1017/S0022112077000111.

Efremenkov, V. M. 1989. "Radioactive Waste Management at Nuclear Power Plants." *International Atomic Energy Agency Bulletin*, March 1989, 37–42.

Ethier, C., and R. Kamm. 1989a. "Flow through Partially Gel-Filled Channels." *Physicochemical Hydrodynamics* 11 (2): 219–227.

Ethier, C., and R. Kamm. 1989b. "Mass-Transfer During Rate-Limited Langmuir Adsorption in a Pore." *Physicochemical Hydrodynamics* 11 (2): 205–217.

Fane, A. G., C. Tang, and R. Wang. 2011. "Membrane Technology for Water: Microfiltration, Ultrafiltration, Nanofiltration, and Reverse Osmosis." In *Treatise on Water Science*, 301. Amsterdam, the Netherlands: Elsevier Science.

Fradin, B., and R. W. Field. 1999. "Crossflow Microfiltration of Magnesium Hydroxide Suspensions: Determination of Critical Fluxes, Measurement and Modelling of Fouling." *Separation and Purification Technology* 16 (1): 25–45. doi:10.1016/S1383-5866(98)00098-7.

Foust, H., and M. Ghosehajra. 2010. "Sizing an Ultrafiltration Process That Will Treat Radioactive Waste." *Separation Science and Technology* 45 (8): 1025–1032. doi:10.1080/01496391003688563.

Foust, H., L. Holton, and L. Demick. 2005. "Maximizing Production Capacity from an Ultrafiltration Process at the Hanford Department of Energy Waste Treatment Facility." *Separation Science and Technology* 40 (16): 3323–3337. doi:10.1080/01496390500423698.

Fowler, V. L., and J. D. Hewitt. 1989. *Filtration of Oak Ridge National Laboratory Simulated Liquid Low-Level Waste* (Oak Ridge, TN: Oak Ridge National Laboratory). ORNL/TM-10653.

Geeting, J. G. H., and R. T. Hallen. 2005. "Filtration, Washing, and Caustic Leaching of Hanford Tank AZ-101 Sludge." *Separation Science and Technology* 40 (1–3): 1–15. doi:10.1081/Ss-200041752.

Geeting, J. G. H., R. T. Hallen, L. K. Jagoda, A. P. Poloski, R. D. Scheele, and D. R. Weier. 2003. *Filtration, Washing, and Caustic Leaching of Hanford Tank AZ-101 Sludge* (Richland, WA: Battelle, Pacific Northwest Division). PNWD-3206 Rev 1; WTP-RPT-043 Rev 1.

Goossen, W. R. A., G. G. Eichholz, and D. W. Tedder. 1991. *Treatment of Gaseous Effluents at Nuclear Facilities* (United States: Harwood Academic Publishers).

Grenier, A., M. Meireles, P. Aimar, and P. Carvin. 2008. "Analysing Flux Decline in Dead-End Filtration." *Chemical Engineering Research & Design* 86 (11a): 1281–1293. doi:10.1016/j.cherd.2008.06.005.

Herman, D. T. 2016. *Evaluation of No-MST Operations in the SRS ARP for Hanford LAWPS* (Aiken, SC: Savannah River National Laboratory).

Herman, D. T., M. R. Poirier, and S. D. Fink. 2006. *Testing and Evaluation of the Modified Design of the 25-Disk Rotary Microfilter* (Aiken, SC: Savannah River National Laboratory). SRNL-STI-2006-00073.

Herman, D. T., M. R. Poirier, D. T. Hobbs, and S. D. Fink. 2003. *Testing of the Spin Tek Rotary Microfilter Using Actual Waste* (Aiken, SC: Westinghouse Savannah River Company). WSRC-TR-2003-00030.

Herman, D. T., D. B. Stefanko, M. R. Poirier, and S. D. Fink. 2009. *Testing of a Full-Scale Rotary Microfilter for the Enhanced Process for Radionuclides Removal* (Aiken, SC: Savannah River National Laboratory). SRNL-STI-2009-00183.

Hermia, J. 1982. "Constant Pressure Blocking Filtration Law: Application to Power Law Non-Newtonian Fluids." *Transactions of the Institution of Chemical Engineers* 60: 183–187.

Hildred, K. L., P. S. Townson, G. V. Hutson, and R. A. Williams. 2000. "Characterisation of Particulates in the BNFL Enhanced Actinide Removal Plant." *Powder Technology* 108 (2–3): 164–172. doi:10.1016/S0032-5910(99)00216-8.

Hlavacek, M., and F. Bouchet. 1993. "Constant Flow-Rate Blocking Laws and an Example of Their Application to Dead-End Microfiltration of Protein Solutions." *Journal of Membrane Science* 82 (3): 285–295. doi:10.1016/0376-7388(93)85193-Z.

Ho, C. C., and A. L. Zydney. 2000. "A Combined Pore Blockage and Cake Filtration Model for Protein Fouling During Microfiltration." *Journal of Colloid and Interface Science* 232 (2): 389–399. doi:10.1006/jcis.2000.7231.

Hobbs, D. T., D. T. Herman, and M. R. Poirier. 2012. *Decontamination Factors and Filtration Flux Impact to ARP at Reduced MST Concentration* (Aiken, SC: Savannah River National Laboratory). SRNL-STI-2012-00299.

International Atomic Energy Agency. *Particulate Filtration in Nuclear Facilities*. Vol. 325 *Technical Reports Series*. Vienna, Austria: International Atomic Energy Agency, 1991.

Iritani, E., and N. Katagiri. 2016. "Developments of Blocking Filtration Model in Membrane Filtration." *Kona Powder and Particle Journal* (33): 179–202. doi:10.14356/kona.2016024.

Iritani, E., N. Katagiri, T. Takenaka, and Y. Yamashita. 2015. "Membrane Pore Blocking During Cake Formation in Constant Pressure and Constant Flux Dead-End Microfiltration of Very Dilute Colloids." *Chemical Engineering Science* 122: 465–473. doi:10.1016/j.ces.2014.09.052.

Kazemi, M. A., M. Soltanieh, and M. Yazdanshenas. 2012. "Modeling of Transient Permeate Flux Decline During Crossflow Microfiltration of Non-Alcoholic Beer with Consideration of Particle Size Distribution." *Journal of Membrane Science* 411: 13–21. doi:10.1016/j.memsci.2012.02.064.

Keefer, M. T., W. P. Mayson III, G. C. Arthur, and T. L. Fellinger. 2017. "Tank Closure Cesium Removal at Savannah River Site." *Waste Management 2017 Conference*, Phoenix, AZ, March 5–9, 2017.

Kim, J., and F. A. DiGiano. 2009. "Fouling Models for Low-Pressure Membrane Systems." *Separation and Purification Technology* 68 (3): 293–304. doi:10.1016/j.seppur.2009.05.018.

Kinoshita, C., K. Fukumoto, and K. Nakai. 1991. "Electron-, Ion- and Neutron-Irradiation Damage in Ceramics." *Annales de Chimie Science des Materiaux (Paris)* 16 (4-6): 379–389.

Kozeny, J. 1927. "Ueber Kapillare Leitung Des Wassers Im Boden." *Sitzungsber Akad. Wiss., Wien* 136 (2(a)): 271–306.

Law, J., B. Mincher, T. G. Garn, M. Greenhalgh, N. Scmitt, and V. Rutledge. 2014. "Development and Testing of an Americium/Lanthanide Separation Flowsheet Using Sodium Bismuthate." ICAPP 2014: International Congress on Advances in Nuclear Power Plants, Charlotte, North Carolina, April 6–9, 2014.

Levy, P. W. 1991. "Radiation Damage Studies on Non-Metals Utilizing Measurements Made During Irradiation." *Journal of Physics and Chemistry of Solids* 52 (1): 319–349. doi:10.1016/0022-3697(91)90070-G.

Lu, W. M., K. J. Hwang, and S. C. Ju. 1993. "Studies on the Mechanism of Cross-Flow Filtration." *Chemical Engineering Science* 48 (5): 863–872. doi:10.1016/0009-2509(93)80325-K.

Mackley, M. R., and N. E. Sherman. 1992. "Cross-Flow Cake Filtration Mechanisms and Kinetics." *Chemical Engineering Science* 47 (12): 3067–3084. doi:10.1016/0009-2509(92)87007-D.

Mackley, M. R., and N. E. Sherman. 1994. "Cross-Flow Filtration with and without Cake Formation." *Chemical Engineering Science* 49 (2): 171–178. doi:10.1016/0009-2509(94)80035-9.

Mann, N. R., R. S. Herbst, T. G. Garn, M. R. Poirier, and S. D. Fink. 2004. *Alternative Ultrafiltration Membrane Testing for the SRS Baseline Process* (Idaho Falls, ID: Idaho National Engineering and Environmental Laboratory). INEEL/EXT-04-01933.

Mann, N. R., and T. A. Todd. 1998. *Evaluation and Testing of the Cells Unit Crossflow Filter on INEEL Dissolved Calcine Slurries* (Idaho Falls, ID: Idaho National Engineering and Environmental Laboratory). INEEL/EXT-98-00749.

Martino, C. J., D. T. Herman, J. A. Pike, and T. B. Peters. 2014. *Actinide Removal Process Sample Analysis, Chemical Modeling, and Filtration Evaluation* (Aiken, SC: Savannah River National Laboratory). SRNL-STI-2003-00700.

Matsumoto, Y., S. Nakao, and S. Kimura. 1988. "Cross-Flow Filtration of Solutions of Polymers Using Ceramic Microfiltration." *International Chemical Engineering* 28 (4): 677–683.

Matthiasson, E., and B. Sivik. 1980. "Concentration Polarization and Fouling." *Desalination* 35 (1–3): 59–103. doi:10.1016/S0011-9164(00)88604-X.

McIntosh, T. W., W. W. Bixby, J. E. Krauss, and D. R. Leap. *An Overview of Waste Management Systems at the West Valley Demonstration Project* (Tucson, AZ: University of Arizona Nuclear Engineering Dept, 1988).

Milner, T. N., and W. T. Watters. 2002. "Development of the Decontamination Approach for the West Valley Demonstration Project Decontamination Project Plan." *Waste Management 2002 Conference*, Tucson, AZ, February 24–28, 2002.

Nazzal, F. F., and M. R. Wiesner. 1994. "Ph and Ionic-Strength Effects on the Performance of Ceramic Membranes in Water Filtration." *Journal of Membrane Science* 93 (1): 91–103. doi:10.1016/0376-7388(94)85019-4.

Nuclear Air Cleaning Handbook. Edited by Jan K. Fretthold James W. Slawski, Matt R. Hargan, Roger W. Zavadoski. Vol. DOE-HDBK-1169-2003 *DOE Technical Standards*, 2003.

Nuclear Regulatory Commission. 2012. "Liquid Radioactive Wastes." In *Radioactive Waste Management Self Study*. United States Nuclear Regulatory Commission.

Oliveira, E. E. D., C. C. R. Barbosa, and J. C. Afonso. 2013. "Stability of a Nanofiltration Membrane after Contact with a Low-Level Liquid Radioactive Waste." *Quimica Nova* 36 (9): 1434–1440.

Palecek, S. P., S. Mochizuki, and A. L. Zydney. 1993. "Effect of Ionic Environment on BSA Filtration and the Properties of BSA Deposits." *Desalination* 90 (1–3): 147–159. doi:10.1016/0011-9164(93)80172-J.

Patel, R., D. Shah, B. G. Prajapti, and M. Patel. 2010. "Overview of Industrial Filtration Technology and Its Applications." *Indian Journal of Science and Technology* 3 (10). http://www.indjst.org/index.php/indjst/article/view/29843.

Peng, H., and A. Y. Tremblay. 2008. "Membrane Regeneration and Filtration Modeling in Treating Oily Wastewaters." *Journal of Membrane Science* 324 (1–2): 59–66. doi:10.1016/j.memsci.2008.06.062.

Peterson, R. A., E. C. Buck, J. Chun, R. C. Daniel, D. L. Herting, E. S. Ilton, G. J. Lumetta, and S. B. Clark. 2018. "Review of the Scientific Understanding of Radioactive Waste at the U.S. DOE Hanford Site." *Environmental Science & Technology* 52 (2): 381–396. doi:10.1021/acs.est.7b04077.

Petzny, W. J., and J. A. Quinn. 1969. "Calibrated Membranes with Coated Pore Walls." *Science* 166 (3906): 751–753. doi:10.1126/science.166.3906.751.

Poirier, M. R., D. T. Herman, F. F. Fondeur, S. D. Fink, R. Haggard, T. Deal, C. Stork, and V. Van Brunt. 2004. *Pilot-Scale Testing of a Rotary Microfilter with Irradiated Filter Disks and Simulated SRS Waste* (Aiken, SC: Westinghouse Savannah River Company). WSRC-MS-2004-00194.

Poloski, A. P., A. W. Etchells, J. Chun, H. E. Adkins, A. M. Casella, M. J. Minette, and S. T. Yokuda. 2010. "A Pipeline Transport Correlation for Slurries with Small but Dense Particles." *Canadian Journal of Chemical Engineering* 88 (2): 182–189. doi:10.1002/cjce.20260.

Poloski, A. P., H. E. Adkins, J. Abrefah, A. M. Casella, R. E. Hohimer, F. Nigel, M. J. Minette, J. J. Toth, J. M. Tingey, and S. T. Yokuda. 2009. *Deposition Velocities of Newtonian and Non-Newtonian Slurries in Pipelines* (Richland, WA: Pacific Northwest National Laboratory). PNNL-17639, WTP-RPT-175.

Poncelet, F.-J., J.-P. Moulin, N. Hubert, M.-F. Debreuille, and S. Runge. 2001. "Continuous Precipitation and Filtration, a Process Technology for Liquid Waste Management." *Waste Management 2001 Conference*, Tucson, AZ, February 28 to March 1, 2001.

Rahman, R. O. A., H. A. Ibrahium, and Y. T. Hung. 2011. "Liquid Radioactive Wastes Treatment: A Review." *Water* 3 (2): 551–565. doi:10.3390/w3020551.

Rana, D., T. Matsuura, M. A. Kassim, and A. F. Ismail. 2013. "Radioactive Decontamination of Water by Membrane Processes—A Review." *Desalination* 321: 77–92. doi:10.1016/j.desal.2012.11.007.

Ripperger, S., and J. Altmann. 2002. "Crossflow Microfiltration – State of the Art." *Separation and Purification Technology* 26 (1): 19–31. doi:10.1016/S1383-5866(01)00113-7.

Robertson, B. C., and A. L. Zydney. 1990. "Protein Adsorption in Asymmetric Ultrafiltration Membranes with Highly Constricted Pores." *Journal of Colloid and Interface Science* 134 (2): 563–575. doi:10.1016/0021-9797(90)90163-I.

Rosell, A. 2015. "Purification of Radioactive Waste Water Using a Ceramic Membrane." Master's Thesis, Department of Chemical and Biological Engineering, Chalmers University of Technology. http://publications.lib.chalmers.se/records/fulltext/218472/218472.pdf.

Ruth, B. F. 1935a. "Studies Im Filtration... IV. Nature of Fluid Flow through Filter Septa and Its Importance in the Filtration Equation." *Industrial and Engineering Chemistry* 27: 806–816. doi:10.1021/ie50307a014.

Ruth, B. F. 1935b. "Studies in Filtration III. Derivation General Filtration Equations." *Industrial and Engineering Chemistry* 27: 708–723. doi:10.1021/ie50306a024.

Ruth, B. F., G. H. Montillon, and R. E. Montonna. 1933a. "Studies in Filtration I Critical Analysis of Filtration Theory." *Industrial and Engineering Chemistry* 25: 76–82. doi:10.1021/ie50277a018.

Ruth, B. F., G. H. Montillon, and R. E. Montonna. 1933b. "Studies in Filtration II Fundamental Axiom of Constant-Pressure Filtration." *Industrial and Engineering Chemistry* 25: 153–161. doi:10.1021/ie50278a010.

Schonewill, P. P., R. C. Daniel, R. W. Shimskey, C. A. Burns, J. M. Billing, and R. A. Peterson. 2015. "Long-Time Performance of a Stainless Steel Crossflow Filter with Simulated Hanford Tank Waste." *Chemical Engineering Research & Design* 102: 69–79. doi:10.1016/j.cherd.2015.06.016.

Schulz, G., and S. Ripperger. 1989. "Concentration Polarization in Crossflow Microfiltration." *Journal of Membrane Science* 40 (2): 173–187. doi:10.1016/0376-7388(89)89003-9.

Singh, G., and L. F. Song. 2005. "Quantifying the Effect of Ionic Strength on Colloidal Fouling Potential in Membrane Filtration." *Journal of Colloid and Interface Science* 284 (2): 630–638. doi:10.1016/j.jcis.2004.10.030.

Song, L. F. 1998a. "Flux Decline in Crossflow Microfiltration and Ultrafiltration: Mechanisms and Modeling of Membrane Fouling." *Journal of Membrane Science* 139 (2): 183–200. doi:10.1016/S0376-7388(97)00263-9.

Song, L. F. 1998b. "A New Model for the Calculation of the Limiting Flux in Ultrafiltration." *Journal of Membrane Science* 144 (1–2): 173–185. doi:10.1016/S0376-7388(98)00057-X.

Stewart, T. L., Frey J.A., Geiser D.W., Manke K.L. 1998. "Overview of U.S. Radioactive Tank Problem." In *Science and Technology for Disposal of Radioactive Tank Wastes*, edited by W. W. Schultz, Lombardo N.J., 3–13. Boston, MA: Springer.

Sun, S. D., Y. L. Yue, X. H. Huang, and D. Y. Meng. 2003. "Protein Adsorption on Blood-Contact Membranes." *Journal of Membrane Science* 222 (1–2): 3–18. doi:10.1016/S0376-7388(03)00313-2.

Van Der Bruggen, B., C. Vandecasteele, T. Van Gestel, W. Doyen, and R. Leysen. 2004. "A Review of Pressure-Driven Membrane Processes in Wastewater Treatment and Drinking Water Production." *Environmental Progress* 22 (1): 46–56. doi:/10.1002/ep.670220116.

Vendel, J., J. C. Laborde, N. Michielsen, and F. Gensdarmes. 2009. "Lessons Learnt over 30 Years of Air Filtration in the Nuclear Industry." *Journal of Physics: Conference Series* 170 (1): 012026. http://stacks.iop.org/1742-6596/170/i=1/a=012026.

Vincent-Vela, M. C., B. Cuartas-Uribe, S. Alvarez-Blanco, and J. Lora-Garcia. 2012. "Analysis of an Ultrafiltration Model: Influence of Operational Conditions." *Desalination* 284: 14–21. doi:10.1016/j.desal.2011.08.030.

Walker, B. W., and D. J. McCabe. 1997. *Oak Ridge National Laboratory Melton Valley Storage Tanks Waste Filtration Process Evaluation* (Aiken, SC: Westinghouse Savannah River Company).

Walker Jr., J. F., and T. E. Kent. 2001. *Wastewater Triad Project: Final Summary Report* (Oak Ridge, TN: Oak Ridge National Laboratory). ORNL/TM-2001/129.

Walker Jr., J. F., P. A. Taylor, R. L. Cummins, B. S. Evans, S. D. Heath, J. D. Hewitt, R. D. Hunt, et al. 1998. *Cesium Removal Demonstration Utilizing Crystalline Silicotitanate Sorbent for Processing Melton Valley Storage Tank Supernate: Final Report* (Oak Ridge, TN: Oak Ridge National Laboratory). ORNL/TM-13503.

Wang, F. Y., M. J. Zhang, W. Peng, Y. M. He, H. J. Lin, J. R. Chen, H. C. Hong, A. J. Wang, and H. Y. Yu. 2014. "Effects of Ionic Strength on Membrane Fouling in a Membrane Bioreactor." *Bioresource Technology* 156: 35–41. doi:10.1016/j.biortech.2014.01.014.

Weatherill, J. S. 2017. "Iron Oxyhydroxide Formation in the Enhanced Actinide Removal Plant." PhD diss., University of Manchester.

Welch, T. D. 2001. *Tank Waste Transport, Pipeline Plugging, and the Prospects for Reducing the Risk of Waste Transfers* (Oak Ridge, TN: Oak Ridge National Laboratory). ORNL/TM-2001/157.

Wilmarth, W. R., G. J. Lumetta, M. E. Johnson, M. R. Poirier, M. C. Thompson, P. C. Suggs, and N. P. Machara. 2011. "Review: Waste-Pretreatment Technologies for Remediation of Legacy Defense Nuclear Wastes." *Solvent Extraction and Ion Exchange* 29 (1): 1–48. doi:10.1080/07366299.2011.539134.

Yee, K. W. K., D. E. Wiley, and J. Bao. 2009. "A Unified Model of the Time Dependence of Flux Decline for the Long-Term Ultrafiltration of Whey." *Journal of Membrane Science* 332 (1-2): 69–80. doi:10.1016/j.memsci.2009.01.041.

Zakrzewska-Trznadel, G. 2006. "Membrane Processes for Environmental Protection: Applications in Nuclear Technology." *Nukleonika* 51, no. 1: S101–S111.

Zakrzewska-Trznadel, G. 2013. "Advances in Membrane Technologies for the Treatment of Liquid Radioactive Waste." *Desalination* 321: 119–130. doi:10.1016/j.desal.2013.02.022.

Zakrzewska-Trznadel, G., M. Harasimowicz, and A. G. Chmielewski. 2001. "Membrane Processes in Nuclear Technology—Application for Liquid Radioactive Waste Treatment." *Separation and Purification Technology* 22-3 (1–3): 617–625. doi:10.1016/S1383-5866(00)00167-2.

Zhang, M. M., and L. F. Song. 2000. "Mechanisms and Parameters Affecting Flux Decline in Cross-Flow Microfiltration and Ultrafiltration of Colloids." *Environmental Science & Technology* 34 (17): 3767–3773. doi:10.1021/es990475u.

Zinkle, S. J. 1999. "Fundamental Radiation Effects Parameters in Metals and Ceramics." *Radiation Effects and Defects in Solids* 148 (1–4): 447–477. doi:10.1080/10420159908229104.

Zydney, A. L., and C. K. Colton. 1986. "A Concentration Polarization Model for the Filtrate Flux in Cross-Flow Microfiltration of Particulate Suspensions." *Chemical Engineering Communications* 47 (1–3): 1–21. doi:10.1080/00986448608911751.

6 Ion Exchange

Reid A. Peterson, Garrett Brown, and Amy M. Rovira

CONTENTS

6.1 INTRODUCTION TO ION EXCHANGE PROCESSES

Ion exchange is generally treated as a subset of adsorption processes in chemical engineering texts. This is in part due to the limited applications of ion exchange processes. The bulk of ion exchange processes are associated with water treatment. The primary ion exchange process that most people run into on a day-to-day basis is their water softener system. Many homes today contain these systems to remove trace impurities from their drinking water. Besides the home market, ion exchange for water treatment also has a significant market in the production of high-purity water for a number of industrial applications. These applications are generally associated with the treatment of relatively large quantities of water that contain trace components. The objective is to produce a stream that is relatively free of contaminants. In many non-nuclear applications, the contaminants are not of any value and can be discarded. In contrast, ion exchange processes for nuclear separations processes frequently are associated with either purifying products or with isolating radiative materials for further processing. These processes tend to be more intensive, process smaller relative volumes, and also usually require recovery of at least one species

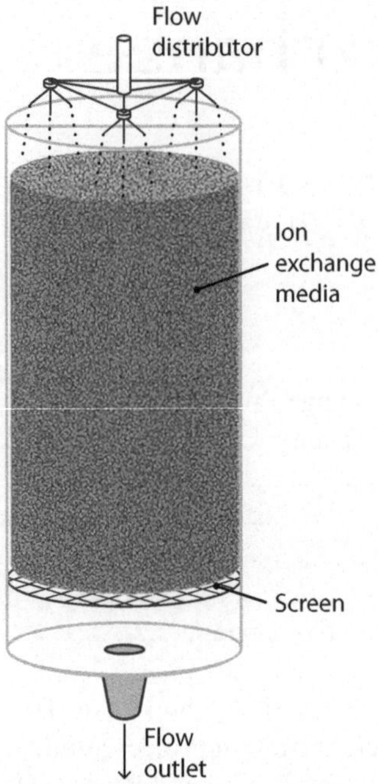

FIGURE 6.1 Configuration of a typical ion exchange column for downflow processing.

of interest from the ion exchange media. A number of examples are discussed, and some of the engineering implications of these processes are highlighted.

For nuclear applications, ion exchange is typically carried out in fixed beds in a vertical configuration (see Figure 6.1). These columns can be processed in either an upflow or downflow configuration, although downflow is the typical operating choice. The ion exchange medium is ideally a spherical material of a relatively uniform size. Typical media range from 80 to 300 μm (16–100 mesh). In general, individual media will have tighter ranges, typically on the order of 100 μm. The ion exchange material is supported on a screen whose mesh size is smaller than the ion exchange media. For a downflow configuration, flow is introduced through a distributor that allows uniform flow through the bed. Channeling, or preferential flow through only a portion of the bed, is one of the challenges to avoid when establishing an ion exchange media bed. Flow rates are generally relatively low; processing rates for non-reactor nuclear separations applications range from a fraction of a gallon per minute to 25 gallons per minute corresponding to superficial velocities that range from fractions of a centimeter per minute to 10 cm/min.

6.2 ION EXCHANGE MEDIA CAPACITY

Ion exchange processes are distinct from normal adsorption processes in that there is exchange of one species for another. Both cationic and anionic species are removed by ion exchange processes; hence, both cationic and anionic ion exchange media are used. A typical cationic ion exchange process for a sodium form ion exchange medium can be expressed as follows:

$$R^{*}Na^{+} + M^{+} \leftrightarrow R^{*}M^{+} + Na^{+} \tag{6.1}$$

where $R^{*}Na^{+}$ is the cationic ion exchange medium in the sodium form, and M^{+} is the soluble metal ion that is to be removed from solution. The first step in developing an ion exchange process is to assess the affinity of the material of choice for the species to be selectively removed. This is typically performed by an equilibrium batch contact. One way to define equilibrium is by measuring the distribution coefficient K_d:

$$K_d = \frac{(C_0 - C_1)}{C_1} \cdot \frac{V}{M} \tag{6.2}$$

where C_0 is the initial concentration of the metal ion of interest, C_1 is the final concentration of the metal ion of interest, V is the volume of the contact solution, and M is the mass of the ion exchange medium. Note that M is the mass of the medium itself and frequently needs to be adjusted for adsorbed water.

In these tests, a fixed volume sample of the target feed solution is contacted with the ion exchange medium and allowed to come to equilibrium. This test solution is generally mixed for a period of at least 24 hours. Figure 6.2 shows typical batch contact data measured for a number of time periods. Typical of this type of test, the batch contact involved 20 mL of solution in contact with 0.2 g of ion exchange medium. These data are for cesium removal from a high sodium matrix associated with high-level nuclear waste. As indicated in Figure 6.2, by 24 hours the system had reached equilibrium to the experimental uncertainty for each of the test temperatures. Note that as the temperature decreases, the measured K_d generally increases, indicating

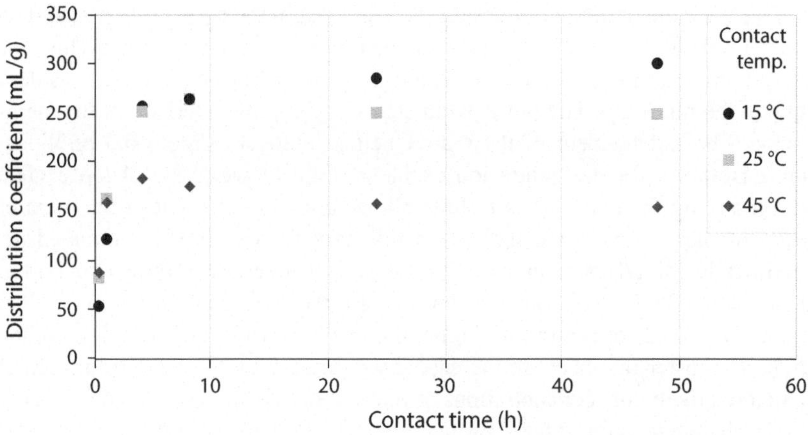

FIGURE 6.2 Cesium batch contact from Tank Waste Simulant. (Adapted from Nash, C.A. et al., *Batch, Kinetics, and Column Data from Spherical Resorcinol-Formaldehyde Resin*, Savannah River National Laboratory, Aiken, SC, 2006.)

that the affinity for the ion exchange medium for the species of interest increases as the temperature decreases. This suggests that operating at lower temperatures has an advantage relative to the capacity of the ion exchange media. There is also typically a trade-off in that the kinetics of the ion exchange generally slow as the temperature decreases (also visible in Figure 6.2) and the potential to drop other species out of high salt solution increases. This last phenomenon is complicated by the high ionic strength generally encountered in processing of nuclear materials.

The K_d measurement is useful in that it can be readily used to assess the capacity of a column. From the K_d measured at the equilibrium feed concentration, the calculation of the capacity—or λ is straightforward:

$$\lambda = K_d \times \rho_b \tag{6.3}$$

where ρ_b is the specific density of the ion exchange bed using the same mass basis for the ion exchange media as was used in determining the K_d. The λ value is also typically expected to represent the 50% breakthrough point for the ion exchange column, assuming that the breakthrough curve is relatively symmetric around this point. This is discussed in greater detail in Section 6.3.

Another way to define equilibrium behavior is through development of an isotherm. Frequently, a Langmuir or Freundlich isotherm is used to describe the capacity of the medium for removal of the selected species. In developing an isotherm, it is important to understand the matrix from which the species of interest is to be removed. The simple Langmuir hybrid isotherm expression is of the form:

$$C_S = \frac{C_{ST} C_A}{(C_A + \beta)} \tag{6.4}$$

where C_S is the concentration of the species of interest loaded onto the solids, C_{ST} is the maximum theoretical loading onto the solids in the absence of any competing ions, C_A is the concentration of the species of interest in the aqueous phase, and the β parameter reflects the selectivity of the media for the species of interest. This results in a two-parameter fit for the isotherm, and only one of those parameters is dependent upon the matrix, β. The other parameter, C_{ST}, is a characteristic of the medium and should be independent of the matrix being deployed. Figure 6.3 provides an example isotherm for plutonium ion exchange onto Dowex® 1 × 4 ion exchange resin. Presenting the data in this fashion allows direct assessment of the maximum capacity of the ion exchange media—in this case the capacity is estimated to be approximately 130 g/L of resin. Also, it is possible to assess something about optimal feed conditions. In this case, any feed above 1 g/L will fully saturate the ion exchange medium. Also, feed concentrations higher than 1 g/mL will saturate the ion exchange medium at a faster rate because the capacity is limited by the equilibrium loading. Note that relatively low concentrations of plutonium that were observed to load onto the ion exchange medium. These results suggest that relatively high decontamination factors would be achievable in column operation. However, it should be noted that this type of batch contact data does not provide information regarding the kinetic behavior of the ion exchange medium.

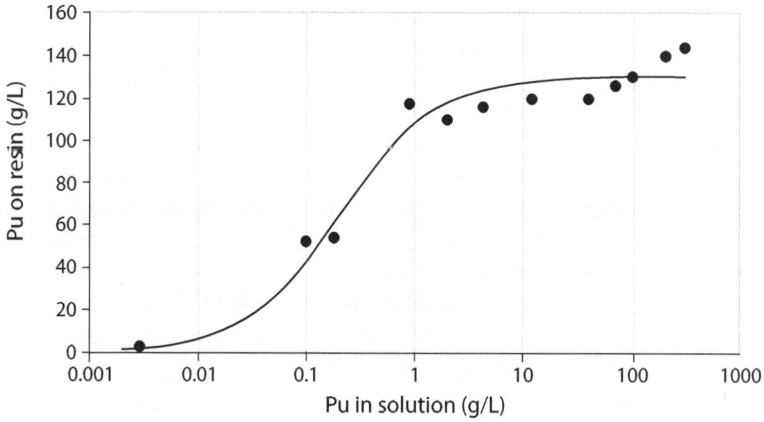

FIGURE 6.3 Plutonium isotherm for anionic sorption on Dowex® resin. (From Bain, P.E., *Solvent Extr. Ion Exc.*, 8, 341–351, 1990.)

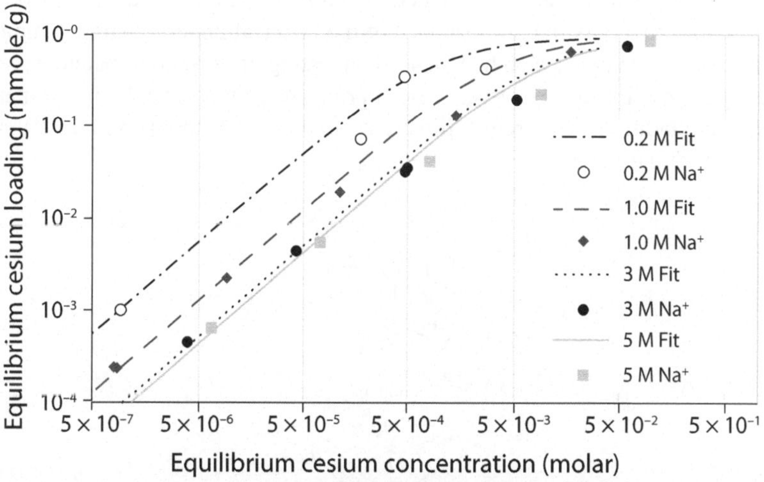

FIGURE 6.4 Cesium isotherm for CS-100 resin at varying sodium molarities. (From Brown, G. N. et al., *Comparison of Organic and Inorganic Ion Exchangers for Removal of Cesium and Strontium from Simulated and Actual Hanford 241-AW-101 DSSF Tank Waste*, Pacific Northwest National Laboratory, Richland, WA, 1996.)

As indicated above, the solutions to be processed by ion exchange for nuclear operations tend to be very complex matrices with many other species present. A good example is tank waste left over after processing materials to produce plutonium and other actinides. These complex solutions contain other ions that may affect the capacity of the ion exchange medium. Figure 6.4 shows an example of a cesium isotherm in the presence of varying sodium concentrations. The data are from cesium batch contacts with Duolite® CS-100 manufactured by Rohm and Haas. For these tests, the ion exchange medium was contacted with varying strengths of cesium- and sodium-bearing solutions. Not surprisingly, sodium dramatically affects the ability of the ion exchange material to exchange sodium for cesium. Note that typical process conditions for this type of feed range from roughly 5×10^{-5} M to just less than 5×10^{-4} M. Typical tank waste contains 5 M sodium or higher.

Over this range of conditions from 0.2 to 5 M sodium, the presence of sodium decreases the cesium loading by roughly an order of magnitude. Also note that the vast majority of the capacity of the ion exchange medium is taken up by the sodium present. Understanding the impact of the competing ions is essential to understanding the efficacy of the ion exchange medium for its intended purpose. Note that even with the significant decrease in cesium loading due to the presence of sodium, the ion exchange medium shown in Figure 6.4 was successfully deployed for separation of cesium from tank waste streams.

6.3 COLUMN PERFORMANCE

Once it has been determined that the ion exchange medium of interest has capacity for the ion of interest and sufficient selectivity compared to the other species present, the next step is to assess the performance of the material under dynamic conditions. Up to this point, the discussion has focused on testing performed in batch processes, frequently at equilibrium. However, typical processing involves columns of material in dynamic conditions. Figure 6.5 illustrates some of the important terms that are part of these dynamics.

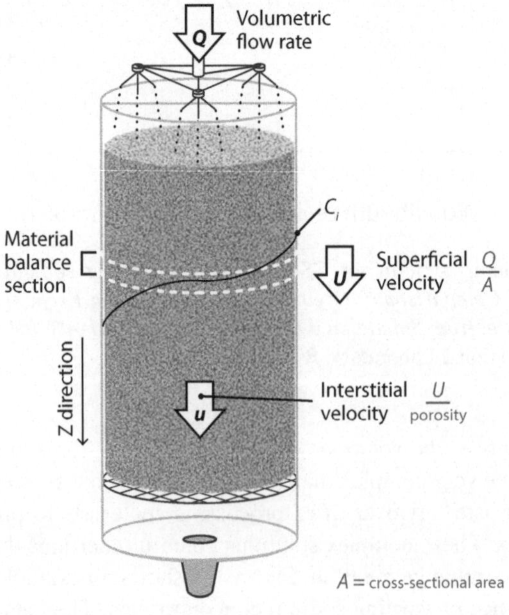

FIGURE 6.5 Material balance over a theoretical column and typical load curve.

Feed is introduced into the column at a volumetric flow rate (Q). This volumetric flow rate can be related to velocity through the column by consideration of the cross-sectional area (A) of the column:

$$U = \frac{Q}{A} \tag{6.5}$$

where U is the superficial velocity through the column. However, the ion exchange medium fills some of the space in the column, so it is also useful to consider the interstitial velocity (u) through the material:

$$u = \frac{U}{\varepsilon_B} \tag{6.6}$$

where ε_B is the porosity of the bed. If a material balance is performed over a section of the ion exchange column, the rate of change in the concentration of the species of interest in solution at that point can be defined, as follows:

$$\varepsilon_B \frac{\partial c_i}{\partial t} = -\varepsilon_B u \frac{\partial c_i}{\partial z} + \varepsilon_B D_{\text{eff},i} \frac{\partial^2 c_i}{\partial z^2} - r_i \tag{6.7}$$

where r_i is the rate of ion exchange for the species of interest. The volume of the segment of the column is adjusted to account for only the available porosity of the bed. The first term is associated with the flow of material into the segment of the column. Figure 6.5 shows a hypothetical concentration profile throughout the column. The concentration profile has a negative slope in the z direction, and the impact of flow through the column is a net increase in the concentration as a function of time. However, this concentration profile also would impose a diffusive gradient. Toward the top of the column, this gradient will cause a decrease in the concentration in the segment of interest because the second derivative with respect to length will be negative. Toward the end of the breakthrough curve, this diffusive effect will cause an increase in the concentration in the segment. The final term is associated with the ion exchange of the species of interest, which will cause a decrease in the concentration of the species of interest.

The last two terms in Equation 6.7 need further assessment. The first of these is the effective diffusivity of the material through the column. Here it is useful to use some experimental data to highlight the dynamics that are occurring in a column. Figure 6.6 provides the breakthrough curve for a cesium ion exchange testing using a 1.25 m tall (L) ion exchange column with a cross-sectional area of 9,000 cm^2 feed at a rate (Q) of 45.5 L/min. The total bed volume (BV) of this ion exchange bed is 1,125 L. It is also useful to define the flow rate in terms of the number of BVs processed per hour, as follows:

$$\frac{BV}{h} = \frac{Q}{BV} \cdot 60 \tag{6.8}$$

Figure 6.6 shows the breakthrough curve for this test in terms of BVs processed. These data can also be used to project the concentration of the species of interest as

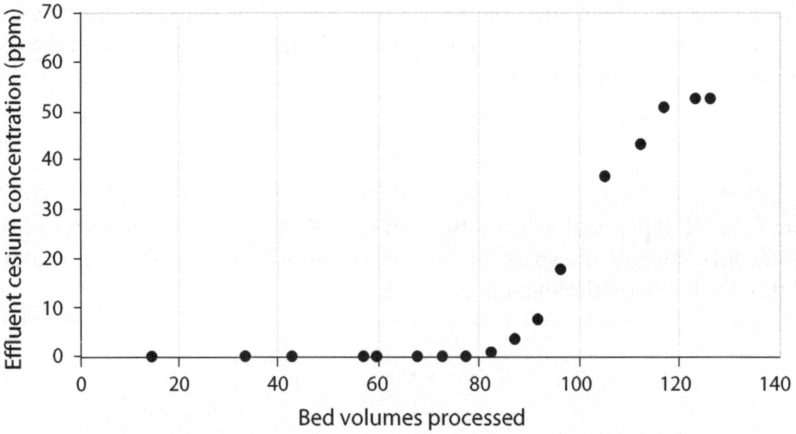

FIGURE 6.6 Breakthrough curve for 1.25 m tall column. (Adapted from Fiskum, S.K. et al., *Sep. Sci. Technol.*, 1–10, 2019.)

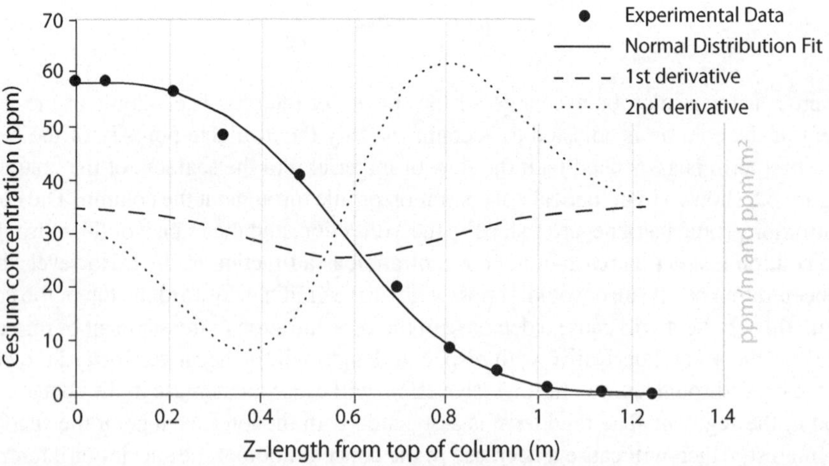

FIGURE 6.7 Estimated column concentration profile and numerical derivatives. (Data adapted from Fiskum, S.K. et al., *Sep. Sci. Technol.*, 1–10, 2019.)

the column is loaded. This can be done if the breakthrough curve has been taken to the nearly complete breakthrough stage as was done in this case. Figure 6.7 shows an of the estimate of the cesium concentration profile in the column after processing 72 BVs, at which point the column had achieved approximately 0.1% breakthrough. The point z_i at which concentration c_i would be found can be estimated as follows:

$$z_i = L\frac{BV_i - BV_f}{BV_e - BV_f} \tag{6.9}$$

where c_i is the concentration measured at the exit after processing BV_i bed volumes. BV_f is the number of volumes to effectively achieve a 100% breakthrough and BV_e is the target number of bed volumes processed—in this case 72 bed volumes. The data from this test can be expressed as a cumulative normal distribution where the mean value ($\bar{z}$) is 0.58 m and the standard deviation (σ) is 0.2 m. Figure 6.7 also shows the fit to these data using the cumulative normal distribution. See Klinkenberg (1948) for a more detailed development of why this fit would be appropriate. If the diffusion term in Equation 6.7 is small and the rate of adsorption is relatively simple (i.e., it does not involve diffusion within the ion exchange media), then the normal distribution is a reasonable fit for the fully developed load curve. As indicated in Figure 6.7, this fit is reasonable for this particular experiment. The error function term can then be used to assess the validity of the first assumption that the diffusion term is small. The first derivative of the concentration as a function of length is merely the probability density function:

$$\frac{\partial c}{\partial z} = c_f \frac{1}{\sqrt{2\pi}\sigma} e^{-\frac{1}{2\sigma^2}(z-\bar{z})} \tag{6.10}$$

And the second derivative is

$$\frac{\partial^2 c}{\partial z^2} = c_f \frac{\bar{z}-z}{\sigma^2} \frac{1}{\sqrt{2\pi}\sigma} e^{-\frac{1}{2\sigma^2}(z-\bar{z})^2} \tag{6.11}$$

Now to get to the first two terms in Equation 6.7 we need the bed porosity and the effective diffusivity. Chung and Wen (1968) developed an empirical correlation for the effective diffusivity using only eddy diffusion:

$$D_{\text{eff}} = \frac{D_p U}{0.2 + 0.11 Re^{0.48}} \tag{6.12}$$

where

$$Re = \frac{D_p \rho U}{\mu}, \tag{6.13}$$

D_p is the diameter of the particle (450 μm for the example of interest), ρ is the density of the carrier fluid (1.25 g/mL for the example), and μ is the viscosity (3 cp for this example). This gives a Reynolds number of 0.16. Note that for a Reynolds number less than about 1, the Peclet number does not change significantly. Using the Reynolds number, it is possible to assess the impact of advection and dispersion on the rate of the transport of the species of interest down the column. Figure 6.8 shows the rate of change as a function of column length for these two terms in Equation 6.7. It is evident that for this set of conditions, advection dominates the transport of the species of interest through the column, and dispersion accounts for less than 3% of the change at any given point. Note that if the breakthrough curve is sharpened by

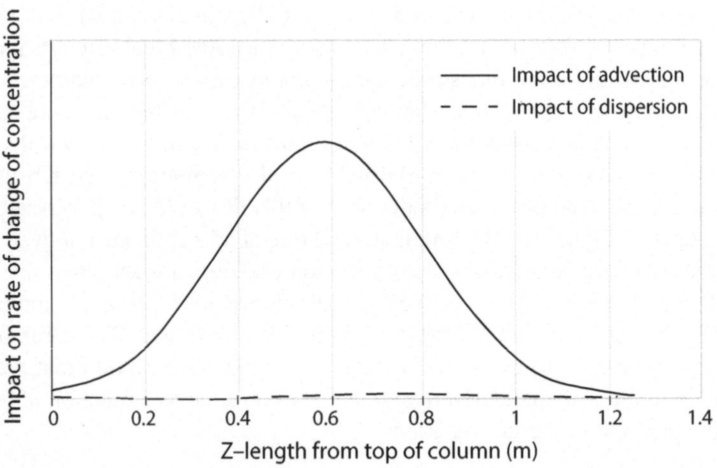

FIGURE 6.8 Impact of advection and axial dispersion on mass transfer through an ion exchange column.

reducing the standard deviation used in Equations 6.10 and 6.11 by half, then the dispersion will account for 10% of the change at the high and low concentration ends of the breakthrough curve and should be considered under these conditions. Decreasing the flow rate for this particular feed would have the effect of sharpening the breakthrough curve.

Now turn to the last term in Equation 6.7, the rate of removal of the species of interest by ion exchange. Five steps control the rate of these processes and can be important to the engineering applications of these processes. The first three steps are associated with the actions of the metal species of interest. These are depicted in Figure 6.9. The first step (1) is transport of the ion of interest from the bulk solution through the boundary layer to the surface of the ion exchange medium. The second step (2) is the transport of the metal ion to within the ion exchange medium itself. Some ion exchange media are characterized by physical pores (as indicated in Figure 6.5) through which the metal ion must diffuse, other ion exchange media are gels through which the metal must be diffuse. The third step (3) involves the ion exchange reaction as indicated in Equation 6.1. The final two steps are the inverse of Steps 2 and 1 for the species that was exchanged; in the example from Equation 6.1, this would be the Na^+ atom. Depending on the processing conditions, the rate-limiting step for the ion exchange process can be any one of these steps. Increasing the flow rate will decrease the boundary layer and increase the kinetics of Step 1. Increasing temperature generally increases the rates of Steps 2 through 4, while increasing the concentration of the metal ion of interest generally increases the rates of Steps 2 and 3. In the example of cationic removal articulated in Equation 6.1, either the sodium content in the feed streams needs to be relatively low, or the ion exchange medium must have a high selectively for the metal ion of interest.

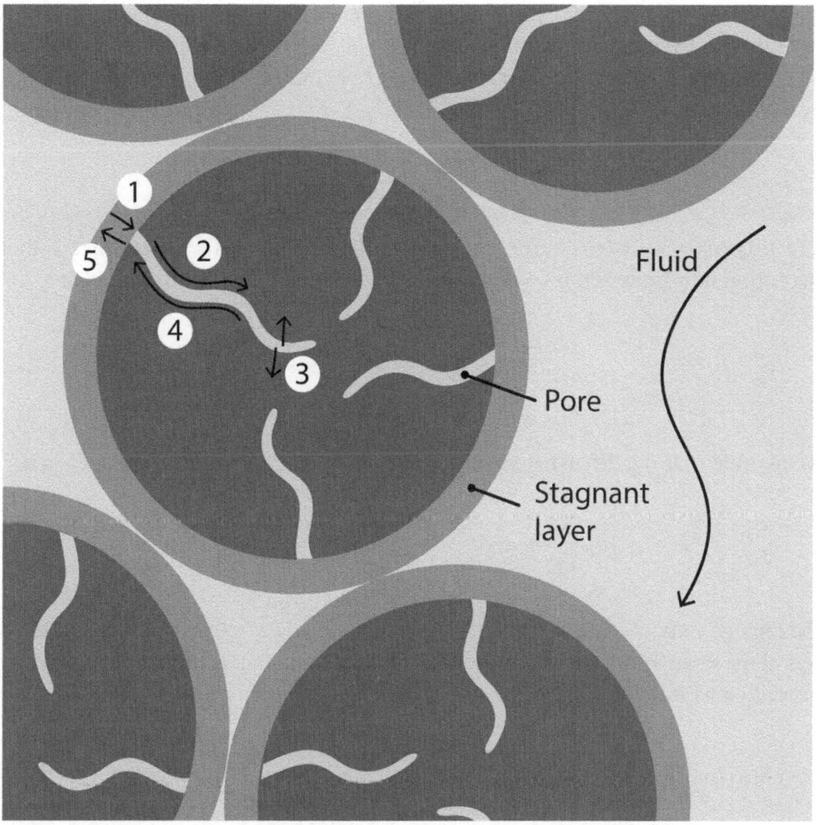

FIGURE 6.9 Representation of the processes that occur around an ion exchange bead.

If we assume the ion exchange medium is a uniform sphere, then the mass transfer through the film can be defined as follows:

$$r_B = 4\pi R_p^2 k(c - c_s) \tag{6.14}$$

where r_B the mass transfer for a single bead, where R_p is the radius of the ion exchange media, k is the mass transfer coefficient, c is the bulk concentration and c_s is the concentration at the surface of the ion exchange medium.

The amount of the medium needs to be normalized by the volume of a bead (V_B) of the ion exchange medium

$$V_B = \frac{4}{3}\pi R_p^3 \tag{6.15}$$

$$r' = \frac{3}{R_p} k(c - c_s) \tag{6.16}$$

where r' is the mass transfer rate for the volume of ion exchange medium. The volume of the ion exchange medium present per unit volume of the column is defined at $1-\varepsilon_B$ so the rate of adsorption for Equation 6.7 becomes

$$r = (1-\varepsilon_B)\frac{3}{R_p}k(c-c_s) \tag{6.17}$$

where this mass transfer coefficient will be dependent on the interstitial velocity. Substitution into Equation 6.7 yields

$$\varepsilon_B\frac{\partial c}{\partial t} = -\varepsilon_B u\frac{\partial c}{\partial z} + \varepsilon_B D_{\text{eff},i}\frac{\partial^2 c}{\partial z^2} - (1-\epsilon_B)\frac{3}{R_P}k(c-c_S) \tag{6.18}$$

If we assume that the diffusion term is small, as was shown above, this becomes:

$$\varepsilon_B\frac{\partial c}{\partial t} = -\varepsilon_B u\frac{\partial c}{\partial z} - (1-\epsilon_B)\frac{3}{R_P}k(c-c_S) \tag{6.19}$$

If the change in concentration with respect to time at any given point is small compared to the change in concentration with respect to distance, then this equation can be rewritten as follows:

$$\varepsilon_B u\frac{\partial c}{\partial z} = -(1-\epsilon_B)\frac{3}{R_P}k(c-c_S) \tag{6.20}$$

A similar material balance can be written around the ion exchange medium:

$$(1-\epsilon_B)\rho_B\frac{\partial c^*}{\partial t} = (1-\epsilon_B)\frac{3}{R_P}k(c-c_S) \tag{6.21}$$

where ρ_B is the density of the ion exchange particles and c^* is the concentration of the ion of interest on the ion exchange medium. If the ion concentration in the bead is uniform (i.e., if the kinetics of transfer within the bead are fast), then it is reasonable to assume that

$$c_s = \frac{c^*}{K_d} \tag{6.22}$$

For the purposes of this example, we will assume that $1/K_d$ is relatively constant over the range of interest. Substitution into Equations 7.19 and 7.20 yields

$$\varepsilon_B u\frac{\partial c}{\partial z} = -(1-\epsilon_B)\frac{3}{R_P}k\left(c-\frac{c^*}{K_d}\right) \tag{6.23}$$

$$\rho_B \frac{\partial c^*}{\partial t} = \frac{3}{R_P} k \left(c - \frac{c^*}{K_d} \right) \tag{6.24}$$

and with some rearrangement:

$$-\frac{\partial c}{\partial z} = \frac{(1-\epsilon_B)}{\varepsilon_B u} \frac{3}{R_P} k \left(c - \frac{c^*}{K_d} \right) \tag{6.25}$$

$$\frac{\partial c^*}{\partial t} = \frac{3}{\rho_B K_d R_P} k \left(K_d c - c^* \right) \tag{6.26}$$

If we define the following three parameters:

$$\alpha = \frac{(1-\epsilon_B)}{\varepsilon_B u} \frac{3k}{R_P} \tag{6.27}$$

$$\beta = \frac{3k}{\rho_B K_d R_P} \tag{6.28}$$

$$\theta = \frac{1}{K_d} \tag{6.29}$$

Equations 6.23 and 6.24 become

$$-\frac{\partial c}{\partial z} = \alpha \left(c - \theta c^* \right) \tag{6.30}$$

$$\frac{\partial c^*}{\partial t} = \beta \left(\frac{c}{\theta} - c^* \right) \tag{6.31}$$

If the initial concentration of ion on the medium is 0 for all values of z and the feed concentration is held constant at C_0, these two equations have the following solution (Hougen and Marshall 1947):

$$\frac{c}{c_0} = 1 - e^{-\beta t} \int_0^{\alpha z} e^{-\alpha z} J_o\left(2i\sqrt{\beta t \alpha z} \right) d(\alpha z) \tag{6.32}$$

$$\frac{c^*}{c_f^*} = 1 - e^{-\alpha z} \int_0^{\beta t} e^{-\beta t} J_o\left(2i\sqrt{\beta t \alpha z} \right) d(\beta t) \tag{6.33}$$

where c_f^* is the equilibrium concentration in equilibrium with the feed for the ion exchange medium, and J_o is the Bessel function of the first kind and zero order. This equation does not have an analytical solution, and so it must be solved either numerically or graphically. However, for sufficiently large length of z, this can be estimated by:

$$\frac{c}{c_0} = \frac{1}{2}\left(1 + \text{erf}\left(\sqrt{\beta t} - \sqrt{\alpha z}\right)\right) \tag{6.34}$$

Figure 6.10 shows the fit for the same configuration described for Figure 6.7, but at varying velocities. Here the film layer mass transfer coefficient is assumed to vary as a function of the Reynolds and Schmidt numbers (Wilson and Geankoplis 1966):

$$k = 1.09 \cdot \frac{u}{\left(Re \cdot Sc\right)^{2/3}} \tag{6.35}$$

so that the mass transfer coefficient increases with velocity to the 1/3 power. Figure 6.10 indicates that this provides a reasonable fit for the expected dependence of the mass transfer on the superficial velocity. Using this expression, it is possible to assess the impact of residence time on the number of BVs required to achieve breakthrough. If we define breakthrough to be C/C_0 of 0.001, we can see a typical plot of the number of BVs required to achieve breakthrough as a function of the throughput, as defined by the number of BVs per hour (Figure 6.11). As one would

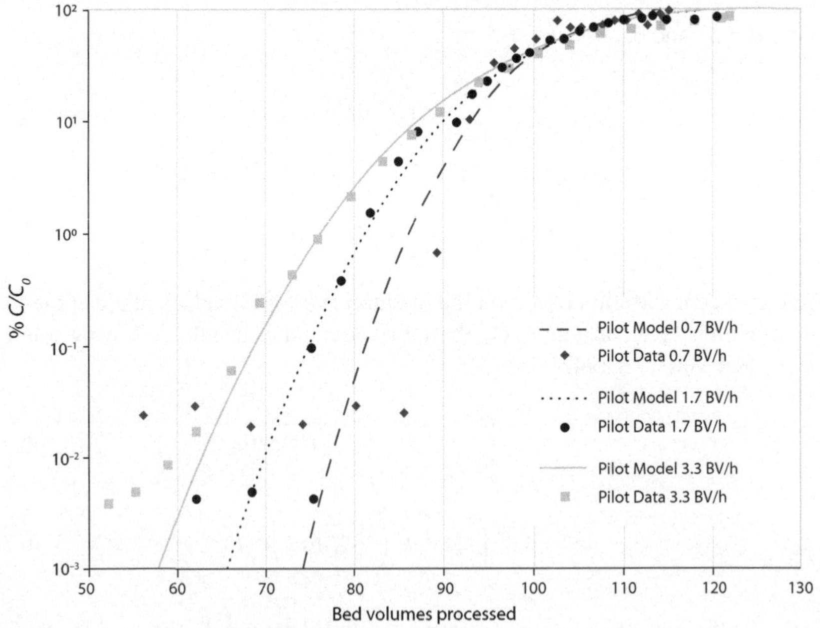

FIGURE 6.10 Impact of increased flow rate on ion exchange column performance. (Adapted from Fiskum, S.K. et al., *Sep. Sci. Technol.*, 1–10, 2019.)

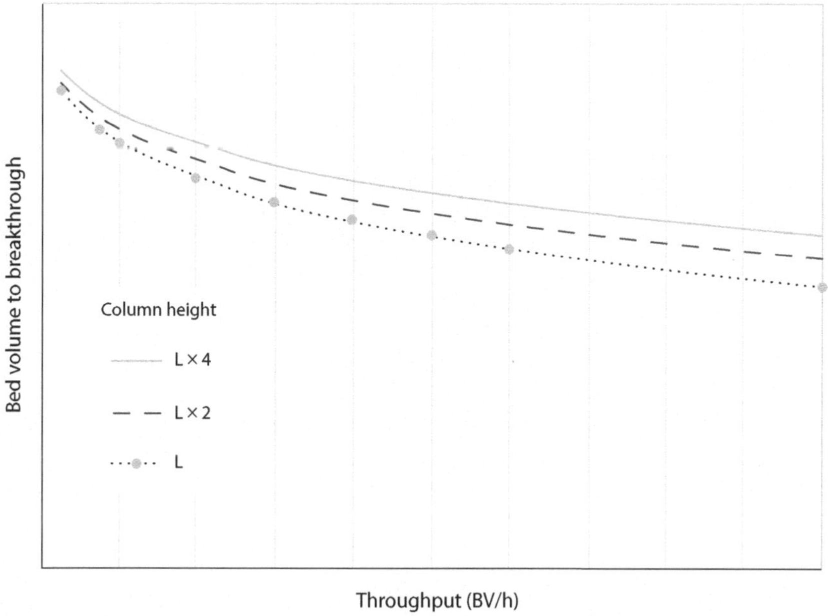

FIGURE 6.11 Impact of increased flow rate on BVs required to achieve breakthrough.

expect, Figure 6.11 indicates that as the throughput increases, the efficacy of the ion exchange medium decreases as indicated by a smaller number of BVs required to achieve breakthrough. This impact is more pronounced in smaller columns due to the slower flow, and thus higher mass transfer resistance associated with smaller columns. Note from a design perspective that increasing the height of the column has about half the impact of decreasing the flowrate, so design trade-offs need to balance these impacts. In addition, increasing the height of the column will be accompanied by increased pressure drops.

Looking back at Figure 6.7, it is apparent that if column operation is stopped when the effluent from the column hits initial breakthrough, the ion exchange material will not be fully used. In the example shown, the ion exchange material usage is roughly 50%. For some systems, this may be desirable, but for many nuclear applications, it is desirable to achieve higher usage of the ion exchange material. Some reasons include the expense of non-eluatable ion exchange material; the degradation of the ion exchange material upon elution (hence the desire to limit the number of elution cycles); throughput concerns associated with frequent elutions; and the desire to produce a concentrated eluant. Therefore, it is useful to evaluate the performance of two columns in series.

Figure 6.12 shows the progression of the loading for three different column conditions, depending upon the value of α. For $\alpha > 100$, the breakthrough curve is relatively sharp, and the mass transfer zone is fully contained within a single column. These conditions create the most efficient use of the ion exchange material. However, to achieve this type of performance, the mass transfer needs to be relatively fast, the

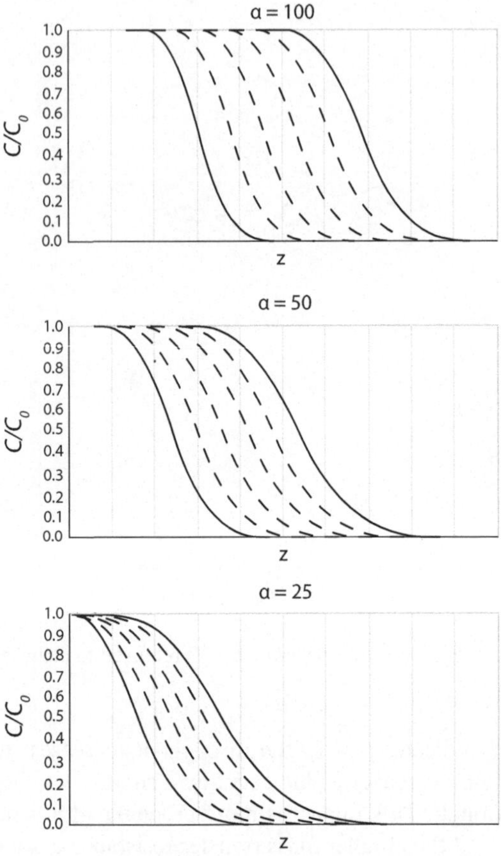

FIGURE 6.12 Impact of the term α on the length of the breakthrough curve.

flow rate needs to be relatively slow, or the particle size needs to be small (yielding a high surface area to volume ratio). If either the flow rate or particle size is increased, then the mass transfer zone spreads out further until, as shown for $\alpha < 25$, the mass transfer zone effectively covers both columns. In cases such as this, it may be warranted to employ three columns in series in order to achieve maximum loading on a column before it is removed from service for either disposition or elution.

6.4 ELUTION PERFORMANCE

As indicated above, a key step in many ion exchange processes for nuclear applications is elution to recover the radionuclide of interest from the ion exchange medium. Typically, this is performed by introducing an eluant solution. The eluant usually has a significantly different composition, such that the ion exchange medium no longer has affinity for the radionuclide of interest. This can be either due to a significant pH change or a drop in the concentration of another ion that draws the radionuclide with it. The objective of this process is to remove the radionuclide of interest in the smallest possible volume and to remove as much of the species of interest as possible so that the ion exchange medium can be used again. In some cases, the radionuclide of interest has some intrinsic value, and recovery is the primary objective. For engineering-scale processes, elution is generally not used to further purify the system. One could imagine a system in which different eluants are used to elute specific radionuclides from the ion exchange material. This approach has not been generally deployed in engineering scale applications.

Because elution generally involves a significant change in chemistry, the first step in elution is to pass a rinse solution through the column. The solution is typically a simple matrix whose key constituents are similar to those of the feed solution to ensure that the radionuclide of interest stays on the ion exchange medium.

Figure 6.13 shows a typically elution profile. The first several BVs of eluant are required to convert the ion exchange material. In this particular case, an acid is being used to elute a sodium form ion exchange material, so the first 3.5 BVs are required to ion exchange H^+ for Na^+ on the ion exchange medium. The subsequent BVs see rapid release of the radionuclide of interest. In this case, the peak concentration of the radionuclide of interest is 60 times greater than the concentration in the feed solution. During this step of the elution process, the eluant is generally well below the equilibrium concentration due to its very high concentration on the ion exchange medium. However, many of the applications for ion exchange of radionuclides require high decontamination factors, and therefore must achieve relatively

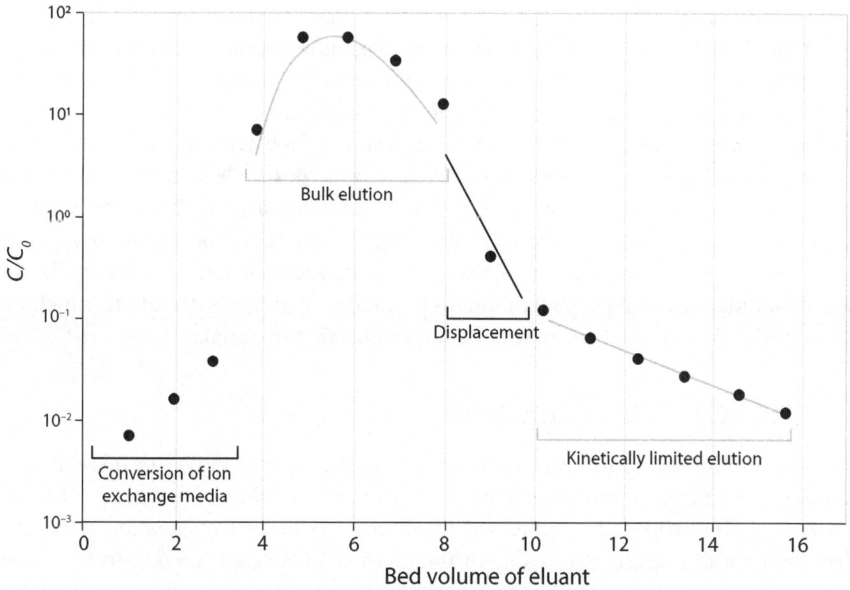

FIGURE 6.13 Typical elution profile for moderate rate of elution. (Adapted from Fiskum, S.K. et al., *Small Column Ion Exchange Testing of SuperLig® 644 for Removal of* ^{137}*Cs from Hanford Waste Tank 241-AP-101 Diluted Feed (Envelope a)*, Pacific Northwest National Laboratory, Richland, WA, 2004b.)

high removal of the radionuclide of interest. This may require extended elution if the kinetics for release of the last few percentages of the radionuclide are slow. This is represented in the final stage of the elution process shown in Figure 6.13. This tail can take up a significant portion of the elution curve because, in this case, approximately half of the elution volume is associated with the elution tail. As can be seen in Figure 6.13, this elution tail follows a simple dilution curve because when plotted on a semi-log plot, the slope of the decrease is linear. This type of behavior is typical for ion exchange media that have slow release kinetics. Note that slowing the flow rate can sharpen the elution peak and reduce the volume of eluant required but may lengthen the overall time required to achieve the desired level of elution.

One consideration for elution is the direction of the introduction of the eluant. As indicated previously, the feed is generally loaded from the top of the ion exchange column. As such, the highest concentration of the radionuclide of interest will be at the top of the column and the lowest concentration will be at the bottom. If elution is performed from the top down, then there is a potential to expose the bottom—cleaner—part of the ion exchange bed to more of the radionuclide that is to be removed. Therefore, from an equilibrium loading point of view, it would be better to introduce the eluant to the bottom of the column, thereby contacting the cleanest eluant with the cleanest portion of the column and ideally using less eluant to achieve the goal of regenerating the column. However, a practical limitation of this approach is associated with the desired throughput for the ion exchange column. The elution flowrate should be less than the flow rate required to fluidize the column, otherwise the efficiency associated with elution from the bottom is lost. Note that elution from the bottom also may require additional valves to be installed, which in radioactive service may be a greater limitation than the additional eluant required to achieve the targeted regeneration.

Temperature can also play an important role in the elution performance of ion exchange media. Increasing the temperature will generally decrease the affinity of the ion exchange medium for the species of interest. However, because much of the volume is associated with slower kinetics, it may be desirable to elute with warmer solutions to speed the elution process. Doing so must also be balanced with the potential for decomposition of the ion exchange medium, a process that is typically accelerated by increasing the temperature. Also, increases in temperature can lead to significant increases in gas generation, in particular hydrogen generation, which can be a significant challenge when processing radioactive materials.

6.5 PLUTONIUM SEPARATION

The first application of ion exchange for plutonium separation was developed as an alternative to the bismuth phosphate and reduction–oxidation reaction (REDOX) processes. The proposed scheme would process spent fuel by loading plutonium (III) onto a cation exchange resin. Uranium would also be loaded onto this resin, as would some of the fission products. While this process was technically viable, it was not able to obtain the fission product decontamination factors that were possible using the plutonium/uranium extraction (PUREX) process. The PUREX process uses solvent extraction to separate uranium and plutonium from spent nuclear fuel (see Chapter 5). The plutonium-bearing product from that stream is destined for a precipitation process (see Chapter 4). The targeted feeds for these precipitation processes range from 50 to 250 g/L of plutonium. The plutonium product from the PUREX process needed to be slightly purified and concentrated prior to being introduced into the precipitation process. This was first accomplished at the Hanford Site through an additional solvent exchange step followed by evaporation. However, the evaporation process resulted in frequent evaporator failures due to the corrosive nature of the process stream. Therefore, these steps were replaced by an ion exchange process; first at Savannah River with a cation exchange process, then at Hanford with an anion exchange process.

6.5.1 Cation Exchange

In the cation exchange process, Pu(III) is removed from the effluent from the PUREX process (Orth 1960). Both Pu(III) and Pu(VI) are present as cations in low-strength nitric acid solutions. Uranium is also present in cation form in the solution and is removed by a cation exchange medium. Dowex® 50W resin was chosen for this process. The feed to this process ranged in concentration from 0.2 to 1.2 g/L of plutonium and approximately 1 mg/L of uranium. The final concentration from the PUREX process was dictated by criticality controls in the solvent extraction process. Figure 6.14 shows the dependence of the Pu(III) K_d on nitric acid concentration using a typical cation exchange resin (Mohapatra et al. 2004). From Figure 6.14, it can also be seen that it was not viable to obtain adequate separation for all other contaminants. The cation exchange medium would have a higher affinity for thorium, and as such would not be expected to produce selectivity relative to that actinide. It should be further noted that Pu(IV) had a significantly higher K_d than Pu(III). However, the choice was made to remove plutonium in the +3 state because of a number of other process considerations. While the resin used had higher affinities for Pu(IV), the actual capacity for Pu(III) was higher, and as such would result in higher ultimate plutonium concentration in the eluate. As shown in Figure 6.14, the K_d for Pu drops

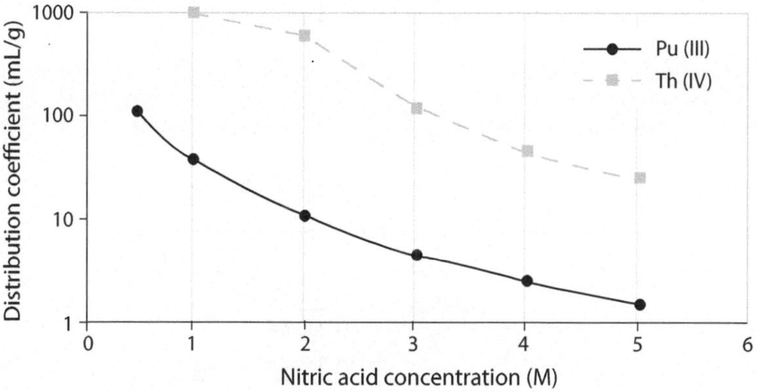

FIGURE 6.14 Distribution coefficient for plutonium(III) on cation resin as a function of nitric acid concentration. (Adapted from Mohapatra, P.K. et al., *Solvent Extr. Ion Exc.*, 22, 267–284, 2004.)

sharply as the nitric acid concentration is increased. The K_d for Pu(VI) does not drop as sharply with acid concentration, and hence more acid would be required to elute the resin if Pu(VI) were adsorbed. In addition, the plutonium arrived at the cation exchange process in the +3 oxidation state, and as such, would not require adjustment prior to processing.

Figure 6.15 depicts a nominal process for plutonium concentration by ion exchange (Navratil 1986). For operations at the Savannah River Site, the cation exchange process was a batch process. The flowsheet shown is for a feed of 0.2 g/L, but as noted above, the plutonium concentration could range up to 1.2 g/L. The stream from the PUREX process contained approximately 0.3 M nitric acid. The plutonium was stabilized in the +3 oxidation state by the addition of 0.025 hydroxylamine sulfate, which prevented the oxidation of the plutonium to the +4 state. The kinetics of the typical ion exchange material (Dowex® 50) were relatively fast. The resin bed was loaded typically at 25–30 BVs/h for a total of 140–200 BVs. The extent of loading is typically monitored by online monitoring of the activity of the stream leaving the ion exchange column. The cationic columns were typically operated so that less than 0.001% of the plutonium left through the effluent stream. At the completion of the loading cycle, the column was washed with 5–20 BVs of 0.025 hydroxylamine sulfate in 0.2 M sulfuric acid to wash off the uranium that was adsorbed by the column.

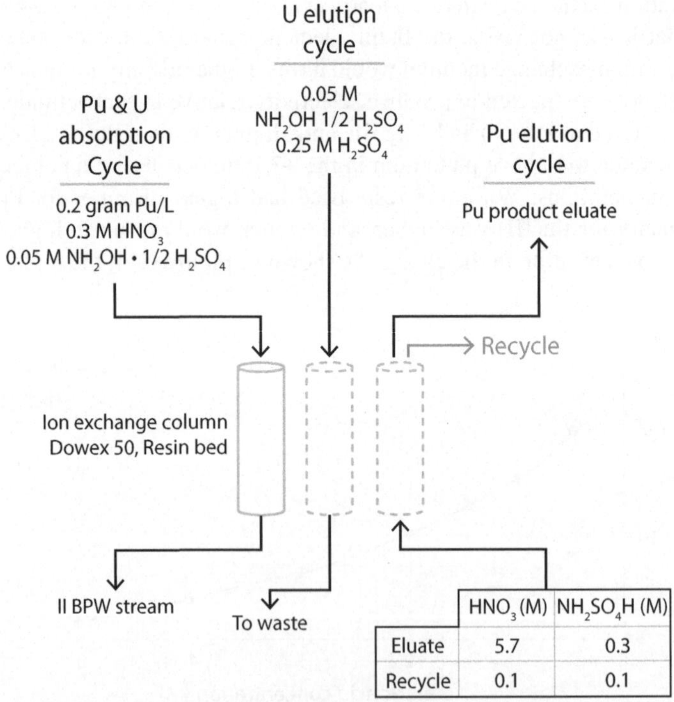

FIGURE 6.15 Nominal process flowsheet for plutonium separation by cation exchange. (Adapted from Navratil, J.D., *Ion Exchange Technology in the Nuclear Fuel Cycle*, International Atomic Energy Agency, Vienna, Austria, 1986.)

This was done at a much slower flow rate—approximately 3 BVs/h. This generally resulted in removal of 99% of the uranium, if the initial uranium concentration was relatively high, and removal of 80%–98% of the zirconium and niobium. Again, only a trace fraction of the plutonium reported to this stream.

At the completion of the washing step, the flow was switched to upflow for elution. The elution was performed with 5.7 M nitric acid, which contained 0.3 M sulfamic acid. This elution was performed at a much slower 0.6 BV/h rate with the goal of achieving a very sharp elution profile. The first BV of eluant was captured for recycle. The second BV of eluant was captured as product and typically contained between 50 and 60 g/L of plutonium. As noted above, the decontamination achieved on select fission products was marginal. For example, roughly 50% of the ruthenium would end up in the eluate product. The volume of eluate recovered was either set by a predetermined amount of eluant or by observation of the color change associated with the decrease in Pu(III) concentration.

The final step was displacement of the concentrated nitric acid with 2.5 BVs of dilute acid (0.1 M nitric acid with 0.05 M hydroxylamine sulfate). This stream contained a significant amount of plutonium, but its concentration was below the targeted concentration for the next precipitation step and was therefore recycled along with the first BV of eluant. At this point, the column was ready to be reloaded with plutonium from the PUREX process (mixed with recycle streams).

The system was relatively robust, and degradation of the resin was relatively slow because the resin would only lose about 25% of its capacity due to irradiation over 500 days. The production of fines was more likely to limit the lifetime of the resin and the resin bed was generally replaced annually because of hydraulic problems.

6.5.2 Plutonium Anion Exchange

At the Hanford Site, there was a desire to replace the evaporation stage that followed the PUREX process to purify plutonium. In the late 1950s, a process was developed based on anion exchange. As the nitrate concentration is increased, plutonium (IV) forms a hexanitrate complex that carries a −2 charge as $Pu(NO_3)_6^{2-}$ (Ryan and Wheelwright 1959). Figure 6.16 shows the distribution coefficient for plutonium at 25°C with Dowex® 1 × 1^{-4} anion exchange resin (Mohapatra et al. 2004; Ryan and Wheelwright 1959; Schiffhauer and Thompson 1993). The maximum loading when there are no competing anions is achieved at around 7 M nitric acid. However, when other anionic species, such as thorium (Navratil 1986) are present, the maximum K_d is a little bit lower because there is competition between plutonium and thorium at higher acid strengths. This process was generally very selective against many of the potential contaminants; only thorium, platinum, gold, palladium, and iridium have an appreciable distribution coefficient (Navratil 1989). As seen in Figure 6.16, a separation factor of 3 is achievable for thorium. The separation factor for uranium is generally greater than 10,000 (Swift 1960) because the uranium K_d is generally less than 10 (Kumaresan et al. 2006). Ryan and Wheelwright (1959) also demonstrated that the capacity of the ion exchange material was sensitive to temperature; the K_d dropped by about a factor of 3 as temperature increased from 25°C to 60°C. However, at the same time, Knoch (1964) demonstrated that the diffusivity of plutonium in

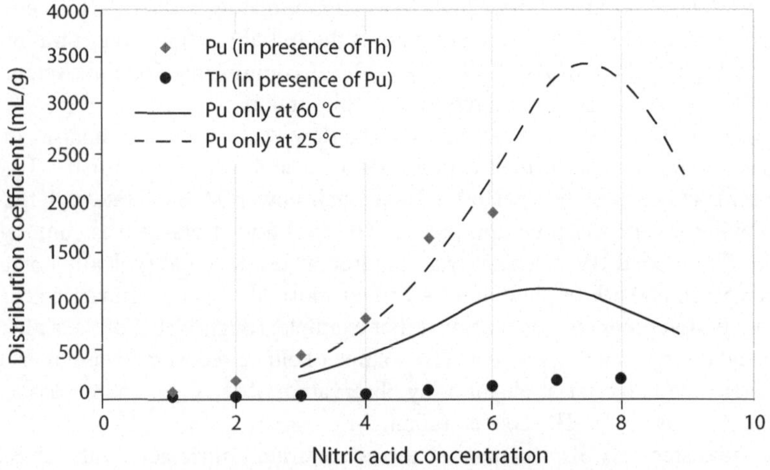

FIGURE 6.16 Anion exchange media capacity for plutonium as a function of nitric acid, with and without thorium present. (Adapted from Ryan, J. L. and Wheelwright, E.J., *Ind. Eng. Chem.*, 51, 60–65, 1959; Mohapatra, P.K. et al., *Solvent Extr. Ion Exc.*, 22, 267–284, 2004.)

concentrated nitric acid increases by a factor of about 6 over the same range of conditions. One of the primary challenges of plutonium ion exchange is to support typical process flow rates while maintaining relatively small columns. The column sizes must be maintained small to avoid criticality concerns, so the column configurations had to be in critically safe geometries. These constraints led to a desire for relatively high-throughput columns. At the relatively high flow rates, the loading kinetics become important, as can be seen in Table 6.1 (Ryan and Wheelwright 1959). Here, the column loading was measured at 50% breakthrough for a feed of 0.55 g/L in 7.2 M of nitric acid. As can be seen, the maximum loading is achieved at approximately 60°C, despite the significantly lower K_d values that were measured. This reflects the faster kinetics that are achieved at higher temperatures.

Based on these characteristics, the last cycle of PUREX solvent extraction and the subsequent evaporation process were replaced at the Hanford Site in 1958 (Swift 1960). A continuous ion exchange process was deployed in which fresh feed was continuously fed to the ion exchange column while the ion exchange was periodically circulated through the system. Figure 6.17 provides an overview of the process.

Feed was introduced halfway through the loading column. The feed flowed downward through the column. The stream from PUREX was 0.5 M nitric acid and 0.05 M sulfuric acid. The plutonium content in the feed was monitored, and 60% nitric acid was added to bring the feed to the targeted nitric acid concentration (6–7 M). The feed was heated to approximately 70°C and then fed into the column. The feed entered at a rate of approximately 10 mL/cm^2/min. The effluent out the bottom of the column was monitored for plutonium content and was found to represent generally less than 0.5% of the feed concentration. Periodically, the resin waste pulsed so that the resin leaving the loading section generally contained 30–50 g/L of plutonium. Thus, the average resin flowrate up through the column ranged from

TABLE 6.1
Plutonium Loading on Dowex® 1 Resin at Varying Temperatures

Loading Temperature (°C)	Pu Loading at 50% Breakthrough
25	56.5
40	72.9
50	93.2
60	130.4
70	123.5
80	80.8

Source: Adapted from Ryan, J.L. and Wheelwright, E.J., *Ind. Eng. Chem.*, 51, 60–65, 1959.

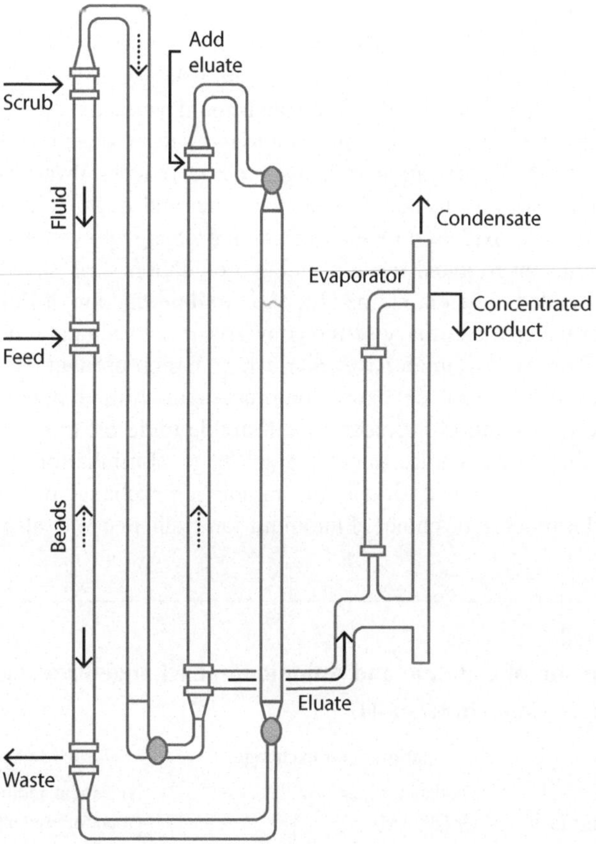

FIGURE 6.17 Process flow diagram for separation of plutonium by anion exchange at the Hanford Site PUREX Plant. (Adapted from Swift, W.H., Plutonium ion exchange processes, U.S.-U.K. Technical Exchange Meeting, Oak Ridge, TN, April 25–27, 1960.)

0.1 to 0.5 mL/cm^2/min. Above the injection point of the plutonium feed was the scrub section, where feed and its contaminants were scrubbed from the resin. At the top of the scrub section, 7 M of nitric acid was introduced to the column also at 10 mL/cm^2/min rate of flow. The net result of this addition at the injection point is that 20 mL/min^2/min of feed was passing through the loading section of the column, effectively reducing the contact time to 5 minutes. The resin left the top of the loading column and free fell and was then transferred to the elution column. Again, the resin was introduced upflow and the eluant passed downflow. The eluant nitric acid concentration ranged from 0.5 to 1.0 M. Eluant flow rates were significantly lower than feed flow rates, resulting in a plutonium concentration in the product stream of 30–50 g/L. The product stream was then subsequently evaporated up to approximately 250 g/L for eventual further precipitation processes. The eluted resin, which contained less than 0.05 g/L of plutonium, was then ready for reintroduction back into the loading column. Because of radiolytic degradation of the resin, it was replaced every few months. The process resulted in excellent decontamination (Navratil 1986) factors for fission products ruthenium (10^4) and zirconium and niobium (10^5).

One of the significant challenges associated with plutonium ion exchange was maintaining the oxidation state of the plutonium. Changes in oxidation state were generally associated with gas production, which could pose safety concerns, but also could lead to channeling in the columns, and thus result in lower than expected performance. Therefore, for the cation exchange process, reductant was added to ensure that the plutonium stayed in the +3 oxidation state, and nitric acid concentrations were limited to avoid oxidation to +4. For anion exchange processes, it was necessary to add reductant to avoid reaching the +6 oxidation state, but the nitric acid concentration needed to be maintained high to avoid production of Pu(III).

Table 6.2 provides a summary of the comparison between cationic and anionic ion exchange. For the Savannah River Site, where fission product decontamination could be achieved by the solvent extraction process, the limited need for nitric acid addition to the feed solution was seen as a desirable trade-off to using cationic ion exchange. At the Hanford Site, where the ability to eliminate one of the solvent extraction steps was seen as desirable, the anionic ion exchange process was determined to be the process of choice. Plutonium ion exchange was also used at Oak

TABLE 6.2
Comparison of Cationic and Anionic Ion Exchange Processes for Plutonium Concentration (1)

	Cationic Ion Exchange	Anionic Ion Exchange
Feed prep	Oxidation adjustment	Nitric acid addition
Decontamination	Good for U	Good for uranium
	Marginal for fission product	Good for fission products
Concentration	50–60 g/L Pu	30–50 g/L Pu
Eluant	5–6 M nitric acid (with reductant)	0.5–1 M nitric acid

Ridge (cationic) and Rocky Flats (Baldwin and Navratil 1983) (anionic) for cleanup of a variety of materials. At Los Alamos, cation exchange was initially used to recover plutonium from slag and crucible dissolver solution. After 1959, anion exchange was generally used at Los Alamos for recovery of plutonium from scrap streams (Baldwin and Navratil 1983).

6.6 CESIUM REMOVAL

Cesium removal has been a focus of waste-processing activities since the 1950s. At that time, the REDOX process was used at the Hanford Site. The heat from the decay of the radiolytic cesium and strontium stored in waste tanks caused a severe temperature rise in the tanks, causing liquid to boil in a significant number of them. For the new PUREX process to be deployed at Hanford, a cesium removal process was needed. At that time, the choice was to use a ferrocyanide ($Na_3Fe(CN)_6$) precipitation process. However, this process was not readily adaptable to the wastes that had already been placed into storage, since the acid-bearing stream had been neutralized. Therefore, in the early 1960s, work began on developing cesium removal processes that used ion exchange. Since the 1960s, a number of ion exchange media have been tested and, in a number of cases, used to process tank waste. Table 6.3 provides a summary of the ion exchange media that have been used in large-scale operations for cesium removal. These processes are discussed in turn, as well as additional ion exchange media that have been developed and tested but not yet deployed at a production scale.

6.6.1 Duolite® for PUREX Sludge Supernate Treatment at Hanford

Duolite® ARC-359 represents an early version of a line of phenolic ion exchange media that were developed for cesium removal. The development of this process, which is summarized in Figure 6.18 (Carlstrom 1977), began in the early 1960s. The process was developed to not only remove the cesium from the waste stream, but to purify the cesium for storage in cesium capsules and for eventual potential use as an irradiation source. The first step in the development of this process was to evaluate several ion exchange materials using synthetic feed materials. The nominal composition of this feed is provided in Table 6.4.

TABLE 6.3
Ion Exchange Media Used in Full-Scale Operations

Media	Facility	Volume Processed	Elutable	Capacity
Duolite® ARC-359	Hanford B-Plant 1968 to 1985	>100 million L	Yes	Low
IONSIV™ IE-96	West Valley 1988 to 1991	2.1 million L	No	Low
CST	Melton Valley Storage Tanks 1996 to 1997	0.1 million L	No	High

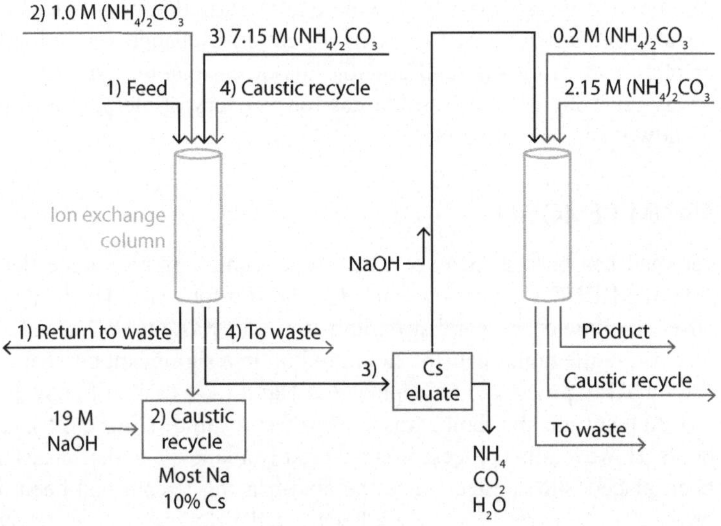

FIGURE 6.18 Process flow diagram for cesium removal at the Hanford B-Plant. (Adapted from Carlstrom, R.F., *Ion Exchange Flowsheet for Recovery of Cesium from PUREX Sludge Supernatant at B Plant*, Atlantic Richfield Hanford Company, Richland, WA, 1977.)

TABLE 6.4
PUREX Sludge Supernate Composition

Component	Molarity
Na	9.0
NO_3^-	2.6
NO_2^-	4.2
CO_3^{--}	1.0
PO_4^{---}	0.028
SO_4^{--}	0.09
Cs-137	1.8 Ci/L

Source: Nelson, J.L. and Mercer, B.W., *Ion Exchange Separation of Cesium from Alkaline Waste Supernatant Solutions*, General Electric, Hanford Atomic Products Operation, Richland, WA, 1963.

A number of potential ion exchange media were assessed using batch equilibrium data (Nelson and Mercer (1963)). From those data, a number of ion exchange media were selected for further study using small laboratory scale 10 mL column volumes. The ion exchange media tested included phenolic resin, an aluminosilicate gel, a natural zeolite, and synthetic zeolites. For relatively low pH values, the synthetic zeolite provided the best performance, yielding initial breakthrough at approximately

20 BVs and a total capacity of approximately 50 BVs. As expected, testing showed that smaller media provided superior kinetics—20–50 mesh provided initial breakthrough at twice the number of BVs as 12–20 mesh. Also, increasing the loading temperature from 25°C to 55°C was shown to significantly lower the capacity of the ion exchange material.

Because the goal of this effort was to produce clean cesium in addition to reducing the heat load in the waste tanks, the process involved a washing step to remove sodium and other cations from the ion exchange material. The process developers (Carlstrom 1977) eventually settled on 1.0 M ammonium carbonate as a wash solution to remove sodium, rubidium, and potassium from the column. The column was then further eluted to remove cesium using 2.15 M ammonium carbonate. Sodium hydroxide was then added to the initial eluant stream (caustic recycle) and this stream was then processed back through the column to convert the ion exchange medium from NH_3 form to Na form. The column was then ready to be reloaded.

The cesium eluant was evaporated to reduce volume and to drive off ammonia and carbonate, which are recovered for production of fresh eluate. Once a sufficient volume of the evaporated cesium stream was gathered, it was caustic adjusted and N-(2-hydroxyethyl)ethylenediaminetriacetic acid (HEDTA) was added to keep iron and aluminum from loading on the ion exchange medium. This feed was then reloaded onto the same ion exchange medium and was then washed with 0.2 M ammonium carbonate. After the extended wash, the elute containing 2.15 M ammonium carbonate becomes the target product, and the column was converted back to sodium form and was ready for a fresh first cycle run.

Some of the considerations for this process included limiting the ammonia concentration to less than 5 M to avoid flammability concerns (this equates to less than 2.5 M of ammonium carbonate). Periodically, the resin bed needed to be fluffed because of the effects of increasing pressure drop. Fluffing was accomplished by performing an upflow water wash with the resin bed in sodium form. The addition of HEDTA had the potential to change the pH of the second feed solution. The ion exchange resin was known to be unstable in acidic conditions, so the pH of the second load solution needed to be monitored to ensure that it remained caustic to avoid resin decomposition.

Overall, the process increased the cesium/sodium ratio by 20,000 and the cesium concentration in the final concentrated eluant was 2,500 higher than in the feed. More than 30 cycles of this process were required to produce sufficient cesium for one cesium capsule.

6.6.2 West Valley Demonstration Project Summary

The West Valley Demonstration Project was a successful example of a large-scale, multistep, high-level radioactive waste (HLW) pretreatment process (Baker and Gessner 1989; Baker et al. 1989; Dalton 1997; Kelly and Meess 1997; Kurath et al. 1995; Ploetz and Leonard 1988; Schiffhauer and Thompson 1993; Schiffhauer and Thompson 1992; Valenti and Elliot 1999). Approximately 2.1 million liters of HLW, originated from neutralization of PUREX wastes, and 190,000 liters of acidic wastes, originating from thorium extraction (THOREX), were pretreated by filtration, ion exchange, sludge washing, and vitrification. To the maximum extent possible, operations were conducted using pre-existing and purpose-built facilities to immobilize long-lived transuranic (TRU) radionuclides with highly radioactive cesium (Cs-137) and strontium (Sr-90) that had been separated from supernatant and sludge-washing liquids. Over 6.4 million liters of liquid HLW were pre-chilled to 10°C–18°C and passed in batches through the ion exchange pretreatment system (3 or 4 columns in series, each 0.93 m i.d. by 4 m long, 1,700 L, 1,300 kg zeolite IONSIV™ IE-96 or TIE-96, loaded downflow at 6 gpm, 0.8 column volume (CV)/h). The mildly caustic, high-sodium PUREX supernatant was processed in nominally 15,000 gallon batches through four IONSIV™ IE-96 columns, while the two PUREX sludge washes (nominally 50,000 and 240,000 gallon batches, respectively) and a final wash of neutralized THOREX sludge that had been mixed with the PUREX waste, were processed through one or two columns of IONSIV™ IE-96 and one or two columns of IONSIV™ TIE-96. As shown in Table 6.5, Kelly Jr. and Meess (1997) estimated 6.6 million Ci of Cs-137 (>99.99%) was captured by 65.3 MT of zeolites, which were then size reduced, blended with the washed PUREX/THOREX sludge, and processed in the HLW Vitrification Facility. The average feed for this process was around 1 Ci/L. The pretreated low-level waste (LLW) liquid effluent from the ion exchange columns was further concentrated by evaporation to 40% total dissolved solids, blended with cement-formers, and solidified into nearly 20,000 individual square 270 L drums that were classified under Resource Conservation and Recovery Act regulations as non-hazardous.

6.6.2.1 Laboratory Testing and Process Definition

An extensive evaluation of process options and laboratory testing (sludge washing for sodium sulfate removal, evaluation of ion exchangers [Duolite® CS-100, Amberlite™ IRC-718, IONSIV™ IE-95, Dowex®-1,X8], and precipitating agents sodium tetraphenyl borate, phosphotungstic acid, and sodium titanate) was completed prior to full-scale operations (Bray et al. 1984; Kurath et al. 1989). The laboratory testing evaluated the materials, phase ratio, composition, concentration, pH, flow rate, and temperature. Results from batch distribution K_d (0.1 g exchanger, 10 mL liquid) and column tests (200 mL, 2.5 cm i.d.) suggested that chilling to 6°C and dilution (2:1) of the supernatant would improve Cs-137 loading by approximately 45% and reduce total ion exchange material needs. Batch K_d values ranged from approximately 45–65 mL/g for the simulated West Valley supernatant (6.9 M Na^+, 0.19 M K^+, 0.0014 M Cs^+, 3.5 M NO_3^-,

TABLE 6.5
Process Volumes Treated by the West Valley Demonstration Project

Stream Processed	Volume (million L)	Cs-137 Removed (MCi)	Zeolite Used (MT)
PUREX supernate	2.34	5.3	45.1
Sludge wash	2.90	1.04	13.0
PUREX/THOREX wash	1.19	0.30	4.9
Total	6.43	6.64	65.3

2.1 M NO_2^-, 0.34 M CO_3^{-2}, 0.30 M SO_4^{-2}, pH 10–11). The zeolite, IONSIV™ IE-95, was shown to remove Cs-137 from the supernatant with a decontamination factor (DF) exceeding 1,000 while being compatible with supernate, sludge washes, and vitrification processes. Kurath et al. (1989) demonstrated remarkable similarity between the cesium ion exchange breakthrough curves from the first two supernatant processing runs to those generated during laboratory testing of simulant solutions. Compensating differences between the laboratory and process-scale temperature (6°C vs. 18°C) and dilution (2:1 vs. 20%) were thought to explain the nearly identical breakthrough curves.

Sludge washing with deionized water and mild caustic was evaluated to reduce the sodium sulfate loading of the HLW glass while leaving TRU components in the sludge. Two washes were chosen to reduce sulfate to acceptable levels, while keeping the volume of diluted wash solutions to a minimum. However, during laboratory testing unacceptable levels of plutonium were found in the wash liquid and additional counter measures were required to meet specifications (Bray and Hara 1993). The pH of the sludge-washing process was increased to 12.5 and a titanate exchanger, TIE-96, was invented (Bray and Burger 1994), which was compatible with the integrated treatment process. Laboratory testing of multiple batches of IONSIV™ TIE-96 under a wide range of conditions demonstrated the simultaneous removal of plutonium, Cs-137, and Sr-90 from simulated and actual sludge-washing solutions.

6.6.2.2 Summary of the Sludge-Washing Process

Direct comparison of laboratory data to process results is inappropriate because of the variable batch processing, chemical composition, actual dilution, and the use of both IE-96 and TIE-96 columns in variable rotating positions in the three-column system. For example, during the first sludge-washing campaign key chemical concentrations varied as follows: 12.4 ± 0.21 pH, 692 ± 130 µCi/mL Cs-137, 0.37 ± 0.48 µCi/mL Sr-90, 0.0149 ± 0.00213 µCi/mL alpha-Pu, 2.68 ± 0.39 M Na, 0.064 ± 0.012 M K. As expected, chemical concentrations were significantly lower during the second sludge washing: 11.8 ± 0.14 pH, 91.7 ± 7.26 µCi/mL Cs-137, 0.035 ± 0.014 µCi/mL Sr-90, 0.0019 ± 0.00096 µCi/mL alpha-Pu, 0.368 ± 0.035 M Na (Dalton 1997). Furthermore, the batch volumes processed varied from 44,100 to 57,600 gal for the first wash and 228,000 to 241,000 gal for the second wash. DFs for Cs-137 varied from 3,100 to 49,000 during the first wash and 8,600 to 40,000 for the second, while those for plutonium varied from 42 to 387 (first wash) and 42 to 253 (second wash). In general, the process results greatly exceeded the design specifications of >1,000 DF Cs-137 and >10 DF Pu, which resulted in significantly lower radiation exposure to personnel in the downstream LLW treatment systems.

6.6.2.3 Process Lessons Learned

The full-scale ion exchange process removal of Cs-137 from West Valley supernatant and sludge-washing liquids was relatively straightforward and directly modeled from simulant-based batch and column laboratory experiments. A number of process challenges encountered required significant problem-solving. Foremost, the plutonium solubility issue during sludge washing required increasing the process pH from 10 to 12.5 and invention of a new ion exchange material. Furthermore, modeling plutonium behavior with simulants was challenging and often incomplete due to the complex chemistry of plutonium in these streams, the low plutonium concentrations concentration, and the ion exchange kinetics of plutonium. During one batch processing cycle, residual cesium-loaded zeolite remained in the emptied ion exchange column. This greatly affected downstream decontamination in the next cycle when fresh zeolite was loaded, and the column was placed in the final position. In another case, a leaking valve at the column exit was responsible for insufficient Cs-137 decontamination, resulting in the batch being returned for additional processing. Due to the number of batches and variability in chemical composition, the ion exchange process used effluent-based criteria to decide when to switch zeolite columns (50% and 5% Cs-137 breakthrough for the lead and middle columns, respectively). The zeolite columns were extremely stable to radiation damage and were chemically compatible with the process streams. Columns partially loaded with Cs-137 from the final supernatant batch were even left for an extended period of time until the first sludge-washing process was initiated. While generally stable in the process radiological and chemical conditions, the IONSIV™ TIE-96 material was shown to generate a mildly acidic effluent during initial rinsing, which required pH adjustment during processing. Extensive testing of both zeolite materials demonstrated sufficient stability under all process conditions, even though aluminum precipitation from the supernatant was observed during laboratory testing. Finally, while the IONSIV™ TIE-96 demonstrated the removal of plutonium and Sr-90 from sludge-washing liquids, the Cs-137 loading was slightly reduced, suggesting that total exchange capacity was split between the radionuclides.

6.6.3 Cesium Removal at Hanford from Tank Waste Supernate

An extensive evaluation of laboratory-scale tests has been performed for ion exchange treatment of the Hanford tank waste supernatant supporting final waste vitrification. Previous studies of organic and inorganic ion exchangers evaluated the capacity of various ion exchange materials in the presence of 5 M Na with a Na/Cs mole ratio that ranged from 100 to 500,000 (Brown et al. 1996). The ion exchange media tested included CS-100, a phenol formaldehyde; resorcinol formaldehyde (RF), an organic polymer; SuperLig® 644, a macrocyclic organic resin; Zeolite TIE-96, an inorganic titanium-treated zeolite; and crystalline silicotitanate (CST), an engineered inorganic zeolite. These media were evaluated for the removal of Cs-137 and other radionuclides from Hanford tank wastes. The nominal composition of the feed is provided in Table 6.6. Cs λ values at a Na/Cs mole ratio of 50,000 are shown in Table 6.7. As noted previously, Cs λ represents the point at which the C/C_0 value is 50% and is a good representation of the capacity of the ion exchange medium.

Full-scale ion exchange process removal of Cs-137 from the supernatant at West Valley Nuclear Waste Site showed that TIE-96 demonstrated the removal of plutonium and Sr-90, which resulted in slightly reduced Cs-137 loading. Cesium batch contacts from tank waste simulant were performed with Duolite® CS-100 and showed the presence of the high molarity sodium dramatically affected the ability of the ion exchange material to exchange sodium for cesium and decreased the cesium loading by roughly an order of magnitude due to competition for the ion exchange sites.

TABLE 6.6
Feed Composition at a Na/Cs Mole Ratio of 50,000

Component	Concentration (M)
Na	5.0
K	0.48
OH (free)	2.1
Sr	7.14E-08
Ca	2.41E-04
Al	0.5
Balance	Nitrate>nitrite>Al>Carbonate>other anions

TABLE 6.7
Performance of Various Ion Exchangers

Ion Exchanger	Elutable?	Cs λ
Zeolite TIE-96	No	23
Duolite® CS-100	Yes	23
SL-644	Yes	211
RF	Yes	157
CST	No	937

From those data, SuperLig® 644 (SL-644), RF, and engineered CST were considered for use within the ion exchange treatment process for Hanford waste and were further studied with small laboratory scale volumes.

The initial River Protection Project-Waste Treatment Plant (RPP-WTP) baseline process for Cs-137 removal included the use of SL-644. The process flowsheet encompassed a four-column carousel design with three continuously operating columns (lead/lag/polish) and one column rotating for regeneration. Figure 6.19 illustrates one step in the carousel rotation of the ion exchange columns. After each cycle the resin was washed with dilute sodium hydroxide and water and then eluted with 0.5 M nitric acid. At the completion of the elution step, the acid is displaced with water and the resin is regenerated to sodium form with sodium hydroxide before the next process cycle. The position of the columns in the processing cycle is changed by valving (not shown in Figure 6.19). Use of this type of carousel arrangement requires a significant number of valves to enable each column to serve in each of the four positions shown in Figure 6.19.

Successful operation of multi-cycle carousel design is dependent on the balance between loading and restoration times. Preliminary results from laboratory scale ion exchange testing at 3 BV/h indicated saturation of the lead column after 100 BVs of processing feed, equivalent to 33 h. Regeneration time for a fully loaded laboratory

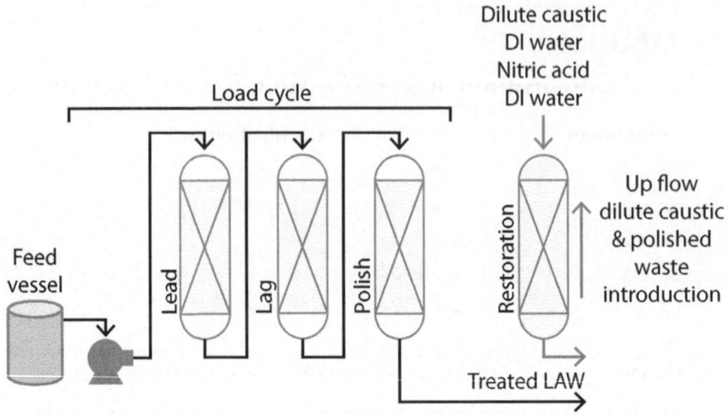

FIGURE 6.19 RPP-WTP carousel rotation of the ion exchange columns.

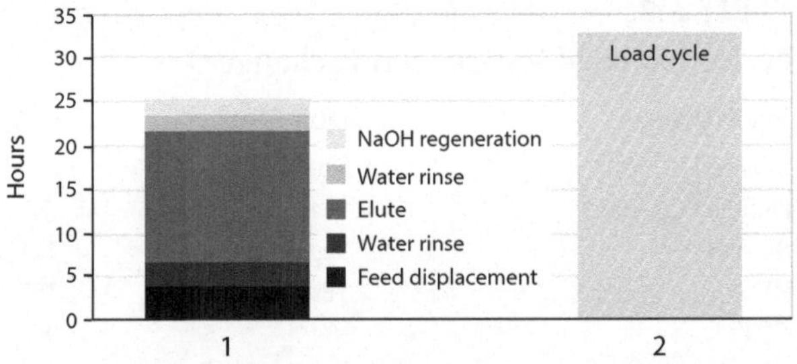

FIGURE 6.20 Loading and restoration balance.

scale column was measured to be 25 hours (Fiskum 2004a). Figure 6.20 shows the breakdown of loading and elution times executed to demonstrate a successful multi-cycle carousel design.

Actual waste testing with SL-644 was evaluated using waste from five different Hanford tanks (Hassan et al. 2000, 2003; King et al. 2001; Fiskum et al. 2004a, 2004b). Table 6.8 provides the 50% breakthrough data for the five Hanford tank wastes with varying potassium concentrations. Breakthrough results indicated that

TABLE 6.8
BT Data for the Five Hanford Tank Wastes

Hanford Tank	BV to 50%	Cs (Ci/L)	K+	Na	OH	References
AN-103	400	0.194	0.121	5.25	2.2	Hassan et al. (2000)
AN-102	250	0.140	0.027	4.8	0.2	King et al. (2001)
AW-101	220	0.175	0.468	5.09	1.8	Hassan et al. (2003)
AP-101	190	0.126	0.744	4.97	2.4	Fiskum et al. (2004c)
AZ-101	120	0.649	0.11	4.87	1.1	Fiskum et al. (2004b)

potassium is not a major competitor of Cs for SL-644, and that sodium concentration plays a bigger role in the Cs capacity of the medium.

Additional testing evaluated elution performance to determine the relative effects of temperature, eluant concentration, and flow rate (Burgeson et al. 2004a). The values of each parameter and the micrograms of Cs present on the resin after processing with 40 BVs of eluant were entered into a statistical software package to facilitate analysis. Flow rate and temperature showed the most statistical significance to the determination of the residual Cs on the resin, as seen by their "p-values." Table 6.9 shows the calculated residual Cs retained by the resin using the following correlation:

$$Cs_R = 2.685 - 0.519[H^+] - 0.0400\ T + 0.835\ FR + 0.00947[Cs] \quad (6.36)$$

where Cs_R is the residual Cs on the resin, μg per total resin sample; $[H^+]$ is the HNO_3 concentration, M; T is the elution temperature, °C; F_R is the flow rate of eluant, BV/h; [Cs] is the Cs concentration of feed, μg/mL.

As seen in Table 6.9, the modeling results indicated that to minimize residual Cs on the resin for a given total elution volume (40 BVs), the temperature should be high, the acid strength should be high, and the flow rate should be minimized.

One of the operational challenges with SL-644 resin is its change in volume as a function of the solution pH and ionic strength (Hassan et al. 2000). Multi-cycle testing showed swelling of 40%–50% when converted from H-form to Na-form (Fiskum et al. 2002). In addition to swelling challenges, the process lifetime of the SL-644 was uncertain. One of the key benefits of the carousel arrangement shown in Figure 6.19 is the ability to reuse the ion exchange medium for an extended period of time. Prior nuclear applications that used eluatable ion exchange media could operate for up to a year without replacement. However, testing had indicated that the SL-644 was susceptible to

TABLE 6.9
Model Residual Resin Cesium Value Predictions at 40 BVs of Processed Eluant

$[HNO_3]$, M	Temperature, °C	Flow Rate, BV/h	Residual Cesium on Resin, μg/g	Trend of Residual Cesium within Modeling Set	Variable
0.5	25	1	2.31	+35%	Flow Rate
0.5	25	1.5	2.73		
0.5	25	2	3.14		
0.5	25	1	2.31	−11%	H^+
0.75	25	1	2.18		
1.0	25	1	2.05		
0.5	25	1	2.31	+35%	Temp
0.5	35	1	1.91		
0.5	45	1	1.51		

Note: All conditions used a loading feed Cs concentration of 5 μg/mL.

chemical degradation. To help resolve this issue, a test was performed to assess the performance of the ion exchange material over multiple load/elute cycles (Hassan et al. 2003). Six cycles of loading, elution, and regeneration were performed to remove Cs from Hanford Tank 241-AW-101 using SL-644 resin. The resin showed good loading and elution performances for all six cycles. The percent Cs removed was greater than 99.99% and the DFs achieved were above 10,000. The resin exhibited sufficient radiation stability in actual Hanford waste solutions for practical applications. After completing six loading cycles, the resin in the primary column absorbed 1.99×10^7 rad with 20% loss of ion exchange capacity at 50% breakthrough (Hassan et al. 2003). While this performance did not live up to the year-long performance seen in prior applications, it was consistent with the process goal of 10 load/elute cycles before replacement.

A second, more significant issue associated with SL-644 was observed during pilot-scale tests, which revealed that converting to Na-form in the column resulted in a significant increase in pressure drop, effectively rendering the bed impermeable. In addition, when introducing the high-density feed to the column fluid head after regeneration, the feed solution would drop through the lower density fluid freeboard with sufficient momentum to churn the resin at the top of the column, resulting in a sloped, uniform bed surface. Due to concerns about poor hydraulic performance and the inability to maintain a homogeneously packed resin bed for efficient removal of cesium, the WTP staff evaluated additional ion exchange media as alternatives to SL-644. Based on an extensive review of available ion exchange materials at the time, RF resin was chosen because it has a high loading capacity per unit volume and was commercially available in granular form. An effort was undertaken to convert that granular form to a spherical bead to ameliorate hydraulic problems.

RF ion exchange resin was first developed at Savannah River National Laboratory in the 1980s to remove cesium from alkaline high-level radioactive liquid waste (Bibler et al. 1990; Bibler and Wallace 1992). The initial form of RF cesium ion exchange resin was alkaline polymerized ground gel (GG). The product was produced by curing the polymer, crushing the material, then sieving it to obtain the targeted particle size distribution. This process produced irregular particles that look like shards of broken glass (see Figure 6.21). The RF resin is marked by a distinct color change upon conversion from the H-form shown in Figure 6.21 to a much darker color when in the Na-form.

The RF was modified to include a polystyrene backbone (Berge et al. 1993). The acid polymerized spherical-bead form of RF resulted in a much tighter particle size, and thus improved hydraulic performance due to its uniform, spherical particle size. Figure 6.22 shows the spherical resorcinol formaldehyde (sRF) product.

Similar to SL-644, RF resin shrinks and swells significantly when cycled between the acidic (H-form) and basic (Na-form) conditions used to operate the ion exchange column. Hydraulic performance properties for SL-644 (as-received and sieved), RF, and sRF were compared to assess the effects of their shrink and swell behavior on particle size (Arm and Blanchard Jr. 2004; Fiskum et al. 2004c). Table 6.10 shows the hydraulic properties for the three resins. The SL-644 resulted in the highest BV swings where the granular and sRF resins were characterized by lower BV changes at nominally 30% change from Na-form to H-form.

Cesium load behavior for SL-644, RF, and sRF were evaluated through column testing using Hanford Tank AZ-102 simulant (Fiskum et al. 2004c). Considerable

FIGURE 6.21 Ground gel resorcinol formaldehyde resin.

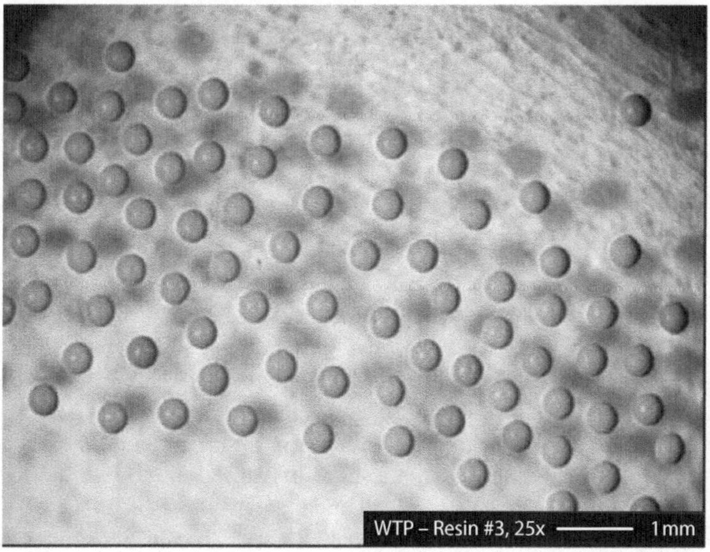

FIGURE 6.22 Spherical resorcinol formaldehyde resin.

variations were observed in the Cs load profiles for the resins, as seen in Figure 6.23. The onset of Cs breakthrough ranged from 90 BVs for sRF to 190 BVs for SL-644, and RF Cs breakthrough fell nominally in between starting at 150 BVs; 50% C/C_0 ranged from 140 BVs for sRF to 275 BVs for SL-644 with RF 50% C/C_0 occurring around 190 BVs.

TABLE 6.10
Hydraulic Properties of the Three Resins

	SL-644 As-received	SL-644 Sieved +20/−30 Mesh	GG RF	sRF
Minimum condition	DI water	Nitric acid	Nitric acid	Nitric acid
Percentage change from minimum	58%	45%	38%	27%
Maximum condition	Dilute hydroxide	Dilute hydroxide	Dilute hydroxide	Dilute hydroxide

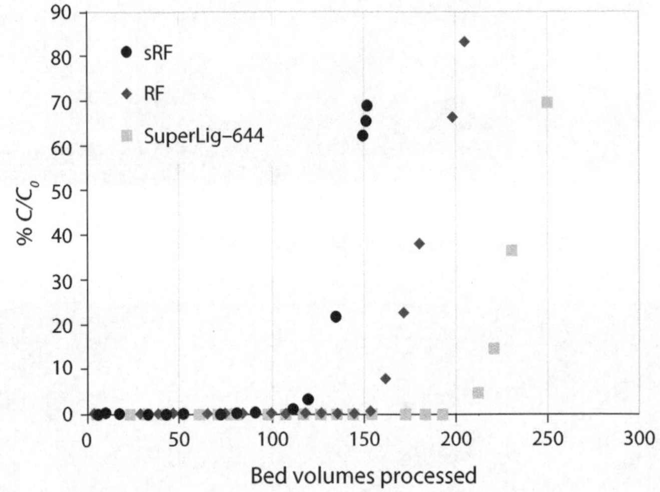

FIGURE 6.23 SuperLig > RF > sRF graph.

F-factor samples were taken to determine the water content in the pretreated resins. The sRF resin contained 25% less mass per unit volume when dried under vacuum to 50°C, whereas the SL-644 resulted saw no mass loss after drying compared to the as-received material.

As with SL-644, the ability to reuse both RF and sRF is important to the viability of these materials for use in this application. Fiskum et al. (2004c) performed an initial assessment of this using three load and elute cycles. The BVs to 50% BT are shown in Table 6.11. The difference between sRF and RF capacity can be attributed to the difference in the mass of the ligand. The average density of sRF was reported to be 0.37 g/mL in the Na^+ form and 0.81 g/mL for RF in the Na^+ form.

To provide further support for the reuse of sRF, six cycles of loading, elution, and regeneration were performed to remove Cs from Hanford Tank 241-AP-105 using sRF resin (Fiskum et al. 2018a). Table 6.12 compares SL-644 (Hassan et al. 2003) and sRF BVs to the 50% BT for each cycle processed.

A 24% reduction in resin loading capacity was observed between cycle 1 and cycle 6 on the SL-644 resin (~4% loss of capacity per cycle). The lead column Cs

TABLE 6.11
Bed Volumes Required to Reach 50% Cs Breakthrough

Cycle	sRF	RF
1	145	184
2	144	174
3	143	169

TABLE 6.12
SL-644 vs sRF BV Processed to 50% BT

Cycle	SL-644 with AW-101	sRF with AP-105
1	222	206
2	208	203
3	195	190
4	155	Not measured
5	180	182
6	166	182

capacity for sRF decreased ~2.8% per cycle. The Cs capacity decrease was attributed to combined chemical, physical, and radiolytic degradation. The cumulative dose exposed to the lead column SL-644 resin was significantly below the target 1.0×10^8 rad to have considerable impact on resin performance. At this low radiation exposure, chemical degradation may have caused most of the observed loss in resin capacity. The primary attack on the sRF resin (and other organic resins) was determined to be by oxygen. Acid elution, water rinse, and 0.1 M NaOH rinses introduce oxygen to the system and are the primary source of resin degradation. Storage of ion exchange media in an inert environment appears to stabilize the ion exchange media, but if in the presence of oxygen, the medium on top will discolor and turn black. Beta decay provides the majority of the radiation dose, and because beta particles have a very short range, they deposit all their energy in a small volume around the point at which the decay takes place.

Changes in the WTP ion exchange process baseline indicated that loading may include a broader range of sodium molarities (0.1–8 M Na). To expand the data range of sRF Cs loading, various sodium concentrations (0.1–5 M) were evaluated at a Cs capacity of 10 ppm Cs (Russell, Rinehart, and Peterson 2013), and the results are shown in Table 6.13. Na concentration has a significant effect on the Cs loading; the more Na present, the less Cs is loaded onto the resin. K^+ concentration was also evaluated and appeared to have nominally no effect on Cs loading at the 10 ppm Cs concentration. Temperature showed a slight effect on Cs loading kinetics; the kinetics increased slightly with increasing temperature.

TABLE 6.13
Impact of Chemistry on sRF Capacity at 10 ppm Cs

Na (M)	K^+ (M)	Temp (°C)	Q (mmol/g)
0.5	0.005	25	0.071
3.0	0.005	25	0.035
5	0.005	25	0.0127
5	0.05	25	0.0126
5	0.005	50	0.006

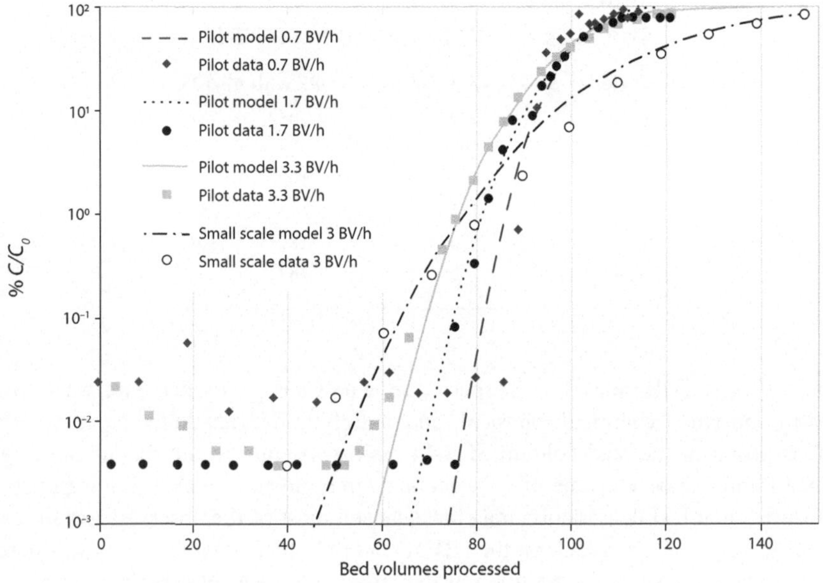

FIGURE 6.24 Small and pilot-scale sRF simulant tests with 10 ppm Cs.

Pilot-scale testing had been conducted with simplified tank waste simulants that provided good Cs load characteristics under nominal processing conditions. A comparison to the small-scale testing with Hanford Tank AP-105 revealed that the model accurately predicted Cs load behavior and validated the model for scaled-up processing, as shown in Figure 6.24. These results showed that small-scale testing can reasonably predict full-scale test results. Parameters such as flowrate, residence time, Cs concentration in the feed, and temperature are needed to be kept consistent when scaling a small-scale test to get equivalent behavior.

Loading testing had demonstrated that the capacity of sRF, while not as high as SL-644, was adequate to meet the processing requirements of greater than 100 BVs (see Table 6.7). The other part of the balance in Figure 6.20 is to ensure that the elution can be performed in an equivalent time period. Therefore, testing was performed

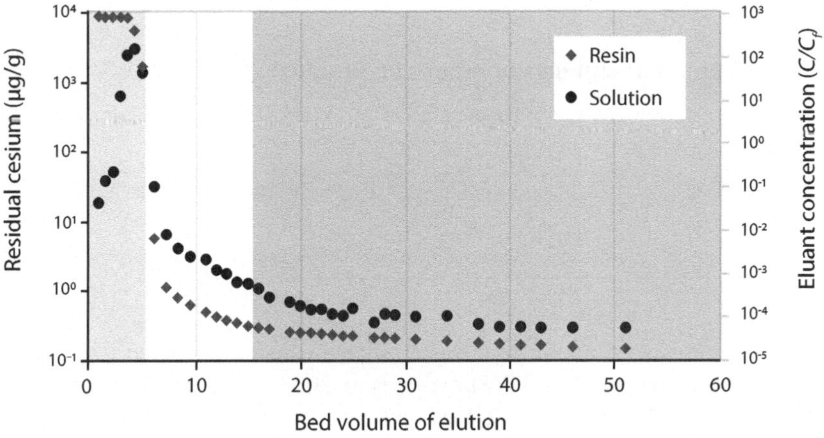

FIGURE 6.25 Elution curve and residual cesium.

to evaluate elution conditions as a function of nitric acid concentration (0.5–1.5 M) using sRF (Burgeson et al. 2004b). Tests were performed using a simulated AN-105 low-activity waste spiked with a $CsNO_3$ Rsolution and traced with Cs-137. The prepared feed solution was analyzed for ^{133}Cs using inductively coupled plasma-mass spectrometry (ICP-MS) to verify the final cesium concentration. Four columns were processed to >90% breakthrough with the AN-105 simulant at 10 μg/mL of Cs. Once the column loading was completed, the feed was displaced, and each column was processed for elution. Figure 6.25 shows the elution curve for the resin and acid eluate solution for 50 BVs of elution. The first 5 BVs of elution are associated with converting the sRF from the Na-form to the H-form neutralizing the interstitial fluid using nitric acid. BVs 5–15 show the cesium that was released upon the H-form conversion as being flushed out. BVs 15–51 show the slow ion exchange of the residual cesium near equilibrium with the acid solution. The spent-resin Cs-137 concentration was used in conjunction with each elution solution C/C_0 value to determine the amount of Cs-137 present on the resin during elution.

Alternate methods for eluting sRF were explored by Oak Ridge National Laboratory, including a staccato flow mode test where 0.5 M nitric acid was eluted for 1 hour at 1 BV/h, followed by 3 hours off (Taylor and Johnson 2009). This method provides time for the eluant to equilibrate, but only showed minimal improvement in elution performance.

GG and sRF elution performance data were collected in an effort to compare the RF Cs elution performance to SL-644 (Fiskum et al. 2004c). Table 6.14 shows the impact of elution conditions on residual cesium at 50 BVs of elution loaded with a 10 μg/mL of feed.

The sRF resin eluted faster and resulted in the least residual Cs on the resin bed after elution. These results suggest that the sRF beads contain pores or other characteristics that allow greater access to active sites within the particles (Fiskum et al. 2004c).

TABLE 6.14
The Impact of Elution Conditions on Residual Cesium

Resin	$[HNO_3]$, M	Temperature, °C	Residual Cesium on Resin, μg/g
sRF	0.5	25	0.15
RF	0.5	45	1.7
RF	1.5	25	3.2
RF	0.5	25	3.9

Elution performance for SL-644 was evaluated using a simulated AN-105 low-activity waste traced with Cs-137 (Burgeson et al. 2004a). Elution was conducted on SL-644 loaded with a 5 μg/mL Cs feed and eluted with 0.5 M HNO_3 at 45°C and 2 BV/h. These elution data were compared to the previously reported sRF data that eluted sRF loaded with 10 μg/mL Cs feed with 0.5 M HNO_3 at 25°C and 1.4 BV/h (Burgeson et al. 2004b). Despite a higher Cs concentration (and thus higher equilibrium cesium loading) and lower temperature, sRF were shown to elute faster than SL-644, as shown in Figure 6.26.

The WTP design basis used an evaporation process to evaporate the cesium eluate. A push to eliminate the evaporator led to studies to evaluate whether the quantity of acid used could be reduced while maintaining adequate decontamination during feed processing. Small-scale ion exchange testing with sRF resin was conducted using an HLW simulant to evaluate the completeness of sRF elution as

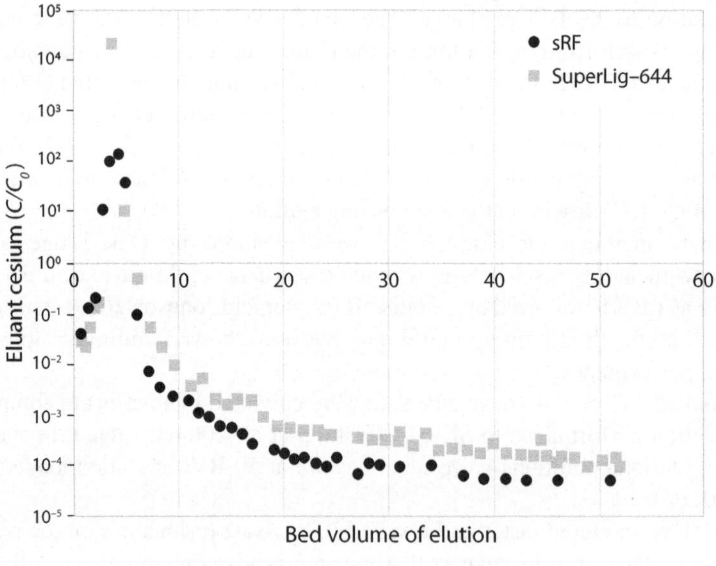

FIGURE 6.26 SL-644 and sRF elution.

a function of nitric acid concentration to help establish acid strength and purity requirements (Brown et al. 2011). Cesium tracer Cs-134 facilitated the monitoring of cesium elution from the column. Columns were loaded with a Cs-137-traced feed and processed to 250 BVs. The columns were then rinsed with 7.5 BVs of 0.1 M NaOH and deionized water then partially neutralized with 3 BVs of 0.5 M HNO_3. Each column was eluted at 25°C with the variable nitric acid concentration (0.02–0.28 M). Columns were then reloaded with Cs-134 tracer to evaluate the source of cesium breakthrough on reloading. Previously, researchers (Fiskum et al. 2004a and 2004c) had estimated that <0.1% C/C_0 (DF > 1,000) would be required to meet waste-loading criteria for Hanford tank wastes. The results from this test, shown in Figure 6.27, indicated nearly all HNO_3 concentrations (0.07, 0.15, 0.23, and 0.28 with the exception of 0.02 M) were sufficient to elute to less than 10^{-3} C/C_0 (<0.1% C/C_0).

To assess the stability of sRF resin, multi-cycle testing using sRF resin on AP-105 Hanford tank waste was conducted in a total of six load elute cycles. The elution with 0.45 M HNO_3 varied for the different process cycles from 16 to 30 BVs. Prior to Test 1 with AP-105, one cycle had been processed with simulant spiked with 60 μg/mL Cs and 0.1 μCi/mL Cs-137 (Fiskum et al. 2018b). The lag column effluent Cs-137 concentration was proportionately low, resulting in low initial C/C_0 values from the lead and lag columns. The Test 2 lead column effluent was the highest of the group; it followed the shortest elution (16 BVs) from Test 1. Figure 6.28 shows the impact of the incomplete elution from Test 1 with initial samples off the lead column being well above the 10% contract limit. Tests 3–5 determined approximately 30 BVs of 0.45 M HNO_3 eluent were required to meet the target DF in the next process cycle.

Due to the timing of the availability of treatment facilities for the eluant and the effluent, there were concerns about the disposition of the cesium-rich eluant. This stream requires treatment in a shielded facility, which was not planned to be

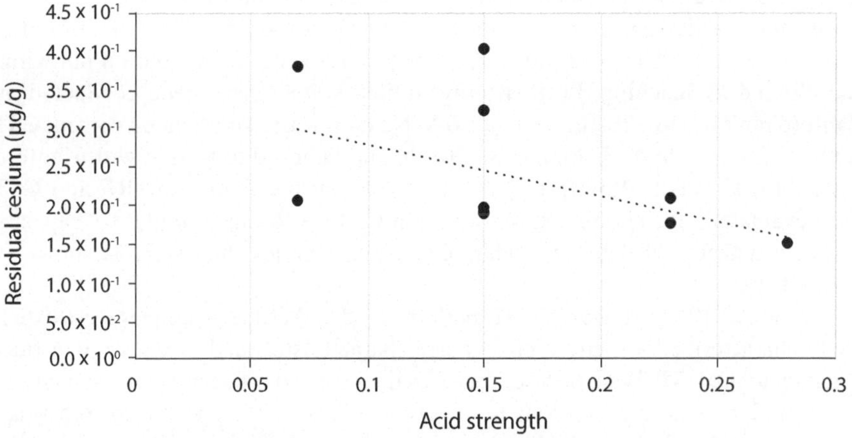

FIGURE 6.27 Residual Cs-134 on sRF resin with varying HNO_3 concentrations.

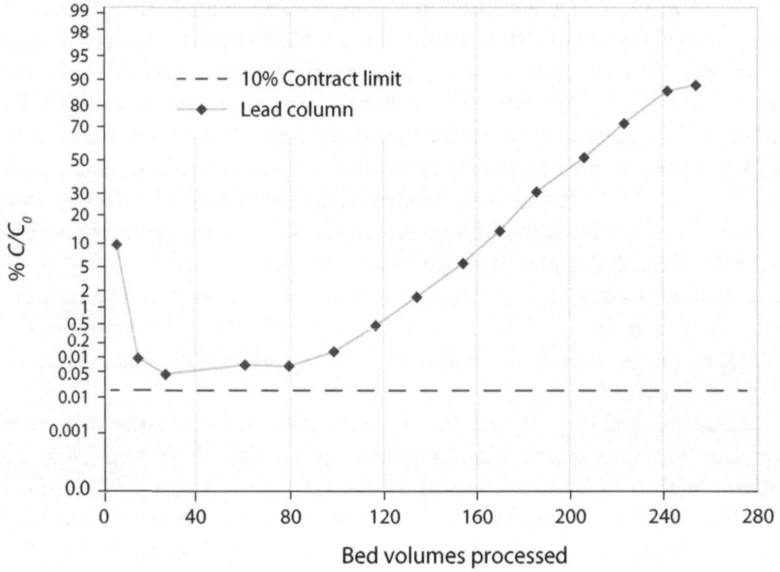

FIGURE 6.28 Test 2 cesium load curve for AP-105 tank waste after 16 BVs of elution.

available at the same time as the low-activity waste disposition path for the effluent. This led to a decision to renew investigation of non-eluatable inorganic CST in July 2018. The use of inorganic ion exchangers offers many advantages over the use of regenerable organic ion exchangers. The inorganic exchangers are generally more resistant to chemical, thermal, and radiation degradation (Pillay 1986; Fiskum et al. 2018a; King 2007; Brown et al. 2011). CST use has been studied for defense waste Cs removal at the Savannah River Site, Melton Valley, and Hanford (King 2007; Walker et al. 1998; Hendrickson, Biyani, and Beck 1996). However, since these earlier tests were conducted, UOP had modified the CST formulation to make the CST more robust relative to fines generation, column plugging, and Nb and Zr leaching. Further, only limited studies have been conducted on Hanford tank wastes. Testing with a 5.6 M Na tank waste simulant on current CST media was conducted (Fiskum et al. 2018b) and later led to testing with Hanford Tank AP-107 waste (Rovira et al. 2018). Loading behaviors for sRF and CST were evaluated, and the results showed that CST has a much higher Cs capacity by about a factor of 4 but considerably slower kinetics than sRF, as shown in Figure 6.29.

During AP-107 testing with CST, both the feed and effluent samples were extensively characterized to better define waste characteristics and assess analyte fractionation to the CST. Table 6.15 summarizes the feed and effluent metals, and anions concentrations and fractionations to the effluent. The CST appeared to sorb small amounts of Cd and also partially remove Ca, Fe, and U; only small percentages of each were reported in the effluent. Small amounts of Th were found in the feed but were not measurable in the effluent analysis.

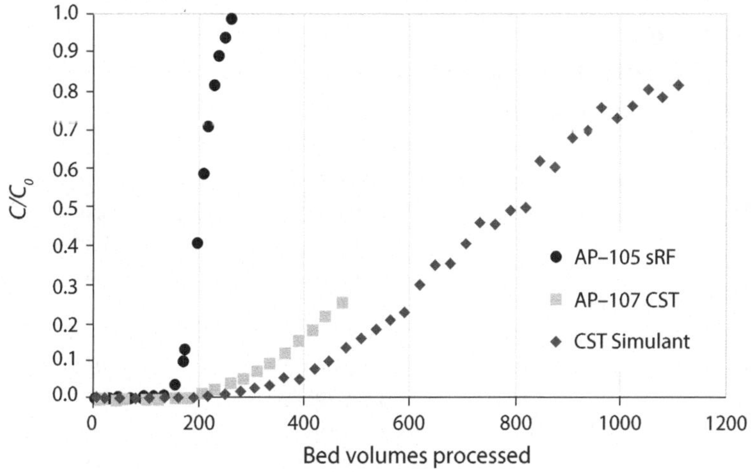

FIGURE 6.29 Cs Loading behavior of sRF and CST.

TABLE 6.15
Analyte Effluent Recovery When Testing AP-107 Waste with CST

Analyte	Feed (ppm)	Effluent (ppm)	Fraction in Effluent (%)
Ca	**34.1**	**17.0**	**49%**
Cd	6.6	5.1	77%
Fe	**19.4**	**12.7**	**65%**
K	3,910	3,800	96%
Th	[12]	–	–
Ti	[0.8]	[0.6]	[77%]

Note: Bracketed values indicate the associated sample results were less than the estimated quantitation limit but greater than the method detection limit.

Table 6.16 summarizes the feed and effluent radioisotope concentrations and fractionations of the effluent. Sr-90 removal is nearly complete and CST appeared to also remove many TRU elements. A large amount of the Pu was assumed to have partitioned to the CST (~70%), and 25%–35% was found in the effluent. The Np-237 results were similar to the Pu results; 31% of the processed Np-237 was accounted for in the effluent, and 69% assumed to have remained on the CST. About 50% of the U fractionated to the CST. The Am and Cm were exceptions. The Cm-243+244 chemistry behavior normally follows that of ^{241}Am; therefore, the Cm-243+244 quantitative recovery (116%) in the effluent was complementary with the 92% recovery for ^{241}Am in the effluent (Rovira et al. 2018).

To assess the scale-up performance of Cs exchange onto the CST, column testing at both the medium (12% of the full bed height) and full height scales were

TABLE 6.16
AP-107 Feed and Effluent Radionuclides

Analyte	Feed (μCi/mL)	Effluent (μCi/mL)	Fraction in Effluent (%)
^{60}Co	<1.9E−3	5.92E−04	–
Cs-137	1.49E+2	1.04E−01	0.04%
^{237}Np	**3.93E−05**	**1.23E−05**	**30.8%**
^{238}Pu	**1.14E−04**	**2.90E−05**	**25.2%**
$^{239+240}Pu$	**5.43E−04**	**1.90E−04**	**34.6%**
^{241}Am	4.17E−04	3.86E−04	91.6%
$^{243+244}Cm$	1.43E−05	1.68E−05	116%
^{90}Sr	**5.81E−01**	**2.72E−04**	**0.05%**
^{99}Tc	9.20E−02	8.90E−02	95.7%
^{126}Sn	1.76E−01	1.74E−01	98.9%
^{238}U	2.07E+01	1.08E+01	52.3%

conducted with a 5.6 M Na simulant (Fiskum et al. 2019). The medium-scale tests used single columns containing 44 mL of <25 mesh CST in a 1.44 cm diameter column, and 27 cm tall CST bed. The full height columns used lead/lag columns containing 1.15 L unsieved (as-received) CST in 2.54 cm diameter columns, and 226 cm tall CST beds. The CST physical properties were characterized for particle size distribution (PSD). The CST d50 of the <25 mesh fraction was 588 μm and the d50 of the full PSD CST was 708 μm.

Figure 6.30 shows the Cs load profiles for the medium and tall column at 1.8 BV/h. The breakthrough curves are virtually identical from 0 to 400 BVs and then diverge where the 2.2 m column experienced a flowrate reduction followed by a stopflow condition. After the stopflow condition, Cs load curves tracking was re-established, despite the extra Cs loading accomplished with added contact time during low-flow and stopflow conditions. The close tracking of the medium and full height scales indicated that mass transfer limitations appear to be primarily associated with diffusion inside the ion exchange medium and that medium column testing can predict full height test Cs load performance. These results suggest that the film diffusion resistance is small compared to the diffusion of cesium within the beads. As such, the increase in superficial velocity between the two tests has a negligible impact on the loading kinetics of this material. This is in contrast to the sRF tests that showed a significant impact of superficial velocity on ion exchange kinetics.

The impact of residence time (in terms of BV/h) on Cs exchange performance was also explored and further demonstrated that fluid residence time with the CST bed drives Cs removal efficiency. Increased residence time allows more time for Cs diffusion into the CST pores. Figure 6.31 compares the effects of residence time on the loading behavior of the ion exchange media.

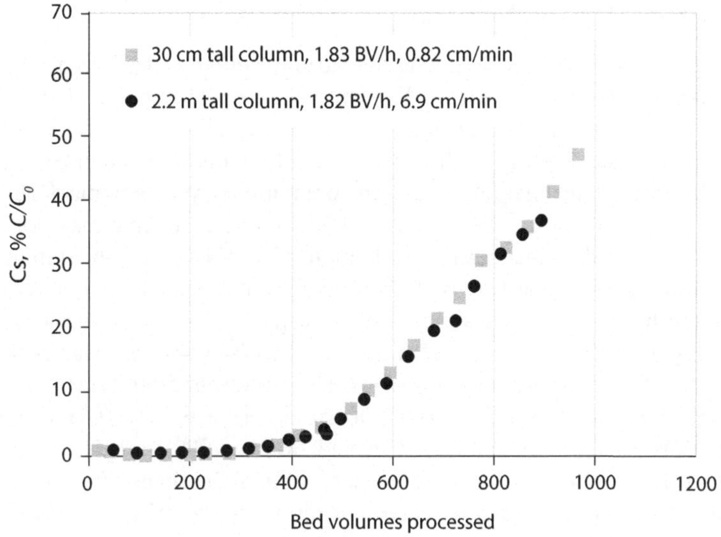

FIGURE 6.30 Cs load profiles at constant residence time for tall and medium height scales.

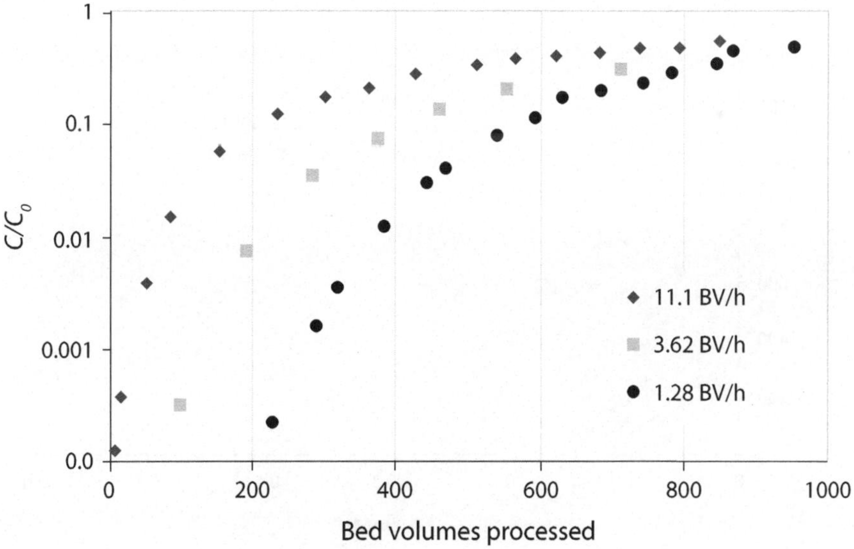

FIGURE 6.31 Cs load profiles at different residence times.

6.7 TECHNETIUM REMOVAL

Technetium is relatively unique among the various radionuclides that must be dealt with coming out of a nuclear reactor. Technetium has a relatively long half-life compared to other species that are also highly soluble in alkaline waste conditions under which defense wastes have typically been stored. In addition, in contrast to cesium, the soluble form of technetium is present predominately in anionic form as the pertechnetate anion (TcO_4^-). IBC Advanced Technologies in American Fork, Utah, has developed a product that has relatively high selectively for technetium in a typical tank waste matrix (SuperLig® 639). Typical K_d values under conditions of interest range from 400 to 1,000, indicating excellent capacity for use as an ion exchange material. Figure 6.32 provides the K_d values for two Hanford tank wastes that have significantly different nitrate concentrations. The nitrate concentration in the AZ-101 (Burgeson, Blanchard Jr, and Deschane 2004b) sample tested was 0.85 M in nitrate, while the AW-101 (Kurath, Blanchard Jr, and Bontha 1999) sample was 2.0 M in nitrate, and the AP-101 (Burgeson, Blanchard Jr, and Deschane 2002) sample was 1.9 M in nitrate. These results indicate that nitrate is the primary competitor for pertechnetate for this ion exchange medium, and that doubling the nitrate concentration roughly halved the capacity of the ion exchange medium. All of these tests were performed between 5 and 6 M Na^+ present. Other results (Rapko, Blanchard Jr, and Carson 2003) demonstrated that the K_d was linear with respect to the sodium concentration. This result suggests that the pertechnetate is removed from tank waste as a $NaTcO_4$ ion pair.

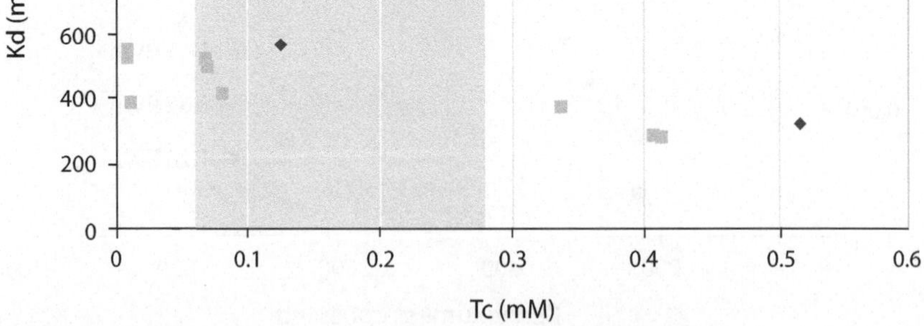

FIGURE 6.32 Batch distribution coefficients for technetium from various tank waste samples. (From Kurath, D.E. et al., *Ion Exchange Distribution Coefficients for* ^{137}*Cs and* ^{99}*Tc Removal from Hanford Tank Supernatants AW-101 (Envelope a) and AN-107 (Envelope C)*, Battelle, Richland, WA, 1999.)

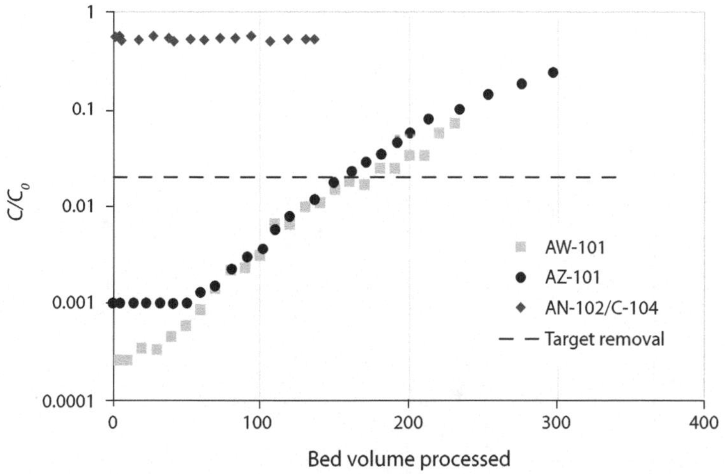

FIGURE 6.33 Breakthrough curves for Tc for tanks containing organics (AN-102) and with relatively low organic loading (AZ-101). (Adapted from Burgeson, I.E. et al., *Small Column Testing of SuperLig® 639 for Removing ^{99}Tc from Hanford Tank Waste Envelope a (Tank 241-AP-101)*, Battelle-Pacific Northwest Division, Richland, WA, 2002; Burgeson, I.E. et al., *Small Column Testing of SuperLig® 639 for Removing ^{99}Tc from Hanford Tank Waste Envelope B (Tank 241-AZ-101)*, Battelle-Pacific Northwest Division, Richland, WA, 2004b.)

The form of technetium present will significantly impact the ability to separate it from tank waste. Figure 6.33 provides the load curves for AZ-101 and AW-101 tank waste along with AN-102/C-104 (Burgeson, Blanchard Jr, and Deschane 2004a). For the first two tanks, the loading behavior at 3 BV/h is as expected for sodium pertechnetate, providing approximately 150 BVs before the target effluent concentration is exceeded (2% C/C_0 in this case). Because the expected capacity is approximately 500–800 BVs for these feeds, it would likely be beneficial to run these feeds longer residence times than tested for a two-column configuration. Note that AZ-101 would be expected to have a higher capacity because of the lower nitrate concentration, but it also has roughly four times as much technetium as the AW-101 sample. The real challenge is represented by the waste from Tank AN-102. This tank received complexant waste from B-Plant during earlier processing to remove strontium and cesium (see Section 6.6). The strontium removal process added strong complexants to keep metals such as iron from contaminating the resultant strontium product. These complexants were subsequently dispositioned to several of the AN tanks. The primary complexants present in these wastes are sodium gluconate, EDTA, and HEDTA. In addition to complexing iron and some actinides, these organic species had the additional undesired effect of reducing the pertechnetate ion to a lower valent soluble complex that could not be removed by this ion exchange material. The data for Tank AN-102 suggest that roughly 50% of the technetium present in that tank had been reduced and was therefore not available for ion exchange.

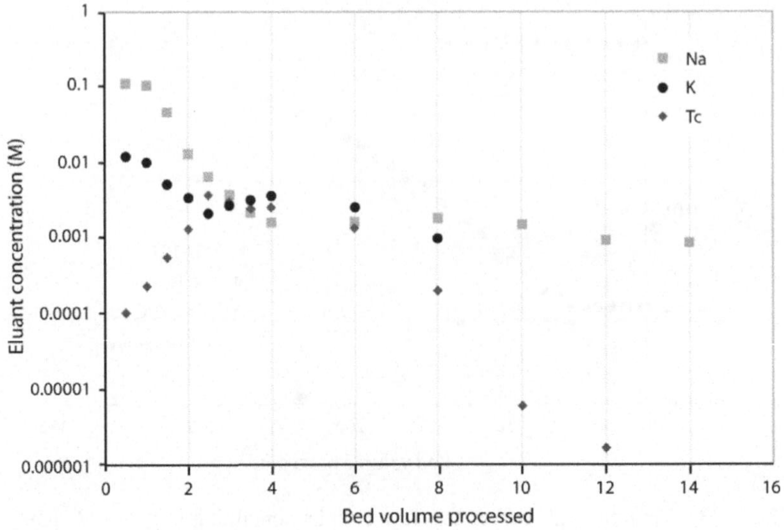

FIGURE 6.34 Elution performance for SL-639 for selected components. (From Hassan, N.M. et al., *Multiple Ion Exchange Column Runs for Cesium and Technetium Removal from AW-101 Waste Sample*, Westinghouse Savannah River Company, Aiken, SC, 2003.)

After loading the technetium onto the SuperLig® 639 (SL-639) resin, the columns were rinsed with 0.1 M NaOH. Then the columns were eluted with hot water. Figure 6.34 shows a typical elution profile (Hassan et al. 2003). The first step in the process is washing out the 0.1 M NaOH that was present, which takes approximately 3 BVs. After that, Tc, K, and Na are eluted off the column. Note that Na continues to elute for an extended period of time. This result confirms the loading results that indicate that the technetium is removed as part of an ion pair. The extended sodium elution is likely attributed to continued elution of sodium and nitrate. Because nitrate is present at a 3 to 4 order higher concentration than pertechnetate, extended elution of nitrate would be expected. Hassan performed five load/elute cycles with SL-639 for technetium removal and saw minimal degradation in performance when 25+ BVs of eluant were used, indicating that this was sufficient to remove the technetium and nitrate from the resin with minimal resin degradation. Table 6.17 shows the feed and eluant composition for AP-101 processing (Burgeson, Blanchard Jr, and Deschane 2002). The technetium is concentrated roughly by a factor of 9 and greater than 99% of the technetium was removed from the feed stream. This ion exchange material is highly selective for technetium over nitrate, by roughly a factor of 10,000. As indicated above, the tail in the elution profile is likely associated with continued elution of small amounts of nitrate, since an order of magnitude more of nitrate has loaded onto the ion exchange medium in comparison to technetium even with the much higher selectivity for technetium.

One of the operational challenges associated with SL-639 is its relatively low density. SL-639 will float in a supernate with a density above 1.26 g/mL

TABLE 6.17
Feed and Eluant Composition for Tank AP-101 Waste

	Feed (M)	Eluant (M)	Concentration Factor
Technetium	2.98E-05	2.70E-04	9.1
Nitrate	1.88	1.65E-03	0.001
Nitrite	0.78	6.74E-04	0.001
Sodium	4.80	1.53E-02	0.003

Source: Burgeson, I.E. et al., *Small Column Testing of SuperLig®639 for Removing ^{99}Tc from Hanford Tank Waste Envelope a (Tank 241-AP-101)*, Battelle-Pacific Northwest Division), Richland, WA, 2002.

(Nash et al. 2015). This limits the normal operating conditions to approximately 5 M sodium because solutions with higher sodium molarities will result in a floating resin bed, which would not provide the same effective column performance because the potential for either channeling or mixing of the resin bed. If higher salt solution concentration were to be processed, the column would need to be designed to constrain the resin bed.

REFERENCES

Arm, S. T., and D. L. Blanchard Jr. 2004. *Pre-Conditioning and Regeneration Requirements of Ground Gel Resorcinol Formaldehyde Ion Exchange Resin* (Richland, WA: Battelle-Pacific Northwest Division). PNWD-3390, WTP-RPT-104.

Bain, P. E. 1990. "A Model Predicting Equilibrium for Plutonium Sorption by Anion-Exchange Resin." *Solvent Extraction and Ion Exchange* 8 (2): 341–351. https://doi.org/10.1080/07366299.1990.11073617.

Baker, M. N., and R. F. Gessner. 1989. *Low-Level Liquid Waste Treatment System Start-Up* (West Valley, NY: West Valley Nuclear Services Company, Inc.). DOE/NE/44139-55. https://www.osti.gov/servlets/purl/10197011.

Baker, M. N., W. E. Mateer, G. H. Metzler, S. R. Reeves, and D. J. Rickettson. 1989. *Design and Construction of the Low-Level Liquid Waste Treatment System* (West Valley, NY: West Valley Nuclear Services Company, Inc.). DOE/NE/44139-56.

Baldwin, C. E., and J. D. Navratil. 1983. "Plutonium Process Chemistry at Rocky Flats." In *Plutonium Chemistry*, edited by W. T. Carnall and G. R. Choppin, 369–380. ACS Symposium Series.

Berge, A., T.-N. Nilsen, J.-O. Bjorgum, and J. Ugelstad. 1993. Process for Preparing a Dispersion and for Preparing Polymer Particles. International Patent WO 93/02112, filed February 4, 1993.

Bibler, J. P., and R. M. Wallace. 1992. Cesium-Specific Phenolic Ion Exchange Resin. Patent Number 5441991. United States.

Bibler, J. P., R. M. Wallace, and L. A. Bray. 1990. "Testing a New Cesium-Specific Ion Exchange Resin for Decontamination of Alkaline High-Activity Waste." Waste Management '90, Tucson, AZ, February 25–March 1, 1990. https://www.osti.gov/servlets/purl/5790227.

Bray, L. A., and L. L. Burger. 1994. Method for Aqueous Radioactive Waste Treatment. Patent Number 5298166. United States.

Bray, L. A., and F. T. Hara. 1993. *Use to Titanium-Treated Zeolite for Plutonium, Strontium, and Cesium Removal from West Valley Alkaline Wastes and Sludge Wash Wastes* (Richland, WA: Pacific Northwest National Laboratory). PNL-SA-21775. http://inis.iaea.org/search/search.aspx?orig_q=RN:25015550.

Bray, L. A., L. K. Holton, T. R. Myers, G. M. Richardson, and B. M. Wise. 1984. *Experimental Data Developed to Support the Selection of a Treatment Process for West Valley Alkaline Supernatant* (Richland, WA: Pacific Northwest National Laboratory). PNL-4969. http://inis.iaea.org/search/search.aspx?orig_q=RN:15047413.

Brown, G. N., L. A. Bray, C. D. Carlson, K. J. Carson, J. R. Deschane, R. J. Elovich, F. V. Hoopes, D. E. Kurath, L. L. Nenninger, and P. K. Tanaka. 1996. *Comparison of Organic and Inorganic Ion Exchangers for Removal of Cesium and Strontium from Simulated and Actual Hanford 241-AW-101 DSSF Tank Waste* (Richland, WA: Pacific Northwest National Laboratory). PNNL-11120, UC-810.

Brown, G. N., J. W. Russell, and R. A. Peterson. 2011. *Small-Column Cesium Ion Exchange Elution Testing of Spherical Resorcinol-Formaldehyde* (Richland, WA: Pacific Northwest National Laboratory). PNNL-20603, WTP-RPT-210.

Burgeson, I. E., D. L. Blanchard Jr., B. J. Cook, and J. R. Deschane. 2004a. *Elution Testing of Resorcinol-Formaldehyde Resins with AN-105 Simulant* (Richland, WA: Pacific Northwest National Laboratory). PNWD-3388, WTP-RPT-105.

Burgeson, I. E., D. L. Blanchard Jr., B. J. Cook, J. R. Deschane, and D. R. Weier. 2004b. *Elution Testing of SuperLig® 644 with AN-105 Simulant* (Richland, WA: Pacific Northwest National Laboratory). PNWD-3399, WTP-RPT-109.

Burgeson, I. E., D. L. Blanchard Jr., and J. R. Deschane. 2002. *Small Column Testing of SuperLig® 639 for Removing ^{99}Tc from Hanford Tank Waste Envelope a (Tank 241-AP-101)* (Richland, WA: Battelle-Pacific Northwest Division). PNWD-3222, WTP-RPT-030.

Burgeson, I. E., D. L. Blanchard Jr., and J. R. Deschane. 2004b. *Small Column Testing of SuperLig® 639 for Removing ^{99}Tc from Hanford Tank Waste Envelope B (Tank 241-AZ-101)* (Richland, WA: Battelle-Pacific Northwest Division).

Burgeson, I. E., D. L. Blanchard Jr., and J. R. Deschane. 2004c. *Small Column Testing of SuperLig® 639 for Removing ^{99}Tc from Hanford Tank Waste 241-AN-102 Supernate (Envelope C) Mixed with Tank 241-C-104 Solids (Envelope D) Wash and Permeate Solutions* (Richland, WA: Battelle-Pacific Northwest Division). PNWD-3252, Rev 1, WTP-RPT-031, Rev 1.

Carlstrom, R. F. 1977. *Ion Exchange Flowsheet for Recovery of Cesium from PUREX Sludge Supernatant at B Plant* (Richland, WA: Atlantic Richfield Hanford Company). ARH-F-106.

Chung, S. F., and C. Y. Wen. 1968. "Longitudinal Dispersion of Liquid Flowing through Fixed and Fluidized Beds." *Al Ch E Journal* 14 (6): 857–866. https://doi.org/10.1002/aic.690140608.

Dalton, W. J. 1997. *Qualification Testing and Full-Scale Demonstration of Titanium-Treated Zeolite for Sludge Wash Processing* (West Valley, NY: West Valley Nuclear Services Company, Inc.). DOE/NE/44139-72. https://www.osti.gov/biblio/565259.

Fiskum, S. K., J. R. Allred, H. A. Colburn, A. M. Rovira, M. R. Smoot, and R. A. Peterson. 2018. *Multi-Cycle Cesium Ion Exchange Testing Using Spherical Resorcinol-Formaldehyde Resin with Diluted Hanford Tank Waste 241-AP-105* (Richland, WA: Pacific Northwest National Laboratory). PNNL-27432, RPT-DFTP-006.

Fiskum, S. K., S. T. Arm, and D. L. Blanchard Jr. 2002. *Aging Study and Small Column Ion Exchange Testing of SuperLig® 644 for Removal of ^{137}Cs from Simulated AW 101 Hanford Tank Waste* (Richland, WA: Pacific Northwest National Laboratory). PNWD-3195, WTP-RPT-015. https://www.pnnl.gov/rpp-wtp/documents/WTP-RPT-015.pdf.

Fiskum, S. K., S. T. Arm, and D. L. Blanchard Jr. 2004a. *Small Column Ion Exchange Testing of SuperLig® 644 for Removal of ^{137}Cs from Hanford Waste Tank 241-AZ-101 (Envelope B)* (Richland, WA: Pacific Northwest National Laboratory). PNWD-3266, Rev 1, WTP-RPT-053, Rev 1.

Fiskum, S. K., S. T. Arm, D. L. Blanchard Jr., and B. M. Rapko. 2004b. *Small Column Ion Exchange Testing of SuperLig® 644 for Removal of ^{137}Cs from Hanford Waste Tank 241-AP-101 Diluted Feed (Envelope a)* (Richland, WA: Pacific Northwest National Laboratory). PNWD-3198, Rev 1, WTP-RPT-016, Rev 1.

Fiskum, S. K., B. X. Augspurger, K. P. Brooks, W. C. Buchmiller, R. L. Russell, M. J. Schweiger, L. A. Snow et al. 2004c. *Comparison Testing of Multiple Resorcinol-Formaldehyde Resins for the River Protection Project-Waste Treatment Plant* (Richland, WA: Pacific Northwest National Laboratory). PNWD-3387, WTP-RPT-103.

Fiskum, S. K., H. A. Colburn, R. A. Peterson, A. M. Rovira, and M. R. Smoot. 2018. *Cesium Ion Exchange Using Crystalline Silicotitanate with 5.6 M Sodium Simulant* (Richland, WA: Pacific Northwest National Laboratory). PNNL-27587, RPT-DFTP-008.

Fiskum, S. K., R. A. Peterson, H. A. Colburn, M. R. Smoot, M. R. Landon, K. A. Colosi, and R. A. Wilson. 2019. "Small- to Large-Scale Comparisons of Cesium Ion Exchange Performance with Spherical Resorcinol Formaldehyde Resin." *Separation Science and Technology* 1–10. https://doi.org/10.1080/01496395.2019.1570262.

Hassan, N. M., K. Adu-Wusu, and C. A. Nash. 2003. *Multiple Ion Exchange Column Runs for Cesium and Technetium Removal from AW-101 Waste Sample* (Aiken, SC: Westinghouse Savannah River Company). WSRC-TR-2003-00098, SRT-RPP-2003-00026.

Hassan, N. M., D. J. McCabe, and W. D. King. 2000. *Small-Scale Ion Exchange Removal of Cesium and Technetium from Hanford Tank 241-AN-103* (Aiken, SC: Westinghouse Savannah River Company). BNF-003-98-0146, Rev. 1.

Hendrickson, D. W., R. K. Biyani, and M. A. Beck. 1996. *Hanford Tank Waste Supernatant Cesium Removal Test Report* (Richland, WA: Westinghouse Hanford Company). WHC-SD-RE-TRP-018.

Hougen, O. A., and W. R. Marshall. 1947. "Adsorption from a Fluid Stream Flowing through a Stationary Granular Bed." *Chemical Engineering Progress* 43 (4): 197–208.

Kelly Jr., S., and D. C. Meess. 1997. *THOREX Processing and Zeolite Transfer for High-Level Waste Stream Processing Blending* (West Valley, NY: West Valley Nuclear Services Company, Inc.). DOE/NE/44139-82. https://digital.library.unt.edu/ark:/67531/metadc695931/m1/3/.

King, W. D. 2007. *Literature Reviews to Support Ion Exchange Technology Selection for Modular Salt Processing* (Aiken, SC: Washington Savannah River Company). WSRC-STI-2007-00609.

King, W. D., N. M. Hassan, and D. J. McCabe. 2001. *Intermediate-Scale Ion Exchange Removal of Cesium and Technetium from Hanford Tank 241-AN-102* (Aiken, SC: Westinghouse Savannah River Company). WSRC-TR-2000-00420, SRT-RPP-2000-00014, Rev 1.

Klinkenberg, A. 1948. "Numerical Evaluation of Equations Describing Transient Heat and Mass Transfer in Packed Solids." *Industrial and Engineering Chemistry* 40 (10): 1992–1998. https://doi.org/10.1021/ie50466a034.

Knoch, W. 1964. "Some Liquid Phase Diffusion Coefficients of Uranium and Plutonium." *Journal of Chemical & Engineering Data* 9 (1): 30–61. https://doi.org/10.1021/je60020a017.

Kumaresan, R., K. N. Sabharwal, T. G. Srinivasan, and P. R. V. Rao. 2006. "Evaluation of New Anion Exchange Resins for Plutonium Processing." *Solvent Extraction and Ion Exchange* 24 (4): 589–602. https://doi.org/10.1080/07366290600762512.

Kurath, D. E., D. L. Blanchard Jr., and J. R. Bontha. 1999. *Ion Exchange Distribution Coefficients for ^{137}Cs and ^{99}Tc Removal from Hanford Tank Supernatants AW-101 (Envelope a) and AN-107 (Envelope C)* (Richland, WA: Battelle). PNWD-2467, BNFL-RPT-009.

Kurath, D. E., L. A. Bray, J. A. Murphy, R. D. Boardman, L. F. Pincock, and N. Christiansen. 1995. "Wastes at the West Valley Nuclear Site, New York." In *Separation Techniques in Nuclear Waste Management*, edited by Thomas E. Carleson, Nathan A. Chipman and Chien M. Wai. Boca Raton, FL: CRC Press.

Kurath, D. E., L. A. Bray, W. A. Ross, and D. K. Ploetz. 1989. *Correlation of Laboratory Testing and Actual Operations for the West Valley Supernatant Treatment System* (Richland, WA: Pacific Northwest National Laboratory). PNL-SA-16871. https://www.osti.gov/biblio/6348484-correlation-laboratory-testing-actual-operations-west-valley-supernatant-treatment-system.

Mohapatra, P. K., L. B. Kumbhare, P. B. Ruikar, and V. K. Manchanda. 2004. "Ion-Exchange Separation of Pu from Macro Concentration of Th in HNO_3 Medium." *Solvent Extraction and Ion Exchange* 22 (2): 267–284. https://doi.org/10.1081/Sei-120030462.

Nash, C. A., M. R. Duignan, and C. E. Duffey. 2006. *Batch, Kinetics, and Column Data from Spherical Resorcinol-Formaldehyde Resin* (Aiken, SC: Savannah River National Laboratory). WSRC-STI-2006-00071, SRNL-RPP-2006-00024.

Nash, C. A., M. R. Williams, M. L. Restivo, L. L. Hamm, and D. J. McCabe. 2015. *SuperLig® 639 Resin Performance Using 8 Molar LAW Feed* (Aiken, SC: Savannah River National Laboratory). SRNL-STI-2015-00489.

Navratil, J. D. 1986. *Ion Exchange Technology in the Nuclear Fuel Cycle* (Vienna, Austria: International Atomic Energy Agency). IAEA-TECDOC-365.

Navratil, J. D. 1989. "Ion Exchange Technology in Spent Fuel Reprocessing." *Journal of Nuclear Science and Technology* 26 (8): 735–743. https://doi.org/10.1080/18811248.1989.9734377.

Nelson, J. L., and B. W. Mercer. 1963. *Ion Exchange Separation of Cesium from Alkaline Waste Supernatant Solutions* (Richland, WA: General Electric, Hanford Atomic Products Operation). HW-76449.

Orth, D. A. 1960. "Plutonium Ion Exchange Processes." U.S.-U.K. Technical Exchange Meeting, Oak Ridge, TN, April 25–27, 1960.

Pillay, K. K. S. 1986. "A Review of the Radiation Stability of Ion Exchange Materials." *Journal of Radioanalytical and Nuclear Chemistry* 102 (1): 247–268. https://doi.org/10.1007/BF02037966.

Ploetz, D. K., and I. M. Leonard. 1988. *Supernatant Treatment System Design through Testing* (West Valley, NY: West Valley Nuclear Services Company, Inc.). DOE/NE/44139-47.

Rapko, B. M., D. L. Blanchard Jr., and K. J. Carson. 2003. *Equilibrium Batch Contact Testing of SuperLig® 639* (Richland, WA: Battelle, Pacific Northwest Division). PNWD-3251, WTP-RPT-026.

Rovira, A. M., S. K. Fiskum, H. A. Colburn, J. R. Allred, M. R. Smoot, and R. A. Peterson. 2018. *Cesium Ion Exchange Testing Using Crystalline Silicotitanate with Hanford Tank Waste 241-AP-107* (Richland, WA: Pacific Northwest National Laboratory). PNNL-27706, RPT-DFTP-011.

Russell, R. L., D. E. Rinehart, and R. A. Peterson. 2013. *Ion Exchange Testing with SRF Resin FY 2012* (Richland, WA: Pacific Northwest National Laboratory). PNNL-21645, WTP-RPT-223.

Ryan, J. L., and E. J. Wheelwright. 1959. "Recovery and Purification of Plutonium by Anion Exchange." *Industrial and Engineering Chemistry* 51 (1): 60–65. https://doi.org/10.1021/ie50589a038.

Schiffhauer, M. A., and M. C. Thompson. 1993. *Installation Start-up and Operation of a PUREX HLW Sludge Mobilization and Wash System at the West Valley Demonstration Project* (West Valley, NY: West Valley Nuclear Services Company, Inc.). http://archive.wmsym.org/1993/V2/2.pdf.

Schiffhauer, M. A., and S. C. Thompson. 1992. *High-Level Waste Tank Modifications, Installation of Mobilization Equipment/Check Out* (West Valley, NY: West Valley Nuclear Services Company, Inc.). DOE/NE/44139-69. https://www.osti.gov/servlets/purl/10197010.

Swift, W. H. 1960. "Plutonium Ion Exchange Processes." U.S.-U.K. Technical Exchange Meeting, Oak Ridge, TN, April 25–27, 1960.

Taylor, P. A., and H. L. Johnson. 2009. *Alternate Methods for Eluting Cesium from Spherical Resorcinol-Formaldehyde Resin* (Oak Ridge, TN: Oak Ridge National Laboratory). ORNL/TM-2008/194.

Valenti, P. J., and D. I. Elliot. 1999. *Operating Experience During High-Level Waste Vitrification at the West Valley Demonstration Project* (West Valley, NY: West Valley Nuclear Services Company, Inc.). DOE/NE/44139-87. https://www.osti.gov/biblio/754792-operating-experience-during-high-level-waste-vitrification-west-valley-demonstration-project.

Walker, D. D., M. J. Barnes, C. L. Crawford, R. A. Peterson, R. F. Swingle, and S. D. Fink. 1998. "In-Tank Precipitation with Tetraphenylborate: Recent Process and Research Results." In *Science and Technology for Disposal of Radioactive Tank Wastes*, edited by W.W. Schulz and N. L. Lombardo, 219–230. Boston, MA: Springer.

Wilson, E. J., and C. J. Geankoplis. 1966. "Liquid Mass Transfer at Very Low Reynolds Numbers in Packed Beds." *Industrial & Engineering Chemistry Fundamentals* 5 (1): 9–14. https://doi.org/10.1021/i160017a002.

7 Non-aqueous Processing

Michael F. Simpson and Andrew M. Casella

CONTENTS

7.1 INTRODUCTION

The early history of nuclear materials processing was dominated by aqueous chemistry. These processes generally involved dissolving the charge from the reactor in strong acid solutions and then processing these materials through one of the unit operations outlined previously; primarily precipitation, ion exchange, or solvent extraction. Currently, solvent extraction is the dominant method for processing spent reactor charges, and plutonium/uranium extraction (PUREX) is the preferred process of choice. This is a relatively mature process, but it has the inherent problem of potential proliferation. Chapter 4 outlines some of the proposed changes to these processes that are being pursued to address these issues.

However, some next generation processes rely on either proliferation resistant flowsheets or support for reactor concepts that rely upon unique fuel charges. Since the early 1960s, researchers have been pursuing alternative nuclear fuel cycles that would rely upon non-aqueous reprocessing unit operations. This chapter will address some non-aqueous processes and their outlook for the future.

7.2 PYROPROCESSING

The terms *pyroprocessing* and *pyrochemical* have been applied to dry, high-temperature chemical processing of spent nuclear fuel using molten salt as an electrolyte or flux. This processing can be used for purposes such as reduction of oxides to metals and separation/extraction of actinides from fission products and other constituents in nuclear fuel such as cladding, bond material, fuel assembly hardware, and fuel alloying metals. In this chapter, the origins of spent fuel pyroprocessing will be presented followed by discussion of current applications and prospects for significant future roles in the nuclear fuel cycle.

Many years before the discovery of nuclear fission when uranium was of little interest as a useful metal, its isolation in metallic form was reported as a pyrochemical reaction. In 1842, Péligot reported the reduction of uranium chloride with sodium metal at high temperature (Péligot 1842). The reaction is as follows.

$$UCl_3 + 3Na = U + 3NaCl \tag{7.1}$$

The temperature that this reaction was carried out at is unknown, but it could have been relatively low to melt the sodium and support fast mass transfer. The standard state free energy change for this reaction is −320 kJ/mole at 500°C, so the reaction likely goes to completion after melting the sodium. The reaction is also exothermic, which will cause the temperature to rapidly rise. In 1930, Driggs and Lilliendahl reported electrolytic recovery of uranium metal in molten $NaCl–CaCl_2$ at 775°C (Driggs and Lilliendahl 1930). The salt melts at about 500°C, and this is the first known report of uranium electrorefining in molten salt. Uranium metal dendrites were formed on a molybdenum cathode rod. This was practically the invention of uranium electrorefining, which would be further developed at Argonne National Laboratory (ANL) in the 1950s (Marzano and Noland 1953; Blumenthal 1955) and scaled up to 10 kg U metal batch sizes by French researchers at CEA in the 1960s (Boisde et al. 1961; Chauvin et al. 1964). The molten salt electrolyte was changed to eutectic LiCl–KCl with 3 mol% UCl_3 by CEA, which is an attractive choice because of the low eutectic point (352°C). Decades later, LiCl–KCl would become the standard choice for the electrolyte. Meanwhile, electrorefining in molten salt to purify plutonium metal was reported by Los Alamos National Laboratory in the 1960s (Mullins et al. 1963). In this process, impure plutonium metal forms a liquid anode surrounded by a molten $NaCl–KCl–PuCl_3$ mixture. Purified plutonium metal reduces onto a steel cathode tube, and the molten plutonium drips down to the bottom of the crucible. A concentric ceramic crucible is used for this process that separates the impure liquid plutonium (anode) from the purified liquid plutonium. The electrochemical half-cell reactions are as follows:

$$Pu_{(l,impure)} = Pu^{3+} + 3e^{-}(\text{anode}) \tag{7.2}$$

$$Pu^{3+} + 3e^{-} = Pu_{(l,pure)}(\text{cathode}) \tag{7.3}$$

Pyroprocessing is, therefore, not a new or speculative technology for reducing or separating actinides. Its use even predates the discovery of practical applications for uranium.

7.3 MODERN PYROPROCESSING METHODS

Modern pyroprocessing technology is considered here to refer to developments that started in the 1980s as developed by ANL, initially under the Integral Fast Reactor (IFR) program. Notably, there is a distinctly different pyroprocessing method developed at the Russian Institute of Atomic Reactors (RIAR) that works with oxide fuel as the feed and generates an oxide product (Kofuji et al. 2007). The RIAR process is not covered in this chapter since it is based on much different chemistry than the ANL process.

Development of the ANL process started in the 1980s under the IFR program, which ran from 1984 to 1994 (Kofuji et al. 2007). It was based around uranium or group actinide (U/TRU) deposition on a cathode in eutectic LiCl–KCl electrolyte. The chemistry was initially developed at ANL-East with non-irradiated materials and then demonstrated with irradiated Experimental Breeder Reactor (EBR)-II fuel at the Fuel Cycle Facility (currently called the Fuel Conditioning Facility, or FCF) at ANL-West (currently called the Materials and Fuels Complex at Idaho National Laboratory (INL)).

Table 7.1 gives a series of standard reduction potentials for actinides, fission products, and molten salt constituents. The electrolyte in an electrorefiner (ER) initially contains only LiCl, KCl, and UCl_3. Spent fuel in metallic form is loaded in the anode basket. When current is passed between the anode and cathode, uranium and species with the lowest reduction potentials will oxidize, while uranium is reduced on the cathode. The key for selective deposition of uranium on the cathode is that no metals with higher reduction potentials than uranium oxidize and dissolve in the salt. These reactions are illustrated below as an incomplete set of examples. All properties are based on standard-state free energy at 500°C as calculated by HSC Chemistry (2019). The given potentials are equilibrium potentials. For the uranium to be oxidized at the anode, the potential must be higher than the given equilibrium potential. For the uranium to be reduced at the cathode, the potential must be lower than the given equilibrium potential.

Anode:

$$U = U^{3+} + 3e^- \qquad E^0 = -2.41\,V \text{ vs. } Cl^-/Cl_2 \tag{7.4}$$

$$3Cs + UCl_3 = 3CsCl + U \qquad \Delta G^\circ = -402 \text{ kJ/mol} \tag{7.5}$$

$$1.5Sr + UCl_3 = 1.5SrCl_2 + U \qquad \Delta G^\circ = -362 \text{ kJ/mol} \tag{7.6}$$

$$La + UCl_3 = LaCl_3 + U \qquad \Delta G^\circ = -183 \text{ kJ/mol} \tag{7.7}$$

TABLE 7.1
Standard Reduction Potentials for a Number of Key Metal Ions in Molten Chloride Salt at 500°C

Metal Ion	E° (V vs. Cl^-/Cl_2)	Metal Ion	E° (V vs. Cl^-/Cl_2)	Metal Ion	E° (V vs. Cl^-/Cl_2)
Ru^{3+}	−0.19	Cm^{3+}	−2.65	Nd^{3+}	−2.95
Co^{2+}	−1.07	Pu^{3+}	−2.72	Na^+	−3.52
U^{4+}	−2.06	Am^{3+}	−2.75	Li^+	−3.57
U^{3+}	−2.41	Pm^{3+}	−2.78	K^+	−3.76
Np^{3+}	−2.52	Gd^{3+}	−2.83	Cs^+	−3.80

Source: HSC Chemistry, Outotek (Version 9), https://www.outotec.com/products/digital-solutions/hsc-chemistry/, 2019

Cathode:

$$U^{3+} + 3e^- = U \qquad E^0 = -2.41\ \text{V vs. } Cl^-/Cl_2 \tag{7.8}$$

Provided there is a high enough initial concentration of UCl_3 in the electrolyte, no other metals will reduce onto the cathode. This presents a problem for recovery of transuranic actinides, which can be used to fabricate fast reactor fuel. Plutonium, americium, neptunium, and curium all require lower reduction potentials than uranium for reduction. Uranium recovery will be highly selective until the UCl_3 has been depleted from the salt. If high current density is passed through the cathode, its potential can be reduced enough to cause co-reduction of TRU with U. But upon stopping the current, the TRU metal will spontaneously re-oxidize by reaction with UCl_3. TRU metal is simply not stable in contact with molten salt containing UCl_3 because of the thermodynamic properties of the compounds.

The workaround for recovery of transuranic actinides involves a liquid cadmium cathode (LCC) and was first reported by Burris et al. (1986). Essentially, the steel cathode normally used in an ER is replaced with a ceramic crucible containing liquid cadmium. It was discovered that plutonium will form a $PuCd_6$ intermetallic compound that effectively reduces the activity coefficient of Pu in the metallic phase. A crucible containing liquid cadmium can be immersed in the molten LiCl–KCl–UCl_3 and polarized to cathodically reduce U^{3+} and TRU metal ions simultaneously (Koyama et al. 1997). This process is illustrated in the conceptual LCC cell shown in Figure 7.1.

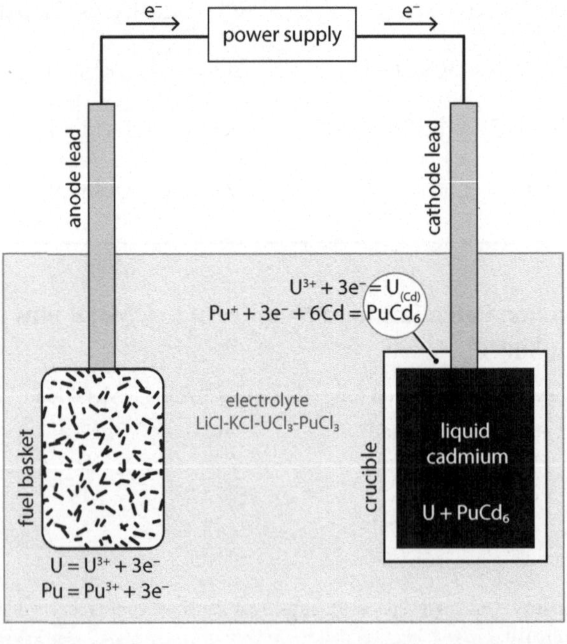

FIGURE 7.1 Illustration of U/TRU recovery on liquid cadmium cathode.

Li and co-workers at Idaho National Laboratory reported separation factors for actinide and rare earth metals obtained using liquid cadmium cathodes in small-scale experiments performed using irradiated fuel from the EBR-II in the Hot Fuel Examination Facility (Li and Simpson 2005). In this study, selective deposition of U, Pu, Np, and Am relative to rare earths was demonstrated. Plutonium's separation factor is only 1.5, meaning that it can readily be co-recovered with uranium on the LCC. Separation factors of rare earths relative to uranium ranged from 6.2 to 59, which implies substantial decontamination of these fission products in the LCC product. Total rare earth concentration in the LCC samples was found to be much less than 5 wt%. While it is assumed that low rare earth concentrations are acceptable in metallic fast reactor fuel, it is unknown whether a limit has been justified. Oxidation-based extraction of rare earth fission products from the U/TRU product has been reported and could be used to polish the LCC product and minimize rare earth fission product contamination (Lee et al. 2011).

Electrorefining is only a single unit operation in the ANL process. Cathode product from the ER is heated to 1,200°C in a cathode processor under vacuum to distill adhered salt and to melt the uranium into ingots (Westphal et al. 2008). The uranium product then has its enrichment lowered via melting with depleted uranium in the casting furnace. This serves the function of converting the uranium into low-enriched uranium, which can be stored relatively economically compared to high-enriched uranium (>20% ^{235}U). The uranium from the driver fuel is down-blended in the casting furnace to just less than 20%, which is now categorized as high-assay low-enriched uranium (HALEU). Preparation of fuel for electrorefining in FCF includes first disassembly of fuel bundles followed by chopping of individual fuel elements. The fuel bundle disassembly occurs in the air cell of FCF, while the chopping occurs in the argon cell of FCF. After each batch of fuel is electrorefined, cladding hulls and undissolved uranium and zirconium remain in the anode basket. The hulls and adhering salt from the anode basket are loaded into a metal waste furnace, blended with zirconium, and heated to 1,600°C to form metal waste forms (Frank et al. 2007). The majority of noble metal fission products such as Mo, Tc, Rh, etc. remain in the hulls and end up being immobilized in the metal waste form (Li and Simpson 2005). The molten LiCl–KCl electrolyte used in both the Mark-IV and Mark-V ERs is progressively contaminated with active metal fission products as chlorides (CsCl, $CeCl_3$, $SrCl_2$, etc.), in addition to NaCl from the bond sodium with each batch of fuel treated. At some time in the future, it will be necessary to remove some of this salt and replace it with clean LiCl–KCl. Various options based on immobilization of the salt in sodalite or glass have been reviewed in the literature (Crum et al. 2009). The original process developed for converting the ER salt into waste forms consisted of absorbing it into zeolite-4A followed by conversion to glass bonded sodalite (Priebe and Bateman 2008). Photographs of simulated glass-bonded sodalite waste forms are shown in Figure 7.2.

Recently a zeolite ion exchange process for dechlorinating the salt as a precursor to waste form fabrication has also been reported and is illustrated in Figure 7.3 (Wasnik et al. 2019). Different than the zeolite-4A based process, the dechlorination process uses H-exchanged zeolite Y. The H^+ ions in the zeolite exchange with metal ions in the salt to form volatile HCl, which is released. The HCl can be disposed of

FIGURE 7.2 Photographs of simulated glass-bonded sodalite waste forms made from surrogate ER salt. Total mass of this waste form is 378 kg.

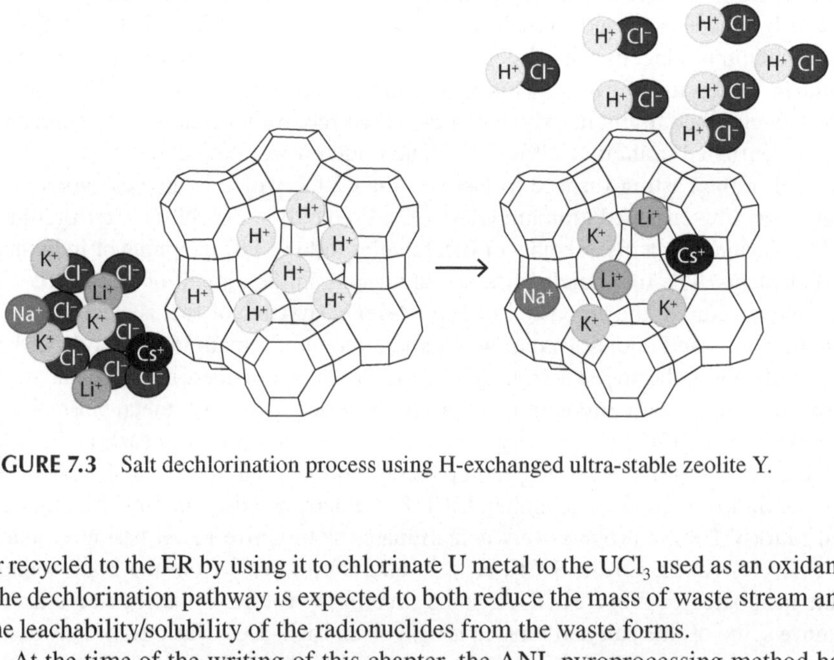

FIGURE 7.3 Salt dechlorination process using H-exchanged ultra-stable zeolite Y.

or recycled to the ER by using it to chlorinate U metal to the UCl_3 used as an oxidant. The dechlorination pathway is expected to both reduce the mass of waste stream and the leachability/solubility of the radionuclides from the waste forms.

At the time of the writing of this chapter, the ANL pyroprocessing method has only been used for processing the irradiated driver fuel and blanket assemblies from EBR-II. This treatment project started in 1996 with the mission to electrorefine both EBR-II driver fuel assemblies and EBR-II blanket assemblies. Currently, INL is focused on driver fuel electrorefining and has approximately 25% of the original 3.35 MT of driver fuel (EBR-II and Fast Flux Test Facility [FFTF]) remaining to be processed based on a recently published Environmental Assessment (DOE 2019). Since the IFR program was shut down in 1994, there is no longer a mission to refabricate fuel from electrorefining U and U/TRU. The electrorefining of EBR-II fuel is motivated by the need to safely and permanently dispose of the waste. EBR-II fuel and blanket assemblies were bonded with sodium metal, which makes untreated, irradiated fuel too reactive to dispose of directly. Sodium reacts vigorously and exothermically with water to form explosive hydrogen gas. The HALEU product from EBR-II may eventually be purchased by a commercial entity for making commercial reactor fuel (DOE 2019), but its transuranic concentration is too high to be accepted into commercial fuel fabrication plants. Despite its successful demonstration, there is currently no need to recover TRU actinides from the Mark-IV or Mark-V ER salt via the LCC. It is expected that TRU actinides will be left in the waste salt, which will in the future be converted into waste forms and disposed of in a geologic repository. A likely site for disposal of waste from pyroprocessing EBR-II fuel is the Waste Isolation Pilot Plant (WIPP) near Carlsbad, New Mexico.

More technical details on the unit operations and fundamental chemistry of spent fuel pyroprocessing can be found elsewhere (Simpson 2018).

7.4 POTENTIAL FUTURE APPLICATIONS FOR PYROPROCESSING

7.4.1 Commercial Fuel Reprocessing

While the modern IFR pyroprocessing method has been demonstrated, used to recover tons of uranium from spent fuel, and extensively modernized and optimized, it is difficult to imagine how it may be used for industrial-scale processing in the near future. The U.S. is arguably in the best position to promote the industrialization of pyroprocessing both domestically and internationally. The U.S. has a unique combination of a large commercial nuclear industry, world-class nuclear research laboratories, and about 40 degree-granting academic nuclear engineering programs. But there is currently strong opposition to nuclear fuel reprocessing in the U.S. government, and there is little economic driving force for implementation of any reprocessing scheme. The nuclear waste agreement between the U.S. government and nuclear utilities has effectively placed the economic and technical burden for nuclear waste disposal on the government.

If the U.S. government were to undertake the mission of using pyroprocessing to reduce the amount of high-level waste requiring permanent geologic disposal, it would need to first reduce the commercial oxide fuel to metallic form prior to electrorefining. The electrolytic reduction process in molten $LiCl–Li_2O$ has been heavily researched over the last two decades, but has not been scaled up to achieve a practical industrial-scale processing rate (Redey and Gourishankar 2003; Herrmann et al. 2006, 2019). A demonstrated and effective approach to processing commercial oxide fuel is aqueous processing such as PUREX (Veliscek-Carolan 2016). There is approximately 2,000 MT of spent fuel generated in the U.S. each year. The La Hague PUREX plant in France has a published throughput rate of 1,700 MT per year. The highest annual output from ERs ever achieved in the FCF at INL was 0.76 MT in 1999. Thus, there is a vast engineering gap for pyroprocessing to overcome in order to process fuel at the scale needed to keep up with commercial spent fuel generation. Designs have been published for 100 MT/year pyroprocessing plants (Chang et al. 2019), but there would need to be 20 such plants to keep up with the current level of spent fuel generation in the U.S. It is difficult to imagine this number of pyroprocessing plants being built in the foreseeable future. Hybrid processing that utilizes PUREX as a head end step for separating low-enriched uranium and sending the transuranic actinides and fission products to a pyroprocessing plant could be evaluated, however, for reprocessing commercial spent fuel (Laidler et al. 2001). Low-enriched uranium can be separated from TRU actinides and fission products via dissolution in nitric acid followed by separation of a purified stream of uranium. The other stream of TRU actinides and fission products could be calcined to form oxides, reduced to metal, and electrorefined to separate TRU actinides from fission products. Typical spent fuel consists of 95.6 wt% uranium after subtracting out the mass of oxygen bonded to the metals (Wald 2009). Thus, the application of PUREX as a head-end processing step for commercial spent fuel could reduce the feed to pyroprocessing plants to only 88 tons per year, which could be processed in a single plant.

7.4.2 Molten Salt Reactors

The molten salt reactor (MSR) is a Generation-IV reactor concept that currently has gained much interest for commercial development and is on the short list of reactor concepts for which research is currently funded by the U.S. Department of Energy (LeBlanc 2010). Notable startup companies that have recently invested in designing MSRs include TerraPower, Terrestrial Energy, Flibe Energy, Elysium Industries, and TransAtomic Power. All of these concepts are actually derived from the Molten Salt Reactor Experiment (MSRE), which was carried out at Oak Ridge National Laboratory in the 1960s. The MSRE featured a molten fluoride salt as the fuel ($LiF–BeF_2–ZrF_4–UF_4$) for a thermal neutron spectrum reactor (Haubenreich and Engel 1970). In contrast, TerraPower and Elysium Industries have worked for the last several years on design of a molten chloride salt-based fast reactor, while the other mentioned companies have worked on development of thermal neutron spectrum fluoride salt reactors. Universally, the ability to operate at high temperatures but low pressures is lauded by developers as the key benefit of MSRs compared to conventional technology. High temperature operation drives either high efficiency electricity production and/or production of valuable chemical products such as hydrogen. Low pressure should correspond to great improvements in operational safety, as the expense associated with robust reactor containment is lessened. Since the fuel in an MSR is designed to always be in fluid form, there is no equivalent to the meltdown accident. The MSR fuel is contained in pipes and vessels that are designed to safely hold molten fuel, while thin zirconium-alloy fuel cladding is not designed to hold molten fuel. The other great benefit of MSRs that was worked on under the MSRE research program is the ability to continuously remove fission products from the fuel while not stopping reactor operation. This can lead to increased run time for the plants and substantial reduction in the amount of nuclear fuel-based waste that is generated. An MSR can conceivably be operated as a chemical reactor in which nuclear reactions are actively occurring. The reactor can be likened to the molten salt electrolyte in an ER, except that fission is perpetually occurring. Fission serves to increase the redox potential of the salt, increase the fission product content, and consume fissionable actinides. Electrorefining spent fuel in molten salt achieves virtually the same functions, except for increasing the redox potential. Thus, some chemical processes that have been studied for application to electrorefining can be applied to achieving optimal operation of an MSR. This includes fuel salt synthesis from commercial spent fuel, internal recycle of actinides, and concentration of fission products into waste forms. Each of these applications is discussed in more detail below.

Modern molten salt reactor design concepts can be divided largely into two classes. In the first class of designs, the reactor fuel is dissolved into a salt in a molten state during reactor operation. In the second class of designs, a molten salt is used as a coolant and moderator fluid within a more traditional solid-fuel prismatic reactor design. These reactor concepts can be further categorized by the category of salts used with the two most common categories being fluoride and chloride salts. The bulk of the current understanding of technologies associated with molten salts comes from the MSRE that operated from 1965 to 1968 at Oak Ridge National Laboratory (ORNL) (Haubenreich and Engel 1970). As such, the focus of the present discussion will be focused on reactor concepts closely related to the MSRE, which employed a molten fluoride fuel salt.

Much of the current resurgent interest in the molten salt reactor centers around the key feature of the molten/liquid fuel system. This fuel system carries the benefits just mentioned of high temperature, low pressure, and elimination of the issue of fuel meltdown. Additionally, blanket salts containing fertile isotopes can be incorporated into the design (Electric Power Research Institute 2015). Portions of these blanket salts can also be diverted during operation through a similar set of chemical processing unit operations in order to extract the fissionable isotopes that have bred in during irradiation. These fissionable isotopes can then be introduced into the fuel salt to replace the original fuel that has been lost due to the fission process. The goal of this design is a self-sustaining operation in which fission products are continuously removed and fuel is continuously replenished so that the reactor operates continuously without shutdown.

The goal of continuous, self-sustained operation is heavily dependent upon the efficiency of the separation and reconditioning operations that are applied to the salts in the reactor. An additional concern is that the separation processes for fluoride salts can be highly corrosive, and the separation processes need to be designed to minimize corrosive attack on the materials of construction of the vessels and piping associated with the separation units.

7.4.2.1 Chloride Fuel Salt Processing

In a molten chloride salt fast reactor (MCFR), fuel salt consists of enriched uranium or transuranic actinides in the form of anhydrous chloride salts. Sources of enriched uranium may include spent fuel, enrichment plants, or decommissioned nuclear weapons. The associated chemical forms for the fuel source include metal (U), oxide (UO_2 or U_3O_8), and fluoride (UF_6). In a MCFR, the uranium would need to be present as UCl_3 and/or UCl_4. Either compound can be synthesized using pyrochemical methods. Thus, pyroprocessing can be utilized to produce either UCl_3 or UCl_4 from U metal or oxide. For example, in the Mark-IV ER in FCF, the salt is replenished with UCl_3 by adding $CdCl_2$ and allowing the following reaction to occur.

$$1.5\,CdCl_2 + U = 1.5\,Cd + UCl_3 \qquad \Delta G^\circ = -292\ \text{kJ/mol at } 500^\circ C \tag{7.9}$$

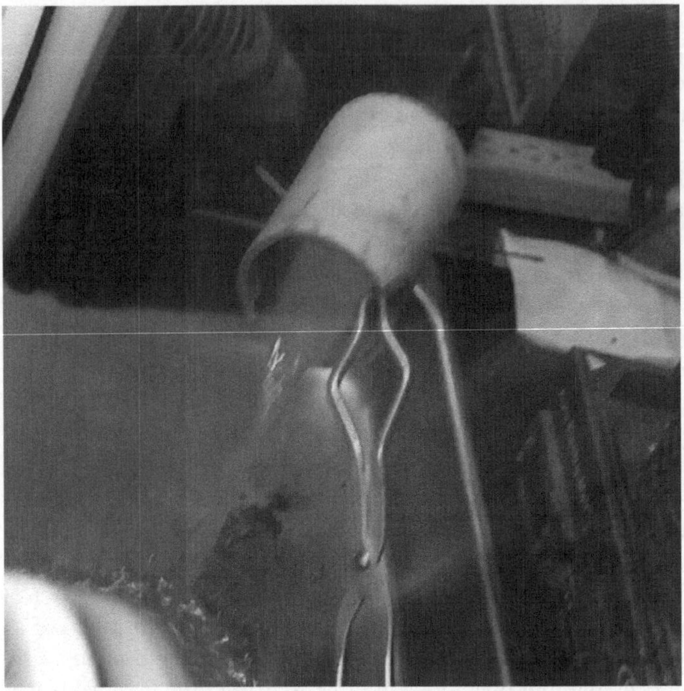

FIGURE 7.4 Sample of molten NaCl–UCl_3 being poured from crucible in a glove box after reaction of U metal with $BiCl_3$ in NaCl.

The problem with using cadmium chloride is that cadmium metal has a high vapor pressure and is a RCRA-listed toxic metal. Thus, virtually all components of the Mark-IV ER are both radioactive and toxic, causing them to be categorized as mixed high-level waste. Alternatively, salts such as $BiCl_3$, $FeCl_2$, and $ZnCl_2$ can also produce UCl_3 from reaction with U metal. Figure 7.4 shows an image of molten NaCl containing UCl_3 being poured in a glove box from an alumina crucible after formation via reaction of U metal with $BiCl_3$.

$$BiCl_3 + U = UCl_3 + Bi \qquad \Delta G° = -468 \text{ kJ/mol at } 500°C \qquad (7.10)$$

$$1.5\ FeCl_2 + U = UCl_3 + 1.5\ Fe \qquad \Delta G° = -333 \text{ kJ/mol at } 500°C \qquad (7.11)$$

$$1.5\ ZnCl_2 + U = UCl_3 + 1.5\ Zn \qquad \Delta G° = -243 \text{ kJ/mol at } 500°C \qquad (7.12)$$

These reactions should be run in the base molten salt of interest, such as Na or NaCl–KCl or NaCl–$CaCl_2$. As can be seen from the NaCl–UCl_3 phase diagram, increasing the UCl_3 concentration has the effect of lowering the liquidus temperature (Sooby et al. 2015).

As uranium is continuously fissioned in a batch of MSR fuel salt, its concentration of fission products will continuously rise. Eventually, solubility limits will be reached, or the operating temperature range of the fuel salt will need to be increased to avoid formation of a solid phase. The obvious workaround to this problem is to add more base salt (NaCl or NaCl–UCl_3), while disposing of salt with a high fission product concentration. But what to do with this salt? As with waste salt from the ERs in FCF, there are many potentially viable solutions (Riley et al. 2019). For the sake of minimizing impact on the geologic repository and maximizing the efficiency of fuel utilization, the actinides should be separated from the fuel salt. Such a process would rely upon pyrochemical reactions such as those listed below.

- Electrolysis:

$$UCl_3 + 3e^- = U + 3Cl^- \tag{7.13}$$

$$3Cl^- = 1.5\,Cl_2 + 2e^- \tag{7.14}$$

- Galvanic reduction (Bagri et al. 2018):

$$UCl_3 + 3e^- = U + 3Cl^- \tag{7.15}$$

$$Gd + 3Cl^- = GdCl_3 + 3e^- \tag{7.16}$$

- Chemical drawdown (Simpson et al. 2012):

$$3Li + UCl_3 = U + 3LiCl \qquad \Delta G^\circ = -336\ \text{kJ/mol at } 500^\circ\text{C} \tag{7.17}$$

A drawing of the galvanic reduction cell is shown in Figure 7.5. A potential difference between the Gd anode on the left and steel cathode on the right drives U metal deposition onto the steel. This process has been demonstrated in laboratory scale experiments using LiCl–KCl–UCl_3 as a surrogate for MSR salt, which resulted in residual U concentration in the salt lower than detection limits (Bagri et al. 2018).

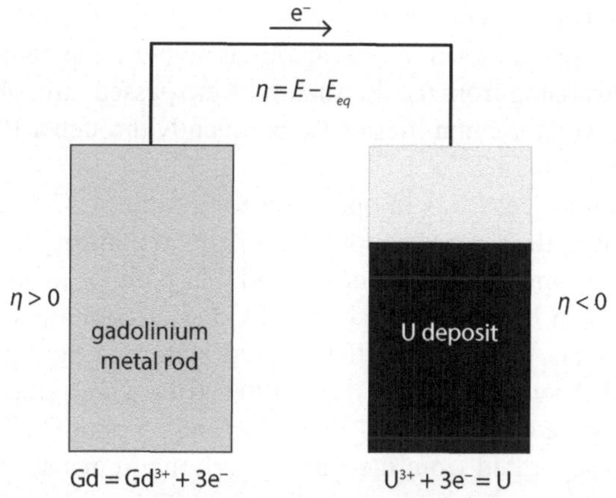

FIGURE 7.5 Drawing of galvanic reduction cell for extraction of U from molten salt. Gd metal is assumed to be the sacrificial anode in this case.

Stabilization of waste salt from MCFRs could be based on the same options that have been surveyed for ER waste salt (Crum et al. 2009). This includes a variety of options for making ceramic or glassy waste forms. The ion exchange process with H-exchanged zeolite Y, previously mentioned, is also applicable to dechlorinating MSR waste salt (Wasnik et al. 2019). Dechlorination is beneficial for minimizing solubility of the waste in water, minimizing waste form volume, and recycling potentially expensive ^{37}Cl to the fuel salt. The last benefit is important, because chloride salt used as an MSR fuel would need to be isotopically enriched in ^{37}Cl to minimize formation of radioactive ^{36}Cl, which would be a major contributor to dose from a geologic repository.

7.4.2.2 Fluoride Fuel Salt Processing

Separation processes for fluoride salts are generally driven by either fluoride volatility or by reductive extraction. The application and effectiveness of these processes depend on the particular salt composition and material of construction in which the process is occurring. In the case of the MSRE, the fuel salt was $^{7}LiF–BeF_2–ZrF_4–UF_4$ (65.0–29.1–5.0–0.9 mole%) and the material of construction was Hastelloy-N (68 Ni–17 Mo–7 Cr–5 Fe). It is important to note that modifications to this salt composition or use of a different material of construction will likely generate results that significantly differ from those discussed in the following paragraphs. The lithium in the fuel salt is specifically enriched to ^{7}Li to prevent the generation of tritium from the neutron irradiation of ^{6}Li.

An early understanding of the effectiveness of the volatile fluoride and reductive extraction separation processes comes from the MSRE itself. The MSRE operated in two stages. During the first stage, the fissionable uranium isotope in the fuel salt was ^{235}U, and during the second stage, the fissionable uranium isotope in the fuel salt was ^{233}U. In between these two stages, it was necessary to remove the ^{235}U that remained in the fuel after irradiation and recharge the fuel salt with ^{233}U. Additional conditioning of the salt by removal of any fission, activation, and corrosion products was also desired. At the end of the first stage, 4,730 kg of fuel was treated to remove the ^{235}U. The first step in the treatment was fluorination in a Hastelloy-N container. Fluorine mixed with helium was bubbled into the salt at a rate of 33 std liters/min through a 1-in dip tube at 450°C. Operation at this temperature limited the corrosion of the Hastelloy-N, which occurs much more rapidly at higher temperatures. The volatile fluoride compounds rising from the fluorination were passed through a 400°C bed of NaF pellets to trap volatile impurities and subsequently through a 100–150°C bed of NaF pellets to capture the UF_6[1].

The fluorination process was in operation for a total of 47 hours over a 6-day period. At this point, the concentration of elements remaining in treated fuel salt was 26 ppm U, 850 ppm Ni, 410 ppm Fe, and 435 ppm Cr as fluorides. The salt was then treated with hydrogen and finely divided zirconium, which caused the fluoride contaminants to reduce to their metal state and agglomerate. The salt was then strained through a fibrous metal filter after which impurity concentrations were found to be 60 ppm Ni, 110 ppm Fe, and 35 ppm Cr. The purity of the uranium that was separated from the salt through the fluorination process is evidenced by a comparison of the dose rate of over 2,000 R/h in the fluorination tank and a dose rate of less than 2 mR/h on a 12 kg container of separated uranium (Haubenreich and Engel 1970).

While the processing steps applied between the two MSRE stages were successful, improvements are necessary in order to advance the technology to the desired state in which continuous online separations can be used. The fuel salt from the first stage of the MSRE was actively fluorinated for 47 hours. Additionally, the fluorination was performed just above the liquidus at ~454°C to minimize corrosion. However, both the Molten Salt Breeder Reactor (MSBR) with a fuel salt mixture of $^7LiF–BeF_2–ThF_4–^{233}UF_4$ (72–16–12–0.3 mole%) (Whatley et al. 1970; Lindauer and Hightower Jr. 1976) and the Flibe Energy, Inc. Liquid Fluoride Thorium Reactor (LFTR) with a fuel salt mixture of $2LiF_2–BeF_2–^{233}UF_4$ and a blanket salt mixture of $2LiF_2–BeF_2–^{232}ThF_4$ (Electric Power Research Institute 2015) are both reportedly designed to operate at 600°C. Additionally, in an online operation, the fluorination will be performed over much longer time periods.

In order to accommodate for increased temperature and exposure time in the fluorination reactor, technologies need to be developed to limit the corrosive attack of the salt on the reaction container materials. One such technology is a frozen salt film that was developed during the MSBR campaign. In this approach, while the salt and fluorinating gases are contacted at operating temperatures (~600°C), the walls of the fluorinator are cooled by external coils to the extent that the salt in contact with the walls freezes. This frozen film of salt limits the corrosive attack of the flowing salt and fluorinating gasses in the center of the fluorinator (Lindauer and Hightower 1976). Another proposed technology is to use a less aggressive fluorinating agent in the fluorinator such as the replacement of F_2 with NF_3 (Sorensen 2018; Scheele et al. 2017).

In the case of the treatment of the fuel from the first stage of the MSRE, the fuel was conditioned with a form of reductive extraction using hydrogen and zirconium metal after the fluorination step. This reductive extraction step can be much more robust, with the molten salt being contacted with a liquid metal that dissolves or reduces fluoride components from the salt. In some designs, the entire fuel-conditioning process is engineered to involve an extensive reductive extraction process complemented subsequently with electrolytic cells and fluorination steps (Whatley et al. 1970). In this design, a stream of the fuel salt from the MSBR would be contacted with liquid bismuth solution in a countercurrent extraction process at 600°C. Initially, this process would remove ^{233}Pr ($^{232}Th + n \rightarrow {}^{233}Th \rightarrow {}^{233}Pr + \beta^-$), allowing it to decay to ^{233}U before being activated by a neutron. The bismuth does not react with the components of the salt but will dissolve components of the salt depending upon the quantity of Li dissolved in the salt according to the equilibrium reaction

$$MF_{n(\text{salt})} + n\,Li_{(\text{Bi})} = M_{(\text{Bi})} + n\,LiF_{(\text{salt})} \tag{7.18}$$

where n is the valence of the metal in the salt phase. Virtually all the zirconium, uranium, plutonium, and protactinium can be extracted from the fuel salt if the bismuth contains 0.1 at% Li. Thorium metal is also dissolved in the liquid bismuth to be used as a reductant in the extraction process.

After the main contact between the liquid bismuth solution and the fluoride salt, the bismuth stream is passed to an electrolytic cell where uranium is oxidized to the

U^{4+} state at the anode, and the thorium and lithium are reduced back to the metal state. Additionally, a portion of the liquid bismuth stream is contacted with HF in the presence of a salt to remove Zr, Zn, Ga, Cd, Sn, Fe, Ni, and Cr. The fluoride salt flowing from the liquid bismuth contactor is subsequently contacted with fluorine to remove uranium, iodine, bromine, and oxygen. This gas stream is separated through passage through NaF pellet beds. A portion of the remaining salt is passed through additional reductive extraction units to remove rare earth elements. Thus, sequencing these steps appropriately was demonstrated to successfully achieve the desired separations. However, there remains ample opportunity to use novel reductive extraction and fluorination techniques to improve the efficiency of the process.

7.5 SUMMARY

The high temperature, dry processes for treating spent nuclear fuel can be traced back to the first reported isolation of uranium metal in the 1800s. Over a nearly 200-year time frame, the processing technology has advanced to accommodate metallic and molten salt-based spent nuclear fuel. Methods for purifying U, concentrating U/TRU, and minimizing fission product waste generation have been developed. The most significant application of pyroprocessing to date has been the operation to treat irradiated driver fuel and blanket assemblies from EBR-II, and it is currently in its third decade of operation. However, looking to the future, there are limited prospects for commercialization of pyroprocessing technology. The 10 MT of HALEU produced from electrorefining EBR-II driver fuel may ultimately be used to fabricate new commercial nuclear fuel. But the process is unlikely to be applied to spent commercial fuel because of the massive quantities in storage and the high annual production rate. PUREX is more suitable for processing large throughputs of light water reactor fuel and may someday be combined with pyroprocessing for maximum efficiency and functionality of fuel reprocessing. It is contended here that the most promising future application of pyroprocessing is to support operation of molten salt reactors—especially molten chloride salt-based reactors. These reactors need pyrochemical operations to synthesize fuel, internally recycle actinides, and produce compact and durable waste forms for permanent disposal of salt waste.

REFERENCES

Bagri, P., J. Ong, C. Zhang, C. Zhang, and M. F. Simpson. 2018. "Galvanic Drawdown of Uranium (III) Chloride and Magnesium (II) Chloride from LiCl-KCl Eutectic Salt Using Gadolinium Sacrificial Anodes." *Journal of Nuclear Materials* 505: 149–158. https://doi.org/10.1016/j.jnucmat.2017.06.007.

Blumenthal, B. 1955. "Melting of High Purity Uranium." *Journal of the Minerals, Metals, & Materials* 7 (11): 1199–1205. https://doi.org/10.1007/BF03379028.

Boisde, G., G. Chauvin, H. Coriou, and J. Hure. 1961. "Mechanisms of Electrorefining of Uranium in Fused Salt Baths." *Electrochimica Acta* 5: 54.

Burris, L., R. K. Steunenberg, and W. E. Miller. 1986. "The Application of Electrorefining for Recovery and Purification of Fuel Discharged from the Integral Fast Reactor." AlChE winter annual meeting, Miami, FL, November 2. https://www.osti.gov/servlets/purl/6840765.

Chang, Y. I., R. W. Benedict, M. D. Bucknor, J. Figueroa, J. E. Herceg, T. R. Johnson, E. R. Koehl et al. 2019. "Conceptual Design of a Pilot-Scale Pyroprocessing Facility." *Nuclear Technology* 205 (5): 708–726. https://doi.org/10.1080/00295450.2018.1513243.

Chauvin, G., H. Coriou, and A. Laroche. 1964. "High-Purity Uranium Production by Electrorefining in Molten Salt Baths." *Journal of Nuclear Materials* 11: 183.

Crum, J. V., S. K. Sundaram, B. J. Riley, J. Matyas, S. A. Arreguin, and J. D. Vienna. 2009. *Alternative Waste Forms for Electro-Chemical Salt Waste* (Richland, WA: Pacific Northwest National Laboratory). PNNL-18930.

DOE. 2019. *Environmental Assessment for Use of DOE-Owned High-Assay Low-Enriched Uranium Stored at Idaho National Laboratory* (Washington, DC: U.S. Department of Energy). DOE/EA-2087.

Driggs, F. H., and W. C. Lilliendahl. 1930. "Preparation of Metal Powders by Electrolysis of Fused Salts: I—Ductile Uranium." *Industrial and Engineering Chemistry* 22: 516–519. https://doi.org/10.1021/ie50245a025.

Electric Power Research Institute. 2015. *Program on Technology Innovation: Technology Assessment of a Molten Salt Reactor Design-the Liquid Fluoride Thorium Reactor (LFTR)* (Palo Alto, CA: EPRI). https://www.epri.com/#/pages/product/3002005460/?lang=en-US

Frank, S. M., D. D. Keiser Jr., and K. C. Marsden. 2007. *Immobilization of Technetium in a Metallic Waste Form* (Idaho Falls, ID: Idaho National Laboratory). INL-CON-07-12883.

Haubenreich, P. N., and J. R. Engel. 1970. "Experience with the Molten-Salt Reactor Experiment." *Nuclear Applications and Technology* 8 (2): 118–136. https://doi.org/10.13182/NT8-2-118.

Herrmann, S. D., S. X. Li, M. F. Simpson, and S. Phongikaroon. 2006. "Electrolytic Reduction of Spent Nuclear Oxide Fuel as Part of an Integral Process to Separate and Recover Actinides from Fission Products." *Separation Science and Technology* 41 (10): 1965–1983. https://doi.org/10.1080/01496390600745602.

Herrmann, S. D., P. K. Tripathy, S. M. Frank, and J. A. King. 2019. "Comparative Study of Monolithic Platinum and Iridium as Oxygen-Evolving Anodes during the Electrolytic Reduction of Uranium Oxide in a Molten LiCl-Li_2O Electrolyte." *Journal of Applied Electrochemistry* 49 (4): 379–388. https://doi.org/10.1007/s10800-019-01287-1.

HSC Chemistry. 2019. Outotek (Version 9). https://www.outotec.com/products/digital-solutions/hsc-chemistry/.

Kofuji, H., F. Sato, M. Myochin, S. Nakanishi, M. V. Kormilitsyn, V. S. Ishunin, and A. V. Bychkov. 2007. "MOX Co-Deposition Tests at RIAR for SF Reprocessing Optimization." *Journal of Nuclear Science and Technology* 44 (3): 349–353. https://doi.org/10.1080/18811248.2007.9711293.

Koyama, T., M. Iizuka, Y. Shoji, R. Fujita, H. Tanaka, T. Kobayashi, and M. Tokiwai. 1997. "An Experimental Study of Molten Salt Electrorefining of Uranium Using Solid Iron Cathode and Liquid Cadmium Cathode for Development of Pyrometallurgical Reprocessing." *Journal of Nuclear Science and Technology* 34 (4): 384–393. https://doi.org/10.1080/18811248.1997.9733678.

Laidler, J. J., L. Burris, E. D. Collins, J. Duguid, R. N. Henry, J. Hill, E. J. Karell et al. 2001. "Chemical Partitioning Technologies for an ATW System." *Progress in Nuclear Energy* 38 (1–2): 65–79. https://doi.org/10.1016/S0149-1970(00)00096-2.

LeBlanc, D. 2010. "Molten Salt Reactors: A New Beginning for an Old Idea." *Nuclear Engineering and Design* 240 (6): 1644–1656. https://doi.org/10.1016/j.nucengdes.2009.12.033.

Lee, H., J.-M. Hur, J.-G. Kim, D.-H. Ahn, Y.-Z. Cho, and S.-W. Paek. 2011. "Korean Pyrochemical Process R&D Activities." *Energy Procedia* 7: 391–395. https://doi.org/10.1016/j.egypro.2011.06.051.

Li, S. X., and M. F. Simpson. 2005. "Anodic Process of Electrorefining Spent Driver Fuel in Molten LiCl-KCl-UCl_3/Cd System." *Mining, Metallurgy & Exploration* 22 (4): 192–198. https://doi.org/10.1007/BF03403322.

Lindauer, R. B., and J. R. Hightower Jr. 1976. *Conceptual Design of a Continuous Fluorinator Experimental Facility (CFEF)* (Oak Ridge, TN: Oak Ridge National Laboratory). ORNL/TM-5253.

Marzano, C., and R. A. Noland. 1953. *The Electrolytic Refining of Uranium.* (Lemont, IL: Argonne National Laboratory). ANL-5102.

Mullins, L. J., J. A. Leary, W. J. Maraman, and A. N. Morgan. 1963. "Plutonium Electrorefining." *Industrial & Engineering Chemistry Process Design and Development* 2 (1): 20–24. https://doi.org/10.1021/i260005a004.

Péligot, E.-M. 1842. "Recherché Sur L'uranium." *Annales de Chimie et de Physique* 5 (5).

Priebe, S., and K. Bateman. 2008. "The Ceramic Waste Form Process at Idaho National Laboratory." *Nuclear Technology* 162 (2): 199–207. https://doi.org/10.13182/Nt08-A3948.

Redey, L. I., and K. Gourishankar. 2003. Direct Electrochemical Reduction of Metal-Oxides. United States Patent 6,540,902.

Riley, B. J., J. McFarlane, G. D. DelCul, J. D. Vienna, C. I. Contescu, and C. W. Forsberg. 2019. "Molten Salt Reactor Waste and Effluent Management Strategies: A Review." *Nuclear Engineering and Design* 345: 94–109. https://doi.org/10.1016/j.nucengdes.2019.02.002.

Scheele, R. D., A. M. Casella, and B. K. McNamara. 2017. "Use of Nitrogen Trifluoride to Purify Molten Salt Reactor Coolant and Heat Transfer Fluoride Salts." *Industrial & Engineering Chemistry Research* 56 (19): 5505–5515. https://doi.org/10.1021/acs.iecr.7b00374.

Simpson, M. F. 2018. "Fundamentals of Spent Nuclear Fuel Pyroprocessing." In *Nuclear Fuel Reprocessing and Waste Management*, edited by J. Zhang, 27–53. Hackensack, NJ: World Scientific Publishing Co.

Simpson, M. F., T. S. Yoo, D. Labrier, M. Lineberry, M. Shaltry, and S. Phongikaroon. 2012. "Selective Reduction of Active Metal Chlorides from Molten LiCl-KCl Using Lithium Drawdown." *Nuclear Engineering and Technology* 44 (7): 767–772. https://doi.org/10.5516/Net.06.2011.010.

Sooby, E. S., A. T. Nelson, J. T. White, and P. M. McIntyre. 2015. "Measurements of the Liquidus Surface and Solidus Transitions of the NaCl-UCl_3 and NaCl-UCl_3-$CeCl_3$ Phase Diagrams." *Journal of Nuclear Materials* 466: 280–285. https://doi.org/10.1016/j.jnucmat.2015.07.050.

Sorensen, K. 2018. "Flibe Energy Has $2.6 Million for Molten Salt Nuclear Research." Molten Salt Reactor Workshop 2018, Oak Ridge, TN, October 4. https://www.nextbigfuture.com/2018/10/flibe-energy-has-2-6-million-for-molten-salt-nuclear-research.html.

Veliscek-Carolan, J. 2016. "Separation of Actinides from Spent Nuclear Fuel: A Review." *Journal of Hazardous Materials* 318: 266–281. https://doi.org/10.1016/j.jhazmat.2016.07.027.

Wald, M. L. 2009. "What Now for Nuclear Waste?" *Scientific American* 301 (2): 46–53. https://doi.org/10.1038/scientificamerican0809-46.

Wasnik, M. S., A. K. Grant, K. Carlson, and M. F. Simpson. 2019. "Dechlorination of Molten Chloride Waste Salt from Electrorefining Via Ion-Exchange Using Pelletized Ultra-Stable H-Y Zeolite in a Fluidized Particle Reactor." *Journal of Radioanalytical and Nuclear Chemistry* 320 (2): 309–322. https://doi.org/10.1007/s10967-019-06476-y.

Westphal, B. R., K. C. Marsden, J. C. Price, and D. V. Laug. 2008. "On the Development of a Distillation Process for the Electrometallurgical Treatment of Irradiated Spent Nuclear Fuel." *Nuclear Engineering and Technology* 40 (3): 163–174. https://doi.org/10.5516/Net.2008.40.3.163.

Whatley, M. E., L. E. Mcneese, W. L. Carter, L. M. Ferris, and E. L. Nicholson. 1970. "Engineering Development of MSBR Fuel Recycle." *Nuclear Applications and Technology* 8 (2): 170–178. https://doi.org/10.13182/Nt70-A28623.

Index

Note: Page numbers in italic and bold refer to figures and tables, respectively.
Page numbers followed by n refer to footnote.